Teaching Science with Hispanic ELLs in K–16 Classrooms

A volume in
Research in Science Education (RISE)
Dennis W. Sunal and Emmett L. Wright, *Series Editors*

Research in Science Education (RISE)

Dennis W. Sunal and Emmett L. Wright, *Series Editors*

Reform in Undergraduate Science Teaching for the 21st Century (2003)
edited by Dennis W. Sunal, Emmett L. Wright, and Jeanelle Bland

The Impact of State and National Standards on K–12 Science Teaching (2006)
edited by Dennis W. Sunal and Emmett L. Wright

The Impact of the Laboratory and Technology on Learning and Teaching Science K–16 (2008)
edited by Dennis W. Sunal and Emmett L. Wright

Teaching Science with Hispanic ELLs in K–16 Classrooms

Edited by

Dennis W. Sunal
University of Alabama

Cynthia S. Sunal
University of Alabama

Emmett L. Wright
Kansas State University

IAP

INFORMATION AGE PUBLISHING, INC.
Charlotte, NC • www.infoagepub.com

Library of Congress Cataloging-in-Publication Data

Teaching science with Hispanic ELLs in K-16 classrooms / edited by Dennis W. Sunal, Cynthia S. Sunal, Emmett L. Wright.
p. cm.
Includes bibliographical references.
ISBN 978-1-61735-047-4 (pbk.) – ISBN 978-1-61735-048-1 (hardcover) – ISBN 978-1-61735-049-8 (e-book)
1. Science–Study and teaching–United States. 2. English language–Study and teaching–United States–Spanish speakers. 3. Hispanic Americans–Education–United States. 4. Science teachers–In-service training–United States. I. Sunal, Dennis W. II. Sunal, Cynthia S. III. Wright, Emmett.
Q183.3.A1T427 2010
507.1–dc22

2010012643

Printed and bound by CPI Group (UK) Ltd, Croydon, CR0 4YY

CONTENTS

Preface to the Series vii

Preface ix

Acknowledgments xi

1 Science Education and Hispanic English Language Learners: The Research Perspective 1
Cynthia Szymanski Sunal and Dennis W. Sunal

2 Fostering Scientific Reasoning as a Strategy to Support Science Learning for English Language Learners 11
Cory Buxton and Okhee Lee

3 Critical Issues in Teaching Science to Hispanic English Language Learners: An Overview 37
Robert D. Leier and Laureen A. Fregeau

4 Promoting Science Understanding and Fluency among Hispanic English Language Learners: Strategies, Explorations, and New Directions 65
Ann M. L. Cavallo and Patricia A. Gomez

5 Synergistic Teaching of Science to English Language Learners: Common Components of Model English Language Learner and Science Instruction 101
Daniel J. Bergman

6 Enhancing Content Instruction for English Language Learners: Learning about Language in Science 135
Luciana C. de Oliveira

7 A Framework for the Effective Science Teaching of English Language Learners in Elementary Schools 151
Trish Stoddart, Jorge Solis, Sara Tolbert, and Marco Bravo

8 Pre-service English Language Learner Science Teacher Preparation in the Southeast United States 183
Teresa J. Kennedy, Jason T. Abbitt, and Michael R. L. Odell

9 Using Inquiry to Teach Inquiry: A Pre-Service Science Education Model with Possibilities for Developing Hispanic English Language Learners' Academic Discourse 201
Elsa Q. Villa and Kerrie Kephart

10 Transformative Professional Development for In-Service Teachers: Enabling Change in Science Teaching to Meet the Needs of Hispanic English Language Learner Students 233
Carla C. Johnson

11 Science as Springboard: Promoting Science Achievement and Aspiration among Latino English Language Learners in the Secondary School 253
Bernadette Musetti and Sara Tolbert

12 A Framework to Support Hispanic Undergraduate Women in STEM Majors 273
Barbara A. Burke and Dennis W. Sunal

About the Authors 313

PREFACE TO THE SERIES

Science education as a professional field has been changing rapidly over the past two decades. Scholars, administrators, practitioners, and students preparing to become teachers of science find it difficult to keep abreast of relevant and applicable knowledge concerning research, leadership, policy, curricula, teaching, and learning that improve science instruction and student science learning. The literature available reports a broad spectrum of diverse science education research, making the search for valid materials on a specific area time-consuming and tedious.

Science education professionals at all levels need to be able to access a comprehensive, timely, and valid source of knowledge about the emerging body of research, theory, policy, and practice in their fields. This body of knowledge would inform researchers about emerging trends in research, research procedures, and technological assistance in key areas of science education. It would inform policy makers in need of information about specific areas in which they make key decisions. It would also help practitioners and students become aware of current research knowledge, policy, and best practice in their fields.

For these reasons, the goal of the book series, *Research in Science Education,* is to provide a comprehensive view of current and emerging knowledge, research strategies, and policy in specific professional fields of science education. This series presents currently unavailable, or difficult to gather, materials from a variety of viewpoints and sources in a usable and organized format.

Teaching Science with Hispanic ELLs in K–16 Classrooms, pages vii–viii

Each volume in the series presents a juried, scholarly, and accessible review of research, theory, and/or policy in a specific field of science education, K–16. Topics covered in each volume are determined by current issues and trends, as well as generative themes related to up-to-date research findings and accepted theory. Published volumes will include empirical studies, policy analysis, literature reviews, and positing of theoretical and conceptual bases.

PREFACE

Research in Science Education (RISE): Teaching Science with Hispanic ELLs in K–16 Classrooms examines research, theory, policy, and practice concerning issues involved in teaching science with Hispanic ELLs in K–16 classrooms and the preparation of teachers for these ELL classroom settings. This RISE Volume addresses pre-service and in-service teachers of science who serve a rapidly growing population of Hispanic students in U.S. classrooms.

The goal of this fourth volume of RISE is to provide a research foundation that demonstrates an agenda for strengthening the preparation and enhancement of teachers of science for regions and states experiencing extensive initial growth of Hispanic ELLs in schools. The goal has been carried out through a series of events that led to the planning and subsequent dissemination of research being conducted by various stakeholders throughout the United States. Researchers were first invited from regions of the country that have had a long history of Hispanic ELLs in classrooms as well as those regions where initial and now extensive growth has occurred only in the past few years. A national conference, Science Teacher Education for Hispanic English Language Learners in the Southeast (SHELLS), funded through the National Science Foundation, was used as one of the dissemination methods to establish and secure commitments from researchers to conduct and report research to strengthen teacher preparation for science. The national call for manuscripts requested the inclusion of major priorities and critical research areas, methodological concerns, and results of implementation of teacher preparation and development programs.

Teaching Science with Hispanic ELLs in K–16 Classrooms, pages ix–x

In developing research manuscripts to be reviewed for *RISE*, Volume 4, researchers were asked to consider the status and effectiveness of pre-service science teacher education programs addressing needs of Hispanic students in classrooms, examining such issues as: what programs exist; differences between programs, especially between regions with long established and more recent and building communities with ELL populations; effectiveness of existing programs in different regions; and differences in the knowledge and skill base required by science teachers who successfully assist Hispanic student populations to accomplish higher science achievement.

This *RISE* volume provides an overview of the research area on preparation and in-service development of teachers of science for Hispanic ELLs, research supported needs and guidelines that should be considered and incorporated in planning professional preparation programs for teaching science in classrooms, and the description of research with example standard and model teacher preparation programs. The chapters outline and describe in detail research-based best practices for a variety of settings.

ACKNOWLEDGMENTS

This fourth volume in the series *Research in Science Education* was an outcome made possible by stakeholders, who gave priority in their professional lives to conduct investigations, write manuscripts, submit their work to the scrutiny of others, and persist through many revisions. Their professional experiences and expertise make this volume possible. Collaboration between the authors developed through discussions and communications that originated at the 2007 national conference, *Science Teacher Education for Hispanic English Language Learners in the Southeast (SHELLS)*, funded through the National Science Foundation. Therefore, we wish to acknowledge the National Science Foundation and individuals attending the conference for providing us with a forum to meet, interact, disseminate, and form professional collaboration communities involving individuals with an interest in improving teaching and learning in science. Special recognition is given to the students, teachers, and administrators who were concerned enough to take part in the investigations and contribute their thoughts to our discussions.

Dennis W. Sunal
Cynthia S. Sunal
Emmett L. Wright
October, 2009

CHAPTER 1

SCIENCE EDUCATION AND HISPANIC ENGLISH LANGUAGE LEARNERS

The Research Perspective

Cynthia Szymanski Sunal and Dennis W. Sunal

ABSTRACT

Developing teachers of science with the knowledge, skills, and dispositions to work effectively with Hispanic English Language Learners (ELLs) is a critical need. To address this need, research, teacher preparation, and classroom practice must be linked. This chapter summarizes the issues that shape the outlines of a research agenda focused on the preparation and professional development of teachers of science (K–16) who can address this need. The origin for the development of this research agenda was a unique forum at a national conference involving a collaborative interdisciplinary team of experienced researchers and practitioners in science education and English language learning. The research agenda outlined in this chapter involves a delineation of major priorities and critical research areas, methodological concerns, and implementation of research needed to focus teacher education and the professional development of teachers on facilitating science instruction in integrated classrooms with Hispanic ELLs.

Teaching Science with Hispanic ELLs in K–16 Classrooms, pages 1–10

INTRODUCTION

The research perspective found in the literature on science education for Hispanic English Language Learners (ELLs) is limited and lacks clarity. The complexity existing among and between Hispanic ELLs, coupled with the need to fully engage these learners in the study of science, is a need still to be fully addressed by researchers. Recognition of real-world complexity involves understanding the wide range of science inquiry-oriented pedagogies and curricula that might build on the strengths Hispanic ELLs bring to the science class. As evidenced in the following chapters in this book, researchers are struggling to identify and investigate critical research areas, key methodological issues, and implementation concerns within a complex educational and social setting. Five interconnecting categories emerge across the chapters in regard to research on science for Hispanic ELLs, (1) the current status of teacher education's preparation and development of teachers of science, (2) the current status of programs for professional development of in-service teachers, (3) the effectiveness of pre- and in-service teacher education and professional development programs, (4) differences found among pre- and in-service teacher preparation and development, and (5) students' meaningful learning of science concepts and the context in which those concepts are understood.

These five categories shape the outlines of a research agenda in which three key questions emerge. These questions underlie the investigations reported here of science education for Hispanic ELLs, although the emphasis on each key question varies among the chapters. The first key question emerging is, "What are important components that must be addressed in research-based pre-service teacher education furthering the development of K–12 teachers of science with Hispanic ELLs?" The second question emerging is, "What are important components that must be addressed in research-based professional development of in-service K–12 teachers of science with Hispanic ELLs?" The third question is, "What components in pre- and in-service education programs are important for supporting the development of K–12 teachers of science with Hispanic students?"

The research agenda described in this chapter and evidenced in this book's chapters is an outcome of a national conference of experienced researchers and practitioners in science education and English language learning. The researchers continued collaboration in interdisciplinary teams over a year following the conference. The working conference was supported by funding from the National Science Foundation (NSF Grant #9729439) and focused on the issue of Science for Hispanic ELLs (SHELLS) in the southeastern U.S.A. The national conference was held April 16–19, 2008 at the University of Alabama in Tuscaloosa, Alabama. These chapters represent the views of the authors, however, not of the National Science

Foundation. The chapters move beyond the southeastern region, identifying and examining connections between research and practice, K–16, found within the nation as a whole. These researchers build on what is known as they investigate current and future directions for research and practice. The chapters take account of regional differences in culture, history, and perspectives while seeking themes, such as inquiry teaching and learning in science, crosscutting regional differences.

Research Questions Associated with the Three Key Questions

Many research questions can be asked in relation to each of the three key questions. As questions are asked and investigated, additional questions arise. The research reported in this book's chapters represents various stages of asking questions, generating hypotheses, investigating hypotheses, and asking more research questions.

Key Question One

The first key question, "What are important components that must be addressed in research-based pre-service teacher education furthering the development of K–12 teachers of science with Hispanic ELLs?" generated several additional research questions that are addressed to varying extents within the chapters. These questions include the following: "How do novice teachers, at the pre-service level and during the early years of in-service teaching, develop their pedagogical content knowledge in science for Hispanic ELLs?"; "To what extent do pre- and in-service teachers receive training from experts in English as a second language (ESL)?"; "How are the standards and strategies of education in ESL integrated into teacher preparation science methods coursework?"; "How can pre- and in-service programs best prepare teachers for working in science with Hispanic ELLs in the range of classroom venues: whole class, sheltered classes, and differentiated middle/high school courses?"; "How can successful structures in science teacher education programs for Hispanic ELLs be disseminated and utilized in various regions of the nation?"; and "How is the effectiveness of the pre- or in-service training for working with Hispanic ELLs in science assessed?"

Key Question Two

The second key question also is addressed in research questions formulated by various authors in this book: "What are important components that must be addressed in research-based professional development of in-service K–12 teachers of science with Hispanic ELLs?" This research question is approached by investigations built around questions found in the following set: "How can we assure fidelity of implementation of the pre-service teacher education model, particularly in an area such as science that is a critical need for often underserved Hispanic ELLs?"; "How can non-traditional class experiences build on home experiences and bridge the space between home and school in science for Hispanic ELLs?"; "What is the nature of appropriate and sufficient ESL preparation of teachers of science?"; "What is the nature of preparation in science, and in science education, for teachers who work with Hispanic ELLs in science?"; "How can teachers be best trained to adapt science curricula to the needs of Hispanic ELLs, particularly when dealing with assessment constraints?"; "What modifications, if any, need to be made within classroom science instruction because Hispanic ELLs' cultural backgrounds may foster different styles or modes of learning?"; "What will be the primary focus of pre-service programs for teachers of science who will have Hispanic ELLs within their classrooms: language development, content translation, teaching strategies, communication skills, writing, reading, and/or experiential learning?"; "Are, or should, pre-service programs preparing teachers to work with Hispanic ELLs be a part of the regular teacher preparation program, or a separate set of courses?"; "What evidence is there that additional training in strategies for teaching learners for whom English is a second language is effective for pre- and in-service teachers?"; and "What factors in the home and family contexts of Hispanic ELLs have an impact on their science learning, K–16?"

Key Question Three

A last set of research questions is found in relation to the third key question, "What components in pre- and in-service education programs are important for supporting the development of K–12 teachers of science with Hispanic students?" These questions guide various studies reported in this book: "How do teacher educators identify and insure coherence of the implementation of a model for preparation of teachers of science to Hispanic ELLs across its components?"; "What structures support the elements of a continuum of teacher development from the pre-service program through the induction period for new teachers?"; "How do we encourage veteran teachers to take on the dual role of mentor for pre-service teachers and of learners of new strategies?"; "What does a community of learners look like if it encompasses the pre-service teacher and the Hispanic ELL?"; "How can pre-service teachers best be involved in practical, real-world experiences with Hispanic ELLs throughout their pre-service program?"; "What does the professional identity of a teacher of science to Hispanic ELLs look like?"; and "How can we best engage pre- and in-service teachers in carrying out investigations into the scholarship of teaching science to Hispanic ELLs?"

The three key questions generate numerous research questions, as indicated above, with additional questions certainly possible. The scholarship in this book identifies some critical research areas, key methodological issues, and implementation concerns and investigates these via a range of approaches. Crosscurrents in the enterprise of education are many, and these become evident across the chapters.

RESEARCH AGENDA

The research agenda described in this book's chapters has the goal of more effective teaching of science to K–12 Hispanic ELLs. To work toward such a goal, researchers investigate the spectrum of K–12 education and extend their work into the undergraduate and graduate levels. Practical outcomes in K–12 and higher education are important to these investigators, as research is used to support best practice. In the view of the researchers in this book, research should be connected to practice, and practice should identify further questions for examination through research.

Hispanic ELLs are found throughout the nation. Some regions, however, are experiencing a rapid increase of Hispanic ELLs, although they have not previously seen significant growth in this population. In the southeastern region, Alabama and Arkansas are examples of states with a rapid increase in Hispanic ELLs. In such regions, the science curriculum is not necessarily aligned with the needs of this growing population of students. Those responsible for curriculum may not yet be aware of how the specific demands of the science curriculum may need adaptation to support the meaningful learning of science by Hispanic ELLs. The research agenda, then, is implemented in a fluid situation and takes into account future directions as well as current needs in science education for Hispanic ELLs.

Research Needs and Efforts

A number of research needs in regard to the science education of Hispanic ELLs, and efforts to address those needs, are described in this book. These needs are briefly discussed below, having been categorized as follows: descriptions of teaching practices and resources, supportive pedagogy, ongoing professional development, integrated lesson planning, and teacher action research.

Descriptions of Teaching Practices and Resources

Efforts are being made to explore current teaching practices in science with Hispanic ELLs. More precise and better description is needed of existing practices and resources at all levels, including K–12 and undergraduate and graduate higher education. Description, however, is particularly needed in school districts, as the diversity and complexity of K–12 education across the nation requires not one portrait of education for Hispanic ELLs, but many. Description is needed of supporting alternative performance-based assessments and tutoring programs, as well as of curricula, pedagogy, and organizational structures. Researchers have an important role in such description and also in analyzing the short- and long-term effects of the components described as the meaningful learning of science by Hispanic ELLs.

Supportive Pedagogy

There is a need to identify and design key elements of pedagogy in science that complement and support the academic and personal development of Hispanic ELLs. Pedagogy specifically designed for appropriate science instruction is needed. Experts exist within school districts and higher education institutions, and are developing research-based practice using effective pedagogy. How such pedagogy can be instituted in school districts and in teacher education programs over the short term and be sustained within these programs over the long term is unclear, and therefore constitutes another research need.

Ongoing Professional Development

On-going professional development must meaningfully engage classroom teachers, administrators, and staff to assist them in integrating Hispanic ELLs and in the teaching of science across the school community. Engagement may include discussion boards, video-conferencing, wikis, modeling, and face-to-face seminars. Adaptation of new interactive technologies as found in social networking options presents opportunities that have yet to be deeply investigated. Engagement, in whatever venue is used, must be designed to relate curriculum and instruction to the research bases in science education and in the education of Hispanic ELLs.

Integrated Lesson Planning

With the high stakes testing agendas seen in U.S. schools, all K–12 teachers, but particularly elementary teachers, are constrained by time and school goals when teaching science. Teacher education science methods courses need to provide integrated lesson plans addressing Hispanic ELLs. For this to occur, knowledgeable ELL faculty must provide assistance to all teacher education faculty, including those in science education. While this book's chapters indicate that ELL researchers have given very limited attention to students' needs in science content learning, such researchers can support science educators as they identify strategies for integrated lessons and modify approaches to better facilitate their students' content learning.

Teacher Action Research

Training is needed by many K–12 pre- and in-service teachers to enable them to develop, implement, and analyze action research studies focusing on the teaching of science to Hispanic ELLs. Research by teachers generates hypotheses that address both emerging and existing needs. Such research impacts specific classrooms directly and can lead to larger and possibly generalizable investigations by teacher researchers as well as higher education researchers.

Threads in the Research Agenda

As needs are identified and hypotheses generated, the research agenda constructed for pre- and in-service science teacher education for Hispanic ELLs is large and complex. Threads running through the research agenda can be described as *people, policy, programs,* and *practice.*

People

People are interpreted in this book's chapters as teacher educators, pre- and in-service teachers, school administrators, K–16 students, and family/community members. Since teacher educators are a critical component of both research efforts and of professional development at the pre- and in-service levels, an important question in need of further investigation is, "How were faculty prepared to teach pre-service teachers about both science content knowledge and facilitating Hispanic ELLs' construction of that knowledge?" Such a question is of particular importance in regions with recent significant growth in Hispanic ELLs, but with little history of serving this population of students.

Policy

Policy recognizes the diverse sets of standards in both science and ESL. These standards exist at the national level, as seen in the National Science Education Standards (National Research Council, 1996) and in the American Association for the Advancement of Science's Benchmarks (American Association for the Advancement of Science, 1993). Individual states also have science standards. The World Class Instructional Design and Assessment Consortium (WIDA, 2008) produced another set of standards for science teaching specifically aimed at ELLs that has been adopted by several states. There are questions to be investigated in relation to how we align WIDA consortium science standards, state standards, and national standards. Such alignment should occur at the K–12 level, but also needs to occur within teacher education programs at both pre- and in-service levels. Beyond investigations of alignment, researchers need to consider whether

Hispanic ELLs and other students more meaningfully learn science when these standards are aligned. Standards are an issue explored throughout this book's chapters with alignments of different sets of standards considered. Assisting teachers in navigating between sets of standards and implementing them in the classroom is difficult, but researchers in this book are describing and analyzing such attempts

Programs

In programs at both pre- and in-service levels, it is important to consider the knowledge and skills teachers currently possess and those which they need to develop, especially in regard to science standards and standards for the instruction of Hispanic ELLs. Since classroom experiences are integral to teacher education programs, researchers also must examine pre-service clinical experiences, considering whether these are exemplary, or even sufficient. Then, researchers must investigate the sorts of experiences pre-service teachers have within their placements in regard to Hispanic ELLs. To which models, for example, are they exposed? Throughout clinical placements, in coursework, and in other components of the teacher education program, researchers should analyze how well the program builds cultural sensitivity.

Practice

Practice is multi-faceted and complex, as it addresses the science learning needs of Hispanic ELLs who themselves are multi-faceted and complex. Researchers need to identify best practices in science education and in ESL education, then identify where they merge. As these are documented, it may be assumed that "holes," or missing components, can be identified and addressed. Practice is a demonstration of pedagogical science content knowledge (PCK). Teachers need to know how to scaffold the learning of Hispanic ELLs in science, and even more specifically, in the various sciences, such as biology, earth science, and physics. Learners must be scaffolded differently in each of the sciences. In-depth case studies of teachers who have both science education backgrounds and training in ESL might help researchers identify pockets of excellence. Teachers who model excellence might enable truth sensing research investigating their PCK and also the examination of effects on student achievement.

Barriers

Researchers encounter barriers to implementation, with methodology, and with identification of critical needs. Although research may not be targeted at identifying barriers, recognizing possible barriers often is an outcome. Barriers of many kinds exist in the teaching of science to Hispanic

ELLs. Several chapters in this book address barriers while presenting the outcomes of the research being reported.

Long-Term Outcomes of the Research Agenda

Hispanic ELLs' access to preparation for participation in the science, technology, engineering, and mathematics (STEM) workforce is critical in the U.S.A. of the 21st century. Participation by Hispanic ELLs currently is limited by the inadequate preparation of pre- and in-service teachers to deliver science instruction to culturally and linguistically diverse populations. Teacher misperceptions of the instructional needs of culturally and linguistically diverse students, such as Hispanic ELLs, hinder their effective use of research-based instructional practices. A major long-term outcome, therefore, of a research agenda focusing on science education for Hispanic ELLs is enabling these students to fully participate in the STEM workforce. Teacher education programs have a role in addressing this problem. In regions of the nation where the numbers of Hispanic ELLs are rapidly growing, but they still do not represent a large percentage of the population, it is important to direct efforts while total numbers are not too large and the scaling up of changes in pre- and in-service teacher education is more manageable. Because the disparity between ELL and non-ELL students continues even after years of schooling, there are barriers to ELLs' successful completion of higher education in science majors. An important question is, "How many ELLs who graduate from high school have an opportunity to participate in the STEM workforce?" For many, the opportunity does not appear to occur. Another question then arises: "What can teacher education programs do to prepare our pre-service teachers to engage Hispanic ELLs in science content learning so that the STEM workforce is impacted?" A concomitant question is, "What can a teacher education program do to prepare pre-service teachers to address equity and opportunity for Hispanic ELLs in STEM areas to enable them to enter the STEM workforce?" Increasing the amount of time Hispanic ELLs have in STEM workforce-related science experiences in their classrooms is one means of addressing questions about facilitating their science learning. Several researchers involve ELLs in such experiences through programs described and analyzed in this book's chapters.

The development of means for identifying higher education faculty members' current status in terms of their ability to prepare pre-service teachers to facilitate the science learning of Hispanic ELLs for the STEM workforce is needed. Such effort would aim to provide information that can be used to assist faculty in identifying their own professional development needs. Once needs are identified, methods of providing faculty with necessary skills can be designed and analyzed. An array of quantitative and qualitative data collecting and analysis techniques might be used to construct instrumentation for needs assessment and for the analysis of faculty professional development approaches.

Assumptions are evident in regard to the preparation of teachers, pre- and in-service, who can enable Hispanic ELLs to become part of the STEM workforce. A primary assumption is that pre-service teachers, and many in-service teachers, are not adequately equipped to teach Hispanic ELLs. Teacher education programs, especially in states with low, but rapidly increasing, Hispanic ELL populations, are struggling to develop adequate programs for preparing pre-service teachers for working with this population of students. Some models or practices for effectively working with Hispanic ELLs in specific settings and environments exist: there is not a vacuum. In many schools, however, a pull-out program continues to be used, based on the view that students must learn language first before they can learn from experiences with science content. These ELLs, then, are held back from learning science content. Finally, policies are not aligned in terms of goals, curriculum, and appropriate pedagogy. This assumption is open to investigation and can inform all stakeholders of the need for continuity and coherence.

REFERENCES

American Association for the Advancement of Science (1993). *Benchmarks for science literacy.* New York: Oxford University Press.

National Research Council. (1996). *National science education standards.* Washington, DC: National Academy Press.

WIDA (World Class Instructional Design and Assessment Consortium) (2008). *Science standards.* Retrieved September 1, 2008 from http://www.wida.us/standards/ELP_StandardLookup.aspx

CHAPTER 2

FOSTERING SCIENTIFIC REASONING AS A STRATEGY TO SUPPORT SCIENCE LEARNING FOR ENGLISH LANGUAGE LEARNERS

Cory Buxton and Okhee Lee

ABSTRACT

We present the conceptual and analytical frameworks for considering the development of scientific reasoning with English language learners (ELLs). We provide examples of the implementation of these frameworks using data on the reasoning complexity of 73 third-grade and 81 4th-grade ELL students as they participated in individual, task-based, interviews on the topics of measurement and transfer of energy in the contexts of school activities and home connections. Our framework builds upon two distinct research traditions: (1) experimental research in developmental and cognitive psychology and (2) research on diverse student groups in science education. Analysis pointed to several consistent patterns in students' reasoning across demographic subgroups of English for speakers of other languages (ESOL) level, home language, and gender. We also highlight the methodological challenge of considering student reasoning in both home and school contexts.

Teaching Science with Hispanic ELLs in K–16 Classrooms, pages 11–36

INTRODUCTION

The challenge of ensuring that all students achieve high academic standards in science has become increasingly complex as the U.S. student population becomes more culturally and linguistically diverse while simultaneously facing increasing demands of high-stakes testing and accountability under the *No Child Left Behind* Act. Teachers of ELL students face the especially challenging task of aiding these students as they struggle to simultaneously develop academic content knowledge and English language and literacy skills (August & Hakuta, 1997; Wong-Fillmore & Snow, 2002).

Reform documents addressing the academic needs of ELL students point to three language learning goals that must be prominent to support school success: (1) to use English to communicate in social settings, (2) to use English to achieve academically in all content areas, and (3) to use English in socially and culturally appropriate ways (Teachers of English to Speakers of Other Languages [TESOL], 1997, 2006). It is the second goal, learning to use English to achieve academically in the content areas, which frequently becomes the greatest struggle for ELL students and their teachers. To succeed academically, ELL students need to develop English language and literacy skills through content area instruction at the same time they are mastering the content goals in an emergent language. Failure to accomplish both of these tasks in unison usually gives rise to a growing achievement gap when ELL students are compared with their fluent English-speaking grade-level peers.

Content area instruction, such as science instruction, has the potential to provide a robust context for meeting learning goals in both academic content and English language and literacy development. For ELL students, however, content area instruction often takes a back seat to language instruction, rather than being viewed as an integral part of language and literacy development. Many teachers continue to assume that ELL students must first master English before learning academic content (Cochran-Smith, 1995). As additional challenges, teachers in the elementary grades are frequently not prepared to teach science content at the level of sophistication required in current state science standards, nor are they sufficiently prepared to meet the learning needs of ELL students (National Center for Education Statistics, 1999). This combination of teacher beliefs, practices, and preparation almost inevitably leads ELL students to fall behind their English-speaking peers in content area learning (August & Hakuta, 1997).

In addition to these instructional challenges, teachers of science face curricular challenges as they are asked to conceptualize and present students with a coherent picture of an ever-growing list of content topics across a broad range of science disciplines. Current science standards documents have been critiqued for lacking a compelling rationale for what content should be taught

(Eisenhart, Finkel, & Marion, 1996), and for fostering a broad coverage model that fails to differentiate standards according to their cognitive complexity (diSessa & Sherin, 1998). More recently, the National Research Council synthesis on learning in the context of school science, *Taking Science to School* (NRC, 2007), has argued that the standards fail to present the core ideas in each discipline in ways that would allow students to engage in sustained investigations at increasingly sophisticated levels of complexity over time.

The current U.S. curricular model, often described as "a mile wide and an inch deep," is detrimental to all students, but can be particularly problematic for ELL students. Students are not given adequate time to explore concepts in depth or from multiple perspectives in ways that allow the practice and application of new academic language skills. All too quickly, instruction has moved on to the next topic, often leaving ELL students floundering both conceptually and linguistically.

In the current climate of assessment and accountability, another significant barrier to supporting the simultaneous development of science content knowledge and academic language is that teachers have tended to become overly focused on the content standards they need to teach, while failing to give adequate attention to how their students are making sense of what is being taught. As is the case for instructional and curricular practices, these assessment practices may be detrimental to all students, but are particularly limiting for ELL students. For example, on the 2005 National Assessment of Educational Progress (NAEP) science results, only 28% of 4th-grade ELL students achieved at the basic level or higher, with only 4% scoring at the proficient level, as compared to 71% of 4th-grade non-ELL students achieving at the basic level or higher, with 31% scoring at the proficient level. Greater awareness of what students are thinking and saying, and particularly an awareness of how students are learning to think and speak scientifically, can help the teacher focus on both the challenges and the progress of ELL students as they work to simultaneously develop science content knowledge and academic language. Additionally, by attending to how students express themselves during science tasks, such as the examples that students use and the questions they ask, teachers can develop a better understanding of the linguistic and cultural resources that ELL students bring to the science classroom, as well as the difficulties students may continue to have with science concepts and scientific language usage even after instruction.

CONCEPTUAL FRAMEWORK: MULTIPLE PERSPECTIVES ON STUDENT REASONING

As we considered strategies for how to work with teachers of ELL students on the complex curricular, instructional, and assessment issues that we have outlined above, we continued to return to the need for teachers to more closely attend to the question of what their students were thinking and the prior knowledge and experiences upon which they were basing those thoughts. This led us to begin conceptualizing a framework of student reasoning grounded in both the learning sciences and research on home language and culture in diverse communities. We realized that the questions we wished to ask about science content and academic language learning of culturally and linguistically diverse students needed to draw upon multiple research traditions. We took as our focus the development of students' scientific reasoning, which we interpreted to mean the logical thought processes used for problem solving during science tasks (Lehrer & Schauble, 2005). We arranged scientific reasoning along a continuum of complexity, ranging from the generation and elaboration of simple statements and assertions at the lower end, to providing causal justifications and explanations of underlying mechanisms at the higher end. We settled upon a framework for analyzing and interpreting the reasoning of ELL students by leveraging the lessons learned from two distinct scholarly traditions: (1) experimental research in developmental and cognitive psychology and (2) research on diverse student groups in science education. Historically, these traditions have largely worked in isolation from one another. If we are to more fully understand the scientific reasoning of culturally and linguistically diverse students, we need to build upon the lessons that have been learned across seemingly disparate paradigms.

Developmental and Cognitive Psychology Perspective on Reasoning

Research on reasoning has its foundations in the highly controlled experimental studies of classical developmental and cognitive psychology. Building on the work of Piaget (Inhelder & Piaget, 1964; Piaget, 1973), reasoning has been construed as the required thinking skills for doing science inquiry through experimentation (Keil & Wilson, 2000), evaluating evidence (Klahr, 2000), and engaging in argumentation in the service of promoting scientific understanding (Kuhn, 1991). From this perspective, reasoning is viewed as an individual cognitive process.

Historically, there have been two main approaches used to study and interpret reasoning, sometimes referred to as "experimentation strategy" and "conceptual change" (Schauble, 1996). The first approach, experimentation strategy, has used a domain-general focus on reasoning and problem solving strategies applicable to a range of science and everyday tasks (Kuhn, 1990; Metz, 1991). The focus has largely been on studying the strategies that individuals develop for generating and interpreting evidence.

These strategies include hypothesis generation, control of variables, and evaluation of evidence. This research has used *knowledge-lean tasks* designed to isolate general skills and strategies in the absence of required domain-specific knowledge. The second approach, conceptual change, has focused on a more domain-specific approach to the development of conceptual knowledge within particular science topics (Carey, 1987; Keil & Wilson, 2000; Smith, Maclin, Houghton, & Hennessey, 2000). This research has used *knowledge-rich tasks* designed to examine the content and structure of students' domain-specific reasoning and problem solving. Recently, some of the researchers within this tradition have shifted their attention from laboratory to classroom settings in order to examine how reasoning develops when students are provided with effective instruction over a sustained period of instructional time in a less controlled school context (Lehrer & Schauble, 2005; Metz, 2004; Toth, Klahr, & Chen, 2000).

Research from developmental and cognitive psychology provides several important lessons for examining the reasoning of ELL students. First, this research points to the challenge for students (and their teachers) of substantially re-crafting their prior ideas and conceptions, and the corresponding importance of identifying students' prior knowledge on science topics (diSessa & Sherin, 1998). Accessing the roots of prior knowledge in meaningful ways must necessarily lead to connections to students' home languages and cultures. Failure to consider the role of home language and culture in prior knowledge formation ignores or even negates critical tools that ELL students have available to construct new understandings of the world.

Second, this research has shown that a focus on generic process skills (e.g., observation, description, prediction, inference) is generally a weak way to promote student reasoning as compared to knowledge-rich activities within a particular domain (NRC, 2007; Zimmerman, 2000). Despite these findings, knowledge-lean science learning activities focusing on the processes of inquiry remain very common in school science instruction at all levels. ELL students are particularly likely to receive knowledge-lean science instruction, due to the misguided belief that ELL students need to be sheltered from more rigorous content knowledge until their English language skills improve. Instead, engaging in complex science tasks is essential for the development of science content knowledge, scientific reasoning skills, and academic English.

Still, such a paucity of research from the developmental and cognitive psychology tradition has been conducted in typical school contexts, let alone with large numbers of ELL students, that many questions remain about how the reasoning of linguistically and culturally diverse students develops in school learning environments. While this tradition brings clarity through highly controlled experimental design, our framework must also account for the unique needs of ELL students in less controlled contexts, such as elementary school learning environments and the students' home and play environments.

Diversity Perspective on Student Reasoning

Unlike the areas of developmental and cognitive psychology, student reasoning has not been a traditional focus of the research on science education with linguistically and culturally diverse student groups. Instead, much of this research has focused on issues such as ensuring equitable learning opportunities (Firestone, Camilli, Yurecko, Monfils, & Mayrowetz, 2000; Tate, 2001), closing the achievement gap (Norman, Ault, Bentz, & Meskimen, 2001; Rodriguez, 2001), or finding ways to make science education more appealing to students who may not readily embrace an identity as a science learner (Brickhouse & Potter, 2001; Seiler, 2001). This research has generally been guided by sociocultural theories borrowed from anthropology and sociology and has rarely focused on questions that are connected to discipline-specific content learning. Still, there are two strands of the research on student diversity in science education that seem most relevant to better understanding the reasoning of ELL students—what we will refer to as research on *continuity* and research on *context.* Some illuminating parallels can be drawn between these issues of continuity and context and the psychological perspectives of knowledge-lean experimentation strategy and knowledge-rich conceptual change.

The question of continuity in classroom research focuses on the connections (or the lack thereof) that learners perceive between the academic tasks they are asked to perform in school and their own developing epistemologies as learners. In other words, how well do normative science practices overlap with students' ways of making sense of the world, and how willing are students to modify their epistemologies to more closely align with normative science practices? The cultural congruence model of learning, for example, argues that students engage more fully in science learning when practices are not seen to be in conflict with their broader worldview (Aikenhead, 2001; Tharp, 1997) and when the communication and interaction patterns in the learning environment do not conflict with culturally accepted norms (Lee & Fradd, 1998). There have been ongoing debates about the degree to which culturally and linguistically diverse students' ways of knowing and talking are continuous or discontinuous with the ways of talking and knowing characteristic of normative science practices (Lee, 2002; Warren, Ballenger, Ogonowski, Rosebery, & Hudicourt-Barnes, 2001). This research on continuity shows some parallels to the research on knowledge-lean experimentation strategy, in that it shares a largely domain-general focus on beliefs and communication patterns that cut across science content areas.

While the research on continuity deals with worldviews and interaction patterns, the research on context addresses the links between classroom learning and knowledge and skills developed beyond the classroom. Much of this work falls under a "funds of knowledge" perspective, highlighting cultural, linguistic, and subject matter knowledge and skills developed and shared by individuals within their communities (González & Moll, 2002; González, Moll, & Amanti, 2005). Understanding students' funds of knowledge requires consideration of a range of learning environments in which students might gain content knowledge or language skills related to a topic, including knowledge constructed in the home (Barton, Drake, Perez, St. Louis, & George, 2004), in the community (Rahm, 2002), and in peer groups (Buxton, 2005), as well as knowledge constructed in the classroom. Research on context shares commonalities with the knowledge-rich research on conceptual change in developmental and cognitive psychology in that the focus tends to be on domain-specific content knowledge as it is developed in unique settings and then potentially mobilized in other settings that share the same domain-specific content.

Research on science learning with linguistically and culturally diverse student groups also provides important lessons when considering the reasoning of ELL students. First, the research on continuity points to the added complexity that must be accounted for when considering the goal of ELL students adopting scientific discourse patterns. In addition to the linguistic challenges of developing the academic language of science, there are epistemological issues involving the degree to which students feel comfortable presenting and representing a normative science worldview (Deyhle & Lecompte, 1994; Hammond, 2001). Teachers need to understand this issue and be willing and able to talk about it explicitly with students who are struggling with questions of belief as well as issues of content and language. Second, the research on context highlights the critical and multifaceted role that prior knowledge plays in developing scientific reasoning. ELL students are often viewed from a deficit perspective due to both their limited English proficiency and the perception (sometimes accurate and sometimes not) that immigrant ELL students bring limited formal education experiences from their countries of origin. Effective teachers of ELL students find ways to move beyond these perceived deficits to make connections between students' lived experiences and grade-appropriate content learning and cognitive complexity (Lee & Fradd, 1998; Warren, Ballenger, Ogonowski, Rosebery & Hudicourt-Barnes, 2001).

In summary, our reading of the research both in developmental and cognitive psychology and in studies of diverse learners points to parallels between these seemingly distant fields and a convergence around the value of teachers attending to their students' emerging scientific reasoning across school and home contexts. While our conceptual and analytical frameworks for fostering reasoning among culturally and linguistically diverse students continues to evolve, we believe that these frameworks will prove fruitful in promoting both science content learning and academic language development for ELL students.

APPLYING OUR REASONING FRAMEWORK IN PRACTICE

Methods

Grounded in our conceptual framework for understanding and supporting the scientific reasoning of culturally and linguistically diverse learners, we developed a series of three elicitation protocols to examine student reasoning about science tasks in both school and home contexts. One protocol was developed for each of the three grade levels (third through fifth) in the larger intervention project. Measurement was the topic selected for third grade, transfer of energy for fourth grade, and changing seasons for fifth grade. Here we will be discussing only the third- and fourth-grade reasoning data.

Third-grade students participated in the measurement elicitation task one-on-one with a member of the research team. The interviews were conducted soon after the students completed the measurement curriculum unit. All interviews were conducted in English, rather than in the students' home language, as English was the language of science instruction in all classes. The task was composed of four parts, with each part addressing four of the topical areas covered in the measurement curriculum unit: length, weight, volume, and temperature (see Appendix A for the elicitation protocol).

The first part of the interview asked students to discuss experiences they had with measurement in the home context. The second part asked students to perform measurement tasks using a ruler, kitchen scale, graduated cylinder, measuring cup, and thermometer. The third part asked students to make estimates about each of the four topical areas of measurement. The final part asked students to discuss experiences they had with measurement in the context of playing with their peers in out-of-school settings. Together, parts one and four were taken to represent the "home" context and parts two and three to represent the "school" context. Each part began with an opening question followed by one or more probes. The task took approximately 30 minutes and was videotaped. The videos were then transcribed for analysis.

The fourth-grade force and motion elicitation followed a parallel procedure, except that there were only three parts: part one asked students to discuss experiences they had with transfer of energy in the home context, part two had students design an experiment on transfer of energy, and part three asked students to discuss experiences they had with transfer of energy in the context of playing with their peers (see Appendix B for the elicitation protocol).

We developed an analytical and coding scheme to categorize student reasoning complexity. We constructed semantic maps of verbal and procedural responses and actions for each of the student interviews. These semantic maps provided an overall verbal and diagrammatic picture of each interview.

Each of the conceptual propositions was then scored using a reasoning complexity rubric along five criteria: (1) generativity (amount and types of topics and assertions brought forth by students); (2) elaboration (supporting details added to ideas within a topic); (3) justification (how well ideas are warranted with evidence and inference); (4) explanation (how underlying causal mechanisms are used to explain assertions); and (5) logical coherence (overall quality of connections between one assertion and other assertions) (see Table 2.1). The first two criteria assess the variety and richness of ideas raised by students, the third and fourth criteria assess the structure of student reasoning, and the final criterion assesses the overall quality of student reasoning. Each of the five complexity criteria was scored on a 5-point scale (0–4), meaning that a score from a zero to four was assigned to each conceptual proposition map for each of the five criteria, and a composite score from a zero to 20 was assigned for overall reasoning complexity.

Finally, tables were constructed to look at cross-case analysis of the five criteria of reasoning complexity (generativity, elaboration, justification, explanation, and logical coherence). For third grade, cross-case analysis was also done for patterns in each of the four measurement topical areas (length, weight, volume, and temperature). In each case, the analysis was conducted with regard to three student demographic variables of ESOL level, home language, and gender. Key findings from analysis of the third-grade data will be presented first, followed by key findings from the fourth-grade data.

Findings

Third Grade

Analysis of overall reasoning complexity for the entire third-grade sample is presented in Table 2.2. The analysis indicates significant differences among the five criteria of reasoning complexity for the entire sample; $F(4) = 581.70$, $p < .001$. Analysis yielded a medium effect magnitude; $\chi^2 = 0.89$. Pairwise comparison *t*-tests indicated that each combination of pairs was statistically significant at $p < .001$. In other words, each reasoning

TABLE 2.1 Student Reasoning Complexity Rubric

Criteria	0	1	2	3	4
Generativity	No observations	One or two observations or simple statements	Three or more observations or simple statements	One or two assertions or conjectures	Three or more assertions or conjectures
Elaboration	No elaboration	One or two elaborations of one idea	One or two elaborations of more than one idea	Three or more elaborations of one idea	Three or more elaborations of more than one idea
Justification	No justifications	One or two justifications of one assertion	One or two justifications of more than one assertion	Three or more justifications of one assertion	Three or more justifications of more than one assertion
Explanation	No explanations	Single mechanism explaining one assertion	Single mechanisms explaining more than one assertion	Multiple or chained mechanisms explaining one assertion	Multiple or chained mechanisms explaining more than one assertion
Logical Coherence	No logical connections or nonsensical connections	Vague connections making superficial sense	Clear and reasonable connections but lack support	Clear and reasonable connections with some specific supports	Clear and reasonable connections with many specific supports

TABLE 2.2 Reasoning Complexity for Entire Sample

n	Generativity *M (SD)*	Elaboration *M (SD)*	Justification *M (SD)*	Explanation *M (SD)*	Logical Coherence *M (SD)*	*df*	*F*	*p*
73	3.7 (0.5)	2.5 (0.6)	1.2 (0.6)	0.6 (0.5)	1.6 (0.4)	4	581.70	.001*

* Significant at $p = .001$.

complexity category (generativity, elaboration, justification, explanation, and logical coherence) is significantly different from the others. The result suggests that students were able to generate topics and examples involving measurement, but were less able to elaborate on those examples with supporting details, still less able to give rational justifications for their ideas, and the least able to give explanations based on underlying mechanisms.

Reasoning data were disaggregated by ESOL level, home language (Spanish or Haitian Creole), and gender. For ESOL level, we followed the school district framework of emergent English speakers (ESOL 1/2), developing English fluency (ESOL 3/4), conversationally fluent but still requiring academic language support (ESOL 5), and non-ESOL. No statistical significance was found for any of the criteria of reasoning complexity by ESOL level. Likewise, there were no significant differences by home language or gender. In other words, the same general pattern of reasoning was seen for third graders no matter how the demographic data were disaggregated.

When third grade reasoning was analyzed across the four topical areas of measurement (length, weight, volume, and temperature), there were no significant differences for the entire sample. However, there were differences when disaggregated by ESOL level. ESOL level 5 students scored significantly higher than ESOL level 1/2 and non-ESOL students for the topics of length and volume ($p = .05$). Third-grade topical area reasoning showed no significant differences by home language or gender.

Fourth Grade

For the fourth grade students, analysis of overall reasoning complexity for the entire sample is shown in Table 2.3. Results are quite similar to those of the third-grade sample, indicating significant differences among the five criteria of reasoning complexity for the entire sample; $F(4) = 245.45$, $p < .001$ and a medium effect magnitude; $\chi^2 = 0.75$. Pairwise comparison *t*-tests indicated that each combination of pairs was statistically significant at $p < .001$, with the exception of elaboration and justification. In other words, each of the reasoning complexity categories (generativity, elaboration, jus-

TABLE 2.3 Reasoning Complexity for Entire Sample

n	Generativity *M (SD)*	Elaboration *M (SD)*	Justification *M (SD)*	Explanation *M (SD)*	Logical Coherence *M (SD)*	*df*	*F*	*p*
81	3.8 (0.4)	2.2 (1.0)	2.2 (0.7)	0.3 (0.6)	2.5 (0.9)	4	245.45	< .001*

* Significant at $p = .001$.

tification, explanation, and logical coherence) was significantly different from the others, except for elaboration and justification. The result suggests that students were able to generate topics and examples involving energy transformations, but were less able to elaborate on those examples with supporting details or to give rational justifications for their ideas, and the least able to give explanations based on underlying causal mechanisms.

The fourth-grade reasoning data were disaggregated by ESOL level, home language (Spanish or Haitian Creole), and gender. Significant differences were observed in reasoning complexity by ESOL level, with ESOL level 5 students significantly outperforming the other ESOL levels, including non-ESOL students, for justification ($p = .05$). No significant differences were found by home language, but when disaggregated by gender, differences in the area of justification were found, with boys outperforming girls ($p = .05$).

DISCUSSION AND IMPLICATIONS FOR TEACHING SCIENCE TO ELL STUDENTS

Discussion

We have presented an overview of the findings from the third and fourth grade reasoning elicitation protocols as examples of how we have enacted our reasoning framework in practice. Our analytic framework for exploring the reasoning of ELL students was derived from our conceptual framework, highlighting the value of developmental and cognitive psychology perspectives as well as cultural and linguistic diversity perspectives to better understand how students think about science tasks.

Our findings proved to be intuitive in some ways and counterintuitive in other ways. For the third-grade students, we had hypothesized that the higher levels of reasoning complexity (justification and explanation) would be more challenging than the lower levels of reasoning complexity (assertions and elaborations). This pattern was clearly seen for the overall sample, with an inverse relationship between increasing level of reasoning complexity and student scores on the rubric.

When we disaggregated the third grade data by demographic subgroups, we hypothesized that the higher the students' ESOL levels were, the better they would score on the reasoning rubric, both because they had more English language ability with which to express themselves and because they would have been better able to comprehend the in-class instruction (primarily done in English) relevant to the elicitation questions. Contrary to our hypothesis, there were fewer and smaller differences than we had anticipated. This finding might be attributable to the fact that our intervention was targeted specifically to support science learning and English language development of ELL students, thus assisting the lower level ESOL students in performing only modestly below their more English proficient peers. Additionally, we were surprised by the finding that ESOL level 5 students outperformed everyone, including the non-ESOL students in topical reasoning. For the demographic subgroups of home language and gender in third grade, we hypothesized that there would be few, if any, significant differences; indeed, none were found.

For the fourth grade students, again we found a combination of intuitive and counterintuitive results. We hypothesized that as students both matured and spent more time in our intervention, the more challenging levels of reasoning complexity would increase. The scores for the lower reasoning levels of assertions and elaborations continued to be high. Fourth grade students did improve in their ability to provide justifications for their assertions. However, results continued to be low for the highest level of reasoning complexity, i.e., explanations of causal mechanisms. While students had studied the concepts that would serve as the underlying mechanisms for explaining transfer of energy, the curriculum provided few opportunities for students to explain these concepts in as loosely guided a fashion as was the case for the elicitation protocol. Students clearly did not interpret the questions in ways that prompted the activation of this prior knowledge.

When we disaggregated the fourth grade data by demographic subgroups, again we hypothesized that the higher the students' ESOL levels were, the better they would score on the reasoning rubric. Contrary to our hypothesis, we found a repeat of the pattern observed with the third-grade sample. Again, ESOL level 5 students outperformed everyone, including the non-ESOL students. While it might be expected that conversationally fluent ESOL level 5 students would outperform ESOL levels 1 to 4 students, we were surprised that the ESOL level 5 students also outperformed the non-ESOL students. The result suggests that ESOL level 5 students were both gaining the academic language of science in English and successfully employing this developing proficiency to execute and express their reasoning about science topics. Thus, an instructional intervention that is primarily in English but with robust supports for academic language development, such as our intervention, could be especially beneficial to students at the level of language proficiency typical of ESOL level 5 (Buxton, Lee, & Santau, 2008).

For the other demographic subgroups by home language and gender in fourth grade, we hypothesized that there would be few, if any, significant differences for the same reasons noted for third grade. For gender groups, we were somewhat surprised that there was a significant difference, with boys outperforming girls in the area of justification. We interpret that this could be, at least in part, a function of the topic of energy transfer. We purposefully selected topics that we believed were gender-neutral when we designed the elicitation task, but the fact remains that fourth-grade boys may be more likely to have experiences taking these things apart (e.g., flashlights, bicycles, soccer balls) and thinking about the processes involved in how they work.

These findings contribute to refining and verifying our conceptual framework for how the reasoning of linguistically and culturally diverse students develops. From the developmental and cognitive psychology perspective, the findings point to the roles of both a domain-generic experimentation strategy and a more domain-specific conceptual change. Part of what might have pushed the fourth-grade students to justify their assertions better than the third-grade students was the ongoing use of the Promoting Science among English Language Learners (P-SELL) curriculum that provided extensive opportunities to practice generating hypotheses, controlling variables and explaining evidence, all hallmarks of experimentation strategy research (Kuhn, 1990). At the same time, part of the challenge of why students so rarely generated examples of causal explanations might stem from the fact that the P-SELL curriculum, because it must cover all of the state science benchmarks, did not provide students with the depth of content knowledge on any given topic (e.g., transfer of energy) needed to promote rich conceptual change (diSessa & Sherin, 1998).

From the perspective of student diversity, these findings point to the important roles of both domain-generic cultural and linguistic continuity and more domain-specific knowledge in context. The fact that so many students seemed content to limit their reasoning to the lower complexity levels of assertions, with occasional elaborations, speaks to the question of whether students' ways of knowing and talking are continuous or discontinuous with the ways of talking and knowing characteristic of normative science practices (Lee, 2002). At the same time, part of what enabled ELL students, even those with emergent levels of English proficiency (i.e., ESOL levels 1 and 2), to generate high scores for the lower reasoning complexity levels might be traced to the focus in the P-SELL curriculum on identifying relevant prior knowledge both through formal science instruction and through funds of knowledge in their communities (González, Moll, & Amanti, 2005).

In short, the findings of this study support the assertion that for ELL students, as for all students, engaging in complex science tasks is essential for the development of scientific reasoning skills, which are supported by science content knowledge and academic English. Understanding and building upon students' prior knowledge, including connections to students' home language and culture, would seem to be an essential part of this approach.

Implications

The conceptual and analytical frameworks we have developed for the investigation of reasoning with culturally and linguistically diverse students are meant to serve as a foundation for our ongoing research and for other related studies. Our focus on ELL students' reasoning raises important questions about relationships between English language proficiency and reasoning complexity, as well as the role that teachers can play in supporting the development of their students' reasoning abilities.

We conceptualized the study of student reasoning as a way to get teachers to think about their students' prior knowledge both from developmental and from sociocultural perspectives. Thus, part of our larger research program examines teacher reasoning, as each teacher watches the videotape of the student from his/her classroom and reflects on what the student was thinking and why. The combination of student reasoning and teacher reasoning will offer multiple avenues for further studies. This approach of promoting teacher reasoning about student reasoning has been advocated in the past (Fennema & Franke, 1992), but has not been a focus of research in science education to date.

The results from this study have also provided new directions for our intervention efforts in teacher professional development and the development of improved curriculum materials for teaching science to ELL students. In our work with teacher professional development we have created multiple sessions on the value of attending to students' reasoning by focusing on the question of what students are thinking and why. We push teachers to think about where students' ideas come from and how these ideas develop from both formal and informal experiences. We see this as a back door approach to encouraging teachers to consider their students' funds of knowledge including their cultural and linguistic resources, areas that many teachers tend to deemphasize and undervalue.

It is clear from our findings that getting students to talk about science is important for the development of their academic language and scientific reasoning. However, not all kinds of science talk are equally valuable, if the goal is to push students towards expressing higher levels of reasoning complexity. Teachers who are consciously monitoring their students' talk for evidence of reasoning complexity can learn to promote science talk that focuses on justifications and causal explanations. While these higher levels of reasoning will remain challenging for elementary grade students, the latest synthesis of research on science learning (NRC, 2007) makes it clear that elementary students are capable of engaging in cognitively complex tasks. In our curriculum development efforts, we continue to build upon our findings from the student reasoning data. For example, in the fifth-grade curriculum, we have included longer instructional sequences focusing on collecting and evaluating evidence, opportunities to engage in model building activities, and supports for engaging in focused scientific conversations. We believe that such experiences will continue to assist ELL students in practicing more complex scientific reasoning.

We expect that our intervention to enhance student reasoning will lead to stronger outcomes in science learning and academic language development. Our ongoing research will look more directly at this relationship. We will also refine our conceptual and analytic frameworks for understanding the progression of reasoning with ELL students by leveraging new thinking from both cognitive and developmental psychology and research on student diversity as well as our own findings. We believe that this conceptual framework for supporting enhanced scientific reasoning for ELL students can aid us in our overarching goal of fostering both high academic achievement and educational equity with diverse student groups in science classrooms.

REFERENCES

Aikenhead, G. (2001). Cultural relevance: Whose culture? What culture? In J. Wallace & W. Louden (Eds.), *Dilemmas of science teaching* (pp. 92–95). New York: Routledge Falmer.

August, D., & Hakuta, K. (Eds.). (1997). *Improving schooling for language-minority children: A research agenda.* Washington, DC: National Academy Press.

Barton, A. C., Drake, C., Perez, J., St. Louis, K., & George, M. (2004). Ecologies of parental engagement in urban education. *Educational Researcher, 33*(4), 3–12.

Brickhouse, N., & Potter, J. (2001). Young women's scientific identity formation in an urban context. *Journal of Research in Science Teaching, 38*(8), 965–980.

Buxton, C. A. (2005). Creating a culture of academic success in an urban science and math magnet high school. *Science Education, 89*(3), 392–417.

Buxton, C., Lee, O., & Santau, A. (2008). Promoting science among English language learners: Professional development for today's culturally and linguistically diverse classrooms. *Journal of Science Teacher Education 19*(5), 495–511.

Carey, S. (1987). *Conceptual change in childhood* (1st MIT Press ed.). Cambridge, MA: MIT Press.

Cochran-Smith, M. (1995). Color blindness and basket making are not the answers: Confronting the dilemmas of race, culture, and language diversity in teacher education. *American EducationalResearch Journal, 32*(3), 493–522.

Deyhle, D., & LeCompte, M. D. (1994). Conflict over child development: Navajo culture and the middle schools. *Theory into Practice, 23*(3), 156–167.

diSessa, A. A., & Sherin, B. L. (1998). What changes in conceptual change? *International Journal of Science Education, 20*(10), 1155–1191.

Eisenhart, M., Finkel, E., & Marion, S. (1996). Creating the conditions for scientific literacy: A re-examination. *American Educational Research Journal, 33*(2), 261–295.

Fennema, E., & Franke, M. L. (1992). Teachers' knowledge and its impact. In Grouws, D. A (Ed.), Handbook of research on mathematics teaching and learning (pp. 147–164). New York: Macmillan Publishing Company.

Firestone, W. A., Camilli, G., Yurecko, M., Monfils, L., & Mayrowetz, D. (2000). State standards and opportunity to learn in New Jersey. *Education Policy Analysis Archives, 8*(35), 1–25.

González, N., & Moll, L. (2002). Cruzando el puente: Building bridges to funds of knowledge. *Educational Policy, 16*(4), 623–641.

González, N., Moll, L. C., & Amanti, C. (2005). *Funds of knowledge: Theorizing practices in households, communities, and classrooms.* Mahwah, NJ: L. Erlbaum Associates.

Hammond, L. (2001). Notes from California: An anthropological approach to urban science education for language minority families. *Journal of Research in Science Teaching 38*(8), 983–999.

Inhelder, B., & Piaget, J. (1964). *The early growth of logic in the child, classification and seriation.* New York: Harper & Row.

Keil, F. C., & Wilson, R. A. (2000). *Explanation and cognition.* Cambridge, MA: MIT Press.

Klahr, D. (2000). *Exploring science: The cognition and development of discovery processes.* Cambridge, MA: MIT Press.

Kuhn, D. (1991). *The skills of argument.* Cambridge: Cambridge University Press.

Kuhn, D. (1990). *Developmental perspectives on teaching and learning thinking skills.* Basel, NY: Karger.

Lehrer, R., & Schauble, L. (2005). Scientific thinking and science literacy. In W. Damon, R. Lerner, K. A., Renninger, & I. E. Sigel, (Eds.), *Child psychology in practice.* Hoboken, NJ: John Wiley & Sons.

Lee, O. (2002). Science inquiry for elementary students from diverse backgrounds. In W. G. Secada (Ed.), *Review of research in education, Vol. 26* (pp. 23–69). Washington, DC: American Educational Research Association.

Lee, O., & Fradd, S. (1998). Science for all, including students from non-English-language backgrounds. *Educational Researcher, 27*(4), 12–21.

Metz, K. E. (2004). Children's understanding of scientific inquiry: Their conceptualization of uncertainty in investigations of their own design. *Cognition and Instruction, 22*(2), 219–290.

Metz, K. E. (1991). Development of explanation: Incremental and fundamental change in children's physics knowledge. *Journal of Research in Science Teaching, 28*(9), 785–797.

National Assessment of Educational Progress (NAEP) 2005. Accessed March 26, 2010 at http://www.nationsreportcard.gov/science_2005.

National Center for Education Statistics. (1999). *Teacher quality: A report on the preparation and qualifications of public school teachers.* Washington, DC: U.S. Department of Education, Office of Educational Research and Improvement.

National Research Council. (2007). *Taking science to school.* Washington, DC: National Academies Press.

Norman, O., Ault, C. R. Jr., Bentz, B., & Meskimen, L. (2001). The black–white "achievement gap" as a perennial challenge of urban science education: A sociocultural and historical overview with implications for research and practice. *Journal of Research in Science Teaching, 38*(10), 1101–1114.

Piaget, J. (1973). *The child and reality: Problems of genetic psychology.* New York: Grossman Publishers.

Rahm, J. (2002). Emergent learning opportunities in an inner-city youth gardening program. *Journal of Research in Science Teaching, 39*(2), 164–184.

Rodriguez, A. (2001). From gap gazing to promising cases: Moving toward equity in urban education reform. *Journal of Research in Science Teaching, 38*(10), 1115–1129.

Schauble, L. (1996). The development of reasoning in knowledge-rich contexts. *Developmental Psychology, 32*(1), 102–119.

Seiler, G. (2001). Reversing the "standard" direction: Science emerging from the lives of African American students. *Journal of Research in Science Teaching, 38*(9), 1000–1014.

Smith, C. L., Maclin, D., Houghton, C., & Hennessey, M. G. (2000). Sixth-grade students' epistemologies of science: The impact of school science experiences on epistemological development. *Cognition and Instruction, 18*(3), 349–422.

Tate, W. (2001). Science education as a civil right: Urban schools and opportunity-to-learn considerations. *Journal of Research in Science Teaching, 38*(9), 1015–1028.

Teachers of English to Speakers of Other Languages. (1997). *ESL standards for pre-K–12 students.* Alexandria, VA: Author.

Teachers of English to Speakers of Other Languages. (2006). *Pre-K–12 English language proficiency standards.* Alexandra, VA: Author.

Tharp, R. (1997). *From at-risk to excellence: Research, theory, and principles for practice.* Santa Cruz, CA: Center for Research on Education, Diversity & Excellence.

Toth, E., Klahr, D., & Chen, Z. (2000). Bridging research and practice: A cognitively based classroom intervention for teaching experimentation skills to elementary school children. *Cognition and Instruction, 18*(4), 423–459.

Warren, B., Ballenger, C., Ogonowski, M., Rosebery, A., & Hudicourt-Barnes, J. (2001). Re-thinking diversity in learning science: The logic of everyday language. *Journal of Research in Science Teaching, 38*(5), 529–552.

Wong-Fillmore, L., & Snow, C. (2002). *What teachers need to know about language.* Washington, DC: Center for Applied Linguistics.

Zimmerman, C. (2000). The development of reasoning skills. *Developmental Review, 20*(1), 99–149.

APPENDIX A
Grade 3 Student Reasoning Elicitation Protocol: Measurement

Opening: Tell the student, "We are going to do a science activity today about measuring things. I'm going to ask you some questions to see what you know about measurement. This isn't a test and I don't expect you to know all the answers. I just want to know how a smart kid like you thinks about these things. Ok? You can ask me any questions about what we are doing whenever you want to. Ok?"

Section 1: Home Context

The purpose of the first section of the interview is to get the student to reflect on ways that he/she has engaged in measurement tasks (or observed others engage in measurement tasks) in the home context.

Initial prompt: We are going to start by talking about times when you measure things at home.

1. Have you ever seen anyone in your family measure how long or how tall something is? What were they measuring? How did they do it? What did they do it with?
   Probe: How tall are you? How do you know?
2. Have you ever seen anyone in your family weigh something to see how heavy it is? What were they weighing? How did they do it? What did they do it with?
   Probe: Who does most of the cooking in your family? Have you ever seen them weigh anything when they are cooking? What did they weigh? How did they weigh it?
   Do you ever go to the food store (market)? Have you ever seen anyone weigh anything there? What did they weigh? How did they weigh it?
3. Have you ever seen anyone in your family find the volume or capacity of something to see how much space it takes up? What were they finding the volume of? How did they do it? What did they do it with?
   Probe: Have you ever needed to take medicine when you were sick? Who gave it to you? How did they know how much to give you? How did they do this?
   Have you ever helped do the laundry? How do you know how much detergent (soap) to put into the washing machine?
4. Have you ever seen anyone in your family take the temperature of anything? What were they taking the temperature of? How did they do it? What did they do it with?

Probe: Have you ever had a fever? Did someone find out what your temperature was? How did they do this?
Have you ever heard people ask about the temperature outside? How do they find out what it is?

Section 2: School Context: Tool Use

Now I'm going to ask you to measure some things for me using some tools. This may be like the kind of activity you are doing in science time here in school.

First, I'd like you to take this plastic cup and fill it about half way up with water from this container.

Is that about halfway full?

1. Now can you tell me the height of the water in the cup? Which tool do you want to use to find out how high the water is in the cup? Ok, go ahead and do it. Can you write your answer in the box for length? (point to place on student sheet)
(*Note:* for all of the measurements, the tools have both metric and traditional systems. Let the student measure in whichever system he/she wishes. If the student asks you which system to use, tell him/her to use the one he/she prefers.)
2. Now I'd like you to weigh the water in the cup. Which tool do you want to use to find out how heavy the water is in the cup? Ok, go ahead and do it. Can you write your answer in the box for weight? (point to place on student sheet)
(*Note:* If student struggles with the units on the scale, prompt with "Do you know how many grams are in a kilogram or how many ounces are in a pound? Can you find 1 kilogram or 1 pound on the scale? What do you think the 200, 400, 600, that go up to 1 kg or the 4, 8, 12 that go up to 1 lb. stand for? Now can you tell me how heavy the water in the cup is?")
3. Now can you tell me the volume or capacity of the water in the cup? (Prompt: Remember that means how much space it takes up.) Which tool do you want to use to find out how much space the water in the cup takes up? Ok, go ahead and do it. Can you write your answer in the box for volume? (point to place on student sheet)
4. Last, I'd like you to tell me the temperature of the water in the cup. Which tool do you want to use to find out how hot or cold the water in the cup is? Ok, go ahead and do it. Can you write your answer in the box for temperature? (point to place on student sheet)

	Length (Height)		Volume (Capacity)	
Cup half-full	centimeter (cm)	inch (in)	milliliter (ml)	ounce (oz)
	Weight		Temperature	
	gram (g)	ounce (oz)	Celsius(°C)	Fahrenheit (°F)

Section 3: School Context: Comparing and Estimating

Now take this second plastic cup just like the first one. This time, I'd like you to fill it completely with water, all the way to the top.

Now I want you to compare the water in the half-filled cup with the water in the full cup.

I have some questions for you to answer. These are sentences that have multiple-choice answers and I want you to tell me which is the best answer. Can you read the first sentence to me?

1. The height of water in the full cup is higher than, lower than, or the same as the height of water in the half-filled cup.

   higher lower same

   Remember when you measured the height of the water in the half cup you said it was [whatever student wrote down]. Now, without measuring, what do you think the height of the water in the full cup is?

   ____

Now can you read me number 2? It's another multiple-choice sentence.

2. The weight of the water in the full cup is heavier than, lighter than, or the same as the weight of water in the half-filled cup.

   heavier lighter same

   Remember when you weighed the water in the half cup you said it was [whatever student wrote down]. Now, without measuring, what do you think the weight of the water in the full cup is?

   ____

Now can you read me number 3.

3. The volume of water in the full cup is larger than, smaller than, or the same as the volume of water in the half-filled cup.

   larger smaller same

   Remember when you measured the volume of the water in the half cup you said it was [whatever student wrote down]. Now, without measuring, what do you think the volume of the water in the full cup is?

   ____

Now can you read me number 4?

4. The temperature of water in the full-cup is warmer than, colder than, or the same as the temperature of water in the half-filled cup.

   warmer colder same

   Remember when you took the temperature of the water in the half cup you said it was [whatever student wrote down]. Now, without measuring, what do you think the temperature of the water in the full cup is?

   ____

Section 4: Connection to Play

So, we talked about measuring and estimating today. We practiced using tools like the ruler and the thermometer. Now, the last thing I want to ask you about is if you ever use measurement when you are playing with your friends or with your brothers or sisters.

1. Can you think of a time playing with your friends that you might need to measure something? Tell me about it.
   Probe: Have you ever wanted to figure out how far something goes? How heavy something is?
2. In the example you were just talking about of [give a student example], do you think it would make a difference if you measured using a tool like the ruler or the scale, or would estimating be good enough? Why do you think so?
3. The tools we used today had two different systems, the metric system (centimeters, grams, milliliters, and Celsius degrees) and the traditional system (inches, ounces, cups, and Fahrenheit degrees). Which of these two systems is easier for you to use? Why do you think so?

APPENDIX B: Grade 4 Student Reasoning Elicitation Protocol: Forms of Energy

Opening: Tell the student, "We are going to do a science activity today about forms of energy and rolling balls. This isn't a test and I don't expect you to know all the answers. I just want to know how a smart kid like you thinks about these things. Ok? You can ask me any questions about what we are doing whenever you want to. Ok?"

Section 1: Home Context

[The purpose of the first section of the interview is to get the student to reflect on ways that he/she has experienced various forms of energy in the home context.]

Initial prompt: We are going to start by talking about things that use energy at home.

1. Can you tell me all the things that use energy in your house?
   Probe: Can you think of other things in your house that use energy? Can you give me any more examples?
2. Sometimes energy is converted from one form to another. Think about a light bulb. What kind of energy does it need to work? What kind of energy is the result?
   Probe: Can you describe what is happening with the light bulb?
3. Think about a flashlight. What kind of energy does it need to work? What kind of energy is the result?
   Probe: Can you describe what is happening with the flashlight?
4. Think about a washing machine. What kind of energy does it need to work? What kind of energy is the result?
   Probe: Can you describe what is happening with the washing machine?

Section 2: School Context: Potential and Kinetic Energy

Tell the student, "Now I'm going to ask you to do a science experiment. This may be like the kind of activity you are doing with your teacher during science here in school. First, I'd like you to look at the inquiry framework in this handout."

Have you used this before in class?
Probe: Tell me about when you used it?
What do you remember about using it?

Today, we are going to use the inquiry framework to think about forms of energy.

Inquiry Framework	
1. Questioning	**This is the question we want to answer today:** *Suppose you roll a ball down a ramp. How does changing the weight of a ball change the amount of energy it has?* (Have the student read the hypotheses. If the student struggles or appears puzzled, the interviewer should re-read the hypotheses to the student.) **Make a hypothesis** The lighter the ball is, the more energy it has. The heavier the ball is, the more energy it has. Changing the weight of the ball does not affect the energy it has. There is not enough information to tell.
2. Planning	**Now we will make a plan:** You have a ramp to roll the balls down and a cup to roll the balls in to. You also have 3 balls of different weights. Remember what you want to find out: *How does changing the weight of the ball change the amount of energy it has?* Tell me your plan for what you will do. *Probe:* Is there something you can measure to help you answer this question? How many times should you roll each ball? If you roll a ball once or 3 times, which will give you a better answer? Why? What if you make a mistake or something goes wrong? Where will you put the ball on the ramp to start it? Where will you place the cup before you roll the ball? (*Ask this question after first trial) Where will you start measuring from? Where will you measure to?

3. Implementing **Gather the materials**

1 ramp
1 meter stick
3 balls (differing weights but all other variables remain the same)
1 box (to set ramp height)
1 cup (cut in half to receive ball)

Remember what you said you would do.
Remember to write down all your measurements in the data table.

	Ball #1 11 grams	Ball #2 22 grams	Ball #3 53 grams
Trial 1			
Trial 2			
Trial 3			

4. Concluding **Draw a conclusion**

What did you find out? Check the most correct statement based on your data.

☐ The lighter the ball was, the farther it pushed the cup.
☐ The heavier the ball was, the farther it pushed the cup.
☐ Changing the weight of the ball did not affect how far the ball pushed the cup.
☐ There is not enough information to tell.

(*Only if student answers "There is not enough information to tell," ask, "What additional information would you need to answer the question?")

What can you conclude?

☐ The lighter the ball was, the more energy it had.
☐ The heavier the ball was, the more energy it had.
☐ Changing the weight of the ball did not affect the energy it had.
☐ There was not enough information to tell.

Compare what you thought would happen with what actually happened. Did the results match your hypothesis?

☐ Yes ☐ No

5. Reporting **Share your results**

What happened when you changed the weight of the ball?

Probe:
How did changing the weight of the ball change the distance the cup moved?

Why do you think this happened?

Probe:
Did the changing distance have anything to do with forms of energy? Why?
Did energy get converted from one form to another?
What kind of energy did the balls have at the top of the ramp?
What kind of energy did the balls have at the bottom of the ramp?

Section 3: Connection to Play

Today we talked about different forms of energy at home. Then we did an experiment about energy and rolling balls. The last thing I want to ask you about is forms of energy when you are playing with your friends.

1. First, tell me all the things you like to play that use energy.
   Probe: Can you think of any more things you play that use energy? Can you think of any more examples?
2. When you slide down a sliding board on a playground, what kind of energy do you need to slide? What kind of energy is the result?
   Probe: Are there similarities between you sliding down the sliding board and the balls rolling down the ramp? Tell me about it.
   Suppose two children are different weights; one is heavy and one is light. When both slide down the sliding board, do they have the same amount of energy? Tell me about it.
   Do they have the same forms of energy? Tell me about it.
3. When you kick a soccer ball, what kind of energy do you need to do it? What kind of energy is the result?
   Probe: Can you describe what is happening when you kick the ball?
4. When you ride a bike, what kind of energy do you need to do it? What kind of energy is the result?
   Probe: Can you describe what is happening when you ride the bike? Do you wear any protection when you bike? Do you think you should? Why?

CHAPTER 3

CRITICAL ISSUES IN TEACHING SCIENCE TO HISPANIC ENGLISH LANGUAGE LEARNERS

An Overview

Robert D. Leier and Laureen A. Fregeau

ABSTRACT

Hispanic English language learners (ELLs) have traditionally underperformed in science education classrooms. This chapter introduces and overviews cultural characteristics, socio-political issues, language acquisition theory, language content issues, science curricular content issues, and pedagogical issues. It presents research on science curricular content in Latin America that K–16 science teachers should consider when teaching science to Hispanic ELLs. Awareness and understanding of these characteristics and issues will allow science teachers to be more effective educators for this growing population of students in the USA.

Teaching Science with Hispanic ELLs in K–16 Classrooms, pages 37–63

INTRODUCTION

The success of Hispanic ELLs in the science classroom is tied to a complex set of issues (Marx, 2008; Gottlieb, Oliver, & Cranley, 2007; Fathman & Crowther, 2006; Calabrese, Goodvin, & Niles, 2005; Zuniga, Olsen, & Winter, 2005; Latino Coalition of Hillsboro County, 2000; Coley, 1999; Cagle, 1997; Peng, Wright, & Hill, 1995). Among the most significant issues are teacher knowledge of, and positive attitude toward, Hispanic cultures and students (Marx, 2008; Rodriquez and Fordham University, 1992); teacher knowledge of student prior science knowledge used to place students in the appropriate level of science classes (Zuniga et al., 2005); teacher employment of culturally responsive pedagogy (Calabrese et al., 2005); and teacher knowledge of English language acquisition and of how to integrate this knowledge into the teaching of science content (Zainuddin, Yahya, Morales-Jones, & Ariza, 2002).

White teachers were found by Marx (2008) to hold negative views of Latina/os, including the belief that Hispanics do not value education. Interacting with this belief was a lack of teacher understanding of Hispanic cultures (Marx, 2008; Rodriquez & Fordham Univ., 1992), resulting in teachers being unable to relate to their Hispanic students. An inability to relate to their Hispanic students was connected by Marx and by Rodriquez & Fordham Univ. to dismal academic success rates for this emerging student population. While the lack of cultural understanding among teachers has been found to negatively affect the academic success of Hispanic students, the Latino Coalition of Hillsboro County (2000) and Cagle (1997) found that positive, close teacher–student relationships were important to Hispanic students' success. Effective teachers of Hispanics, therefore, have been described as culturally responsive, understanding, and employing the cultures of their students in teaching content and strategies (Calabrese et al., 2005).

Negative perceptions and a lack of cultural understanding on the part of teachers are not the only reasons for Hispanic students' lack of success in science classes. Researchers also have found that Hispanic ELLs take fewer science courses and are placed in less demanding science classes than their non-Hispanic peers, and that these placements are not based on student capabilities or knowledge (Zuniga et al., 2005; Coley, 1999; Peng et al., 1995; Klopfenstein, 2004).

CULTURAL CHARACTERISTICS

The information about the cultural characteristics of Hispanic students and socio-political issues pertaining to Hispanic ELLs in this section is intended to assist science teachers toward developing positive attitudes concerning Hispanic cultures and students, as well as to provide a foundation for culturally responsive pedagogy. This information also can facilitate science teachers' ability to relate to the variety of cultures and backgrounds found among their Hispanic ELL students.

People in the United States who originate from Spain and the countries of Hispanic America typically are grouped under the label *Hispanic*. Nationalities include: Argentinean, Bolivian, Chilean, Colombian, Costa Rican, Cuban, Dominican Republican, Ecuadoran, El Salvadoran, Guatemalan, Honduran, Mexican, Nicaraguan, Panamanian, Paraguayan, Peruvian, Uruguayan, Venezuelan, and Puerto Rican (territory of the USA). Under this umbrella also are people from the southwestern United States, Florida, and other parts of the country with historical connections to Spain. Hispanics are a combination of diverse ethnicities, races, languages, and cultural characteristics (Tienda & Mitchell, 2006). This diversity of cultural characteristics found among Hispanic ELLs adds high levels of complexity to teachers' efforts to provide culturally responsive teaching of science to their Hispanic ELLs.

Teachers of Hispanic ELLs must undergo self-examination of their general attitudes and possible stereotypes about Hispanics, understand the specific Hispanic subcultures they are serving, and make appropriate adjustments. This task begins with identifying the Hispanic subcultures among one's students. To assist in this task we provide below an introduction to the diversity that is Hispanic culture. Due to the complexity of cultures representing the Hispanic population in the United States, teachers will need to do further research on the specific Hispanic ELL populations in their respective classrooms.

Students who are identified as Hispanic ELLs may possess one or a combination of Latino and Indigenous cultural characteristics.

Latino

Latinos have a common Spanish cultural heritage and language. Latino immigrants originate from countries, however, that diverged into distinctively different cultures that vary in their unique combination of economic, environmental, and socio-political characteristics. Latino ELLs speak Spanish as their home language. The language variation of Spanish a Latino ELL may speak is characterized by dialects and regionalisms. It should be noted that Latinos inhabited what is now the western and southwestern USA since the late 1500s, though the descendants of these people typically speak English now.

Indigenous

Indigenous people have cultures and languages originating in the Americas prior to the arrival of the Spanish. Their cultures have assimilated differing aspects of Spanish culture while retaining varying degrees of their pre-Spanish culture and language. Indigenous cultures may be interrelated though they are as distinct as any other geographically related group of cultures who are associated together simply due to geography. Many have evolved separately in different regions of Latin America. They generally inhabit the lands their ancestors inhabited before the Spanish arrived, with some living on reservations. Indigenous ELLs speak languages that include: Amerindian, Aymara, Aztec, Carib, Garífuno, Guaraní, Incan, Kuna, Mapuche, Mayan, and Taino. Mexico alone has sixty-two living indigenous languages with over six million speakers. These include Nahuatl (2.50 million speakers), Maya (1.50 million speakers), Diidza/Zapoteca (.80 million speakers), and Savi/Mixteco (.75 million speakers) (Instituto Nacional de Estadística, Geografía e Informática, 2000). Some indigenous people marry Latinos and combine cultural characteristics. Others have adopted Latino culture and language to cope with ethnic cleansing or discrimination against indigenous peoples, while covertly retaining indigenous culture (Stonich, 2001).

SOCIO-POLITICAL ISSUES

The cultural identification issues mentioned above are further complicated by socio-political status. It is critical to recognize the socio-political issues that cause Hispanic ELLs to be categorized into immigrant, migrant, or USA citizenship statuses.

Immigrant

Immigrant Hispanic ELLs are recent arrivals to the USA who may or may not have the intention of becoming permanent residents or citizens. They originate from all the nationalities mentioned above. Since assimilation usually is a priority, second generation immigrants typically become fluent in English and also maintain first language skills, at least oral skills (Suro & Passel, 2003).

Migrant

Migrant ELLs come from families who seek seasonal employment in the USA. They have been migrating between countries or within the borders of the USA for many generations. They may or may not be citizens of the USA.

Many continue using Spanish or an indigenous language as their home language even after many generations have worked and lived in the USA (Farmer & Moon, 2007).

US Citizenship

Hispanic ELLs may have Latino or indigenous cultural backgrounds with ancestors who originated in the southwestern United States, Puerto Rico, or Florida and have been residents or citizens since these areas became part of the USA. These ELLs may retain Spanish as their home language and use Spanish, English, or "Spanglish" (a combination of English and Spanish) for business negotiations (Suro & Passel, 2003).

LANGUAGE ACQUISITION THEORY

Science teachers may question why they should learn about second language acquisition theory rather than be given "recipes" on how to teach ELLs. Understanding language theory allows the teacher to adapt lessons according to "why" they will work and not superficially "how" they will work. Quick-fix recipes are limited, in that they do not work for all cases.

Teachers need not be linguists or learn Spanish to be effective science teachers for Hispanic ELLs, though understanding important features of the Spanish language and learning how language, in general, is structured can be very beneficial in understanding the obstacles ELLs have in acquiring academic content. The field of second language theory is relatively new but has made great strides in understanding principles that guide learning another language (Mantero, 2002). A few of the notable researchers and writers in this field include Stephen Krashen and James Asher. In his work in the early 1980s, Stephen Krashen (1981) provided the discipline with a theoretical insight into how ELLs acquire another language in stages similar to first language acquisition. More in-depth information is available on Krashen's website: http://www.sdkrashen.com. Approximately at the same time, James Asher (1979) was developing teaching approaches for English language instruction and assessment that mimicked first language acquisition development. Asher's Total Physical Response Method is widely used today for ELLs with limited language skills (Asher, 2003). More information is available at http://www.tpr-world.com/Merchant2/merchant.mvc?Screen = CTGY&Category_Code = 200, and http://www.context.org/ICLIB/IC06/Asher.htm.

It is important for science teachers to distinguish the social language skills needed for interpersonal interaction from the academic language skills needed to achieve in school settings. Cummins (1979) termed the former, Basic Interpersonal Communication Skills (BICS) and the latter, Cognitive Academic Language Proficiencies (CALP). Through the seminal research of Jim Cummings (1979), we can now differentiate these different language skills and use them for preparing ELLs to function both within and outside of the classroom. This language acquisition distinction cautions teachers to be wary of assuming that if an ELL speaks English fluently in social situations, then he/she has sufficient English to achieve in the academic science classroom. An ELL with limited oral communication ability, that is, low level listening and/or speaking proficiencies, will not necessarily have a correspondingly low proficiency level in English literacy (Riches & Genesee, 2006). ELLs with formal educations tend to read and write more proficiently than they can communicate orally (August & Shanahan, 2006).

LANGUAGE CONTENT ISSUES

Science teachers need to be aware of the language content issues that may hinder Hispanic ELLs from performing to their potential. Since science teachers teach not only science content to their ELL students but also the use and form of the English language, all science teachers with ELL students are English language teachers, even though many have little or no English language teacher training (Feryok, 2008). It is, therefore, important that science teachers acquire a basic understanding of second language acquisition issues. To acknowledge and act on the necessity to educate science teachers about the needs of ELLs at all levels in developing English language proficiency will contribute to the success of the Hispanic ELL (Feryok, 2008; Dobb, 2004). A couple of the more important language content issues to consider will be discussed in this section: grade appropriate language content standards and maintenance of first language through bilingual materials.

Grade Appropriate Language Content Standards

Science teachers need to know what to expect from Hispanics ELLs in terms of comprehension and production of the English language. Performance indicators exist for the grade-appropriate language content standards at the students' respective language proficiency levels. ELLs who are at the same grade level as their mainstream classmates may understand similar science concepts, but will have different ways of expressing their knowledge, depending on their English language proficiency levels (Echevarria and Short, 2004). Science teachers need to take into consideration the likelihood that ELLs at a specific age or grade level probably will not have the same English language skills as native English speakers. They are likely, however, to have cognitive capabilities for learning science content similar to those of native English speakers (Gottlieb, Oliver, & Cranley, 2007).

The World-class Instructional Design and Assessment Consortium (WIDA, 2007) has developed content specific performance indicators for grade-level clusters (Pre-K–K, 1–2, 3–5, 6–8, 9–12) that provide assistance in determining what ELLs may be able to do in the four language domains of speaking, listening, reading, and writing, according to several language proficiency levels. In grades 6–8, for example, teaching climate and temperature change is a learning objective in science. Beginning ELLs may only be able to point to information on charts or respond in one or two words concerning temperature change. Intermediate ELLs may be able to state or compare differences from charts. More advanced ELLs may be able to summarize, explain, and present information on temperature change from charts.

Maintenance of First Language through Bilingual Materials

It is advantageous for science teachers to provide supporting materials in the ELL's first language when available. These supporting materials should be accessible in both hard copy and electronic forms. Even though the students are learning the content through English, this provides an avenue in which ELLs can check their understanding of what they have been taught. It is more feasible for science teachers to acquire academic materials if the ELL's first language is Spanish, but becomes more difficult if the first language is an indigenous language, since materials in these languages often are not readily available. Such materials may be of limited use to the ELLs who originate from indigenous cultures, however, since these students may have not been taught literacy in their respective first languages. It should be noted that there is a tendency for science teachers to provide Spanish-language academic materials to Hispanic ELLs under the assumption that this will substitute for teaching them science in English. Since science teachers teach both science and the language needed to communicate about science in academic settings, they need to learn strategies to teach English language science to ELLs (Fathman & Crowther, 2006).

SCIENCE CURRICULAR CONTENT ISSUES

Science curricular content issues influence the success of Hispanic ELLs. Grade-appropriate science content standards, academic science language, awareness of Latin American scientific accomplishments, and science content taught in Latin American schools are several of the more important issues to consider when teaching science to Hispanic ELLs.

Grade-Appropriate Science Content Standards

Science teachers need to know how to make the adjustments to science instructional approaches based on the English language proficiency levels of Hispanic ELLs. These students will come to US science classrooms with diverse experiences in science education, depending on the degree of formal science education they received in their home countries. Some Hispanic ELLs will not have been exposed to the grade level concepts to which they are being introduced in the US science classroom. Other Hispanic ELLs will have been exposed to curricula similar to, or more advanced than, grade level science curricula in US school systems. Others will have learned science content that is not traditionally taught in the USA (e.g. prevention of malaria and polio). Science teachers need to be aware of the academic science background each Hispanic ELL brings to the classroom and teach accordingly. Later in this chapter, we provide examples from science curricula of various Latin American countries.

Science Academic Language

Hispanic ELLs need to learn the specific "science language" necessary for participating in science classrooms. Science teachers should identify key science content vocabulary that will be used in instruction and assessment and that may be problematic for Hispanic ELLs. Students may know science content and vocabulary in Spanish, but be unable to explain it to the science teacher in English (Zainuddin, Yahya, Morales-Jones, & Ariza, 2002). Science teachers may address this "language gap" by providing science content in contexts that Hispanic ELLs may already know, making the content comprehensible. The metric system is an example of a context that can be employed in this approach. It is taught for science measurement in the USA and also is the primary measurement system used in Latin American countries. Science concepts involving metric measurement have an element of familiarity for students from Latin America.

Scientific Accomplishments in Latin America

Hispanic ELLs will be more motivated to learn science content if it is made relevant to their cultural community. Hispanic ELLs need to be able to associate successful people from their culture with important scientific accomplishments. Science teachers, therefore, need to become aware of, and be ready to provide examples of, important achievements made by Hispanic scientists during lessons and discussions. ELLs will be able to associate these accomplishments with the subject matter and build on background knowledge. The scientific accomplishments should be both current and historical.

Examples of significant science accomplishments that embrace both indigenous and Latino cultures are:

1. The Mayan culture that comprises parts of southern Mexico and Guatemala contributed several scientific accomplishments to the New World. They constructed hundreds of pyramids and astrological observatories, developed the Mayan calendar to track astrological phenomena, and created a system of hieroglyphic writing to document their scientific findings (Coe, 2002).
2. The Quechua, who have occupied the high plateau country around the Andes Mountains in South America for centuries, have developed over one hundred different varieties of the potato, a vegetable which was unknown in Europe until the arrival of the Spanish in the Americas. The potato is now a basic food around the world. The Quechua also invented their own method of freeze-drying before Spanish/European contact (Landon, 1993).
3. A recent example of a Latino scientific accomplishment is one by Dr. Inés Cifuentes, the 2006 National Hispanic Scientist of the Year. She was the first female graduate of Columbia University's doctoral program in seismology. Dr. Cifuentes has been recognized for improving science programs in public schools and for assisting Central American refugees in the Washington, D.C. area. She currently works at the American Geophysical Union (Hispanic PR Wire, 2007).
4. The history of science and nature should include Latinas and Latinos such as Ellen Ochoa (1958–) the first female Hispanic astronaut, who served on a nine-day mission aboard the shuttle Discovery in 1993. The astronauts were studying the earth's ozone layer. A pioneer of spacecraft technology, she patented an optical system to detect defects in a repeating pattern. At NASA's Ames Research Center, she led a research group working primarily on optical systems for automated space exploration (National Women's History Project, 2008).
5. Mario Molino, born and raised in Mexico City, was awarded the Nobel Prize in chemistry for his research on stratospheric ozone depletion. He became interested in chemistry as a youth, pursuing degrees in chemical engineering and a Ph.D. in Physical Chemistry. Molino became interested in the area of science, technology,

and society (STS). As a post-doctoral researcher, Molino conducted research in chemistry at the University of California-Irvine, choosing to focus on chlorofluorocarbons (CFCs) in the atmosphere leading to the CFC-ozone depletion theory, and then to the recognition of stratospheric ozone depletion as a global concern. Molino moved from Massachusetts Institute of Technology to the University of California-San Diego and then to Mexico City, where he created a new center for strategic studies in energy and environment. He currently is working on air quality and global change issues (Molino, 2006).

PEDAGOGICAL ISSUES

Besides language and science content issues, science teachers need to be aware of the pedagogical issues that may influence the academic performance of Hispanic ELLs. Teaching science can be made more effective by encouraging the teaching of science across the curriculum, implementing teaching approaches and strategies that are culturally appropriate, using authentic assessment approaches and strategies, and involving parents and the community in the educational process. These pedagogical issues will be discussed in the following sections.

Science Across the Curriculum

Science teachers should encourage their non-science colleagues to incorporate grade-related science themes into their language arts, mathematics, and social studies curricula. Such integration is beneficial because it can reinforce what is presented and taught in the science classroom.

An example is provided below using Alabama's Science Standard #7 in Grade 6 to Describe Earth's biomes.

- In language arts, this standard can be reinforced by the language arts teacher having students research and then produce an oral report on the characteristics of different biomes (e.g., rainforests, tundra, grasslands). Bring in schema from immigrant ELL's country (biomes found there).
- In mathematics, this standard can be reinforced by the mathematics teacher having students interpret information from tables to plot line graphs for each land biome's temperature and precipitation throughout a one-year period. (Include biomes from immigrant ELL's country in graph).

- In social studies, this standard can be reinforced by the geography teacher having students identify where the terrestrial biomes exist on each of the different continents (include immigrant ELL's country as an example). Additionally, students select a biome and research the history of how this distinct biome supports human life.

Besides having working relationships with fellow teachers, science teachers should generate an ongoing exchange with the English for Speakers of Other Languages (ESOL) liaison or resource personnel assigned to the school. There should be structured time in which to discuss lingual, cultural, and academic issues concerning instruction and assessment for Hispanic ELLs. Just as the teaching of the English language should be acknowledged across all disciplines, science content should be visible throughout the curriculum, not only in science classrooms (Thorson, 2002). (For additional information see Appendix A: Resources for Implementing Science Across the Curriculum)

Teaching Approaches and Strategies

Science teachers need to understand the importance of hands-on and cooperative approaches for teaching Hispanic ELLs in place of traditional didactic approaches. In hands-on approaches, students perform science as they construct meaning and acquire understanding (Christensen, 1995). Hands-on approaches to science include critical analysis of information. Thinking critically (using a minds-on approach) about science is part of the scientific method. However, for ELLs (and, we believe, most students) minds-on learning without experiential (hands-on) learning is language-rich and decontextualized, and thus can be less comprehensible. ELLs need to use all their senses as a bridge between lower language-level and content learning. The family-oriented culture of Hispanics lends itself to supportive and cooperative approaches to teaching science (Tharp & Gallimore, 1988). Less abstract, low-context, embedded strategies allow ELLs to understand concepts without the confusion of the second language. ELLs can learn new concepts without knowing the specific English language needed for that concept (Padrón, Waxman, & Rivera, 2002). This can be accomplished through the use of visuals, experiments, projects, etcetera.

One teaching strategy effective for ELLs is to incorporate cognates for science vocabulary building and understanding of science concepts. English and Spanish languages have many words that have similar origins and resemble each other in spelling and meaning. Science teachers who can utilize these cognates in instruction assist with the translation of science concepts from one language to the other. Science teachers need not to be fluent in both languages to understand the numerous science cognates in English and Spanish (Kress, 2008). A few examples of English/Spanish cognates in science include: *conservation/conservación*, *ecology/ecología*, and *biology/biología*. A complete list of Spanish/English cognates is available online at http://coe.sdsu.edu/people/jmora/MoraModules/SpEngCognates.htm. A Spanish cognate dictionary can be found at http://www.latinamericalinks.com/spanish_cognates.htm.

Another effective teaching strategy based on the works of Vygotsky (1978) is to build contextual background based on the understanding of the cultural and environmental knowledge of one's Hispanic ELLs. Knowledge of the environment in the region from which the Hispanic ELL originated can be valuable information for the science teacher. Knowing if a Hispanic ELL has an agricultural background, for example, may provide for opportunities to discuss animal husbandry or crop production. Such discussion can lead to investigating various questions and scientific problems in chemistry or biology concerning feed for animals, fertilizer, and crop pesticides. For those students from large urban environments, such as Mexico City or Guatemala City, examples involving overcrowding, environmental pollution, and urban ecological systems are appropriate. These authentic situations provide for high interest in acquiring key concepts in science (Wells, 1999).

Other ways of building contextual background may include teaching about climate and climatic change. Many recently arrived Hispanic ELLs will be familiar with climatic conditions in their respective countries. Since seasons vary in Latin American countries as they do in the USA, examples from South American countries located south of the equator should be included in teaching global patterns and seasons opposite to those found in the northern hemisphere. When teaching about the water cycle, for instance, one can include examples with which Hispanic ELLs may be familiar, such as tropical and subtropical variations by elevations and geographical locations.

Assessment Approaches and Strategies

Assessing Hispanic ELLs in science can be frustrating since traditional paper and pencil tests in English can mask what a student has learned in the science classroom. Students may understand a scientific concept but not understand how to express that knowledge in English (Geisinger & Carlson, 1992). Responses in Spanish are useful only if the teacher is bilingual or has a bilingual aide. Using authentic assessments and appropriate assessment accommodations can be a tool in revealing what Hispanic ELLs have learned (O'Malley & Valdez Pierce, 1996).

Science teachers can incorporate authentic assessment approaches that are less language-oriented and more content-based. Besides checklists, anecdotal records, and project-oriented, process-based, and self and peer assessments, portfolio assessments can provide insights unavailable in other assessment formats (Moya & O'Malley, 1994).

There are numerous ways that science teachers can prepare ELLs to take tests as well as provide appropriate accommodations while they are taking tests (Fregeau & Leier, 2008). In test preparation, Hispanic ELLs may need clarification of the key points that will be assessed and help with scientific vocabulary that may be confusing because of the language. An example is the use of the word *gas* to indicate gasoline that is pumped into vehicles, as opposed to atmospheric gas, such as helium or nitrogen. This difference in the definition of *gas* could cause confusion among Hispanic ELLs unless the explanation of this shortened form of *gasoline* is explained beforehand. While taking tests, ELLs may use appropriate language translation accommodations, such as picture dictionaries or word-to-word translators, and be given extra time if needed for translation. Tests also may be given orally for students with limited literacy experience (Fregeau & Leier, 2008).

Parents, School, and the Community

Hispanic parents generally place high value on education. Science teachers need to be aware of the numerous barriers that may hinder forming partnerships with Hispanic parents. Language is the most obvious barrier. Translators with academic backgrounds in both Spanish and indigenous languages are essential for clear communication. Volunteer indigenous language translators in the community, especially at local churches serving indigenous populations from Latin American countries, may be available. An understanding of the Hispanic family structure and the daily survival needs of the parents, especially those in low-income situations, can provide important information concerning scheduling of parent meetings. As noted earlier, parents can be a great source of information concerning the formal and informal science education their child has received. A more direct line of communication between the science teacher and the parents may reveal the importance the Hispanic family places on education. Finally, Hispanic parents need to understand specific ways in which they can participate in the education of their children. Hispanic parents may feel intimidated in regard to providing any assistance to their child's teacher, since, in their country, the teacher may be the most educated person in the community. Offering assistance or suggestions to the teacher may be considered an insult in some circumstances (Ovando, Combs, & Collier, 2006).

It is critical that science teachers involve the parents of Hispanic ELLs in the education of their children (Garcia, 2002). There are several strategies teachers can use to maximize parent participation and overcome possible barriers. One strategy is to provide science content materials students can use with their parents at home to reinforce scientific concepts, preferably in a bilingual format. Through this strategy, parents learn more about how their children are educated and, at the same time, can participate in the learning/teaching process (Ovando et al., 2006).

Another strategy is to learn how to communicate effectively with parents so they will feel welcomed and not intimidated when associating with the school. Understanding Hispanic proxemics that specifically address personal space and non-verbal communication skills is a start in reaching across cultural barriers. When Anglos communicate with each other, for example, they usually maintain a comfortable spatial distance of 36 to 48 inches, while Hispanics tend to stand closer to each other, at about 18 inches. Drifting far from these culturally acceptable distances can communicate a feeling of alienation and distrust (Burger, 1968). There are other nonverbal communication issues that need to be considered, such as forms of greeting, touching, and gestures, and these may differ between indigenous and Latino cultures (Cruz, 2001; Irujo, 1989; Shuter, 1976).

Even when Hispanic ELLs have received little formal education in their countries, they will have acquired significant science knowledge through informal education provided by parents and extended family members. Science teachers can become aware of informally learned knowledge and of how their Hispanic ELLs learn by incorporating parents into the educational process. Inviting Hispanic parents to school to demonstrate "traditional" approaches to science education can bridge the formal/informal education gap; it also demonstrates respect for Hispanic parents' culture and an acknowledgment of the value of the student's culture, in addition to placing importance on science from Latin American countries. One example of science knowledge learned through informal education is how the Mayans plant squash, beans, and corn together in a mutual relationship.

SCIENCE CURRICULAR CONTENT IN LATIN AMERICA

Science teachers need to know what science- and mathematics-related subjects are taught in the Latin American schools where their Hispanic ELLs began their education. Knowing the grade level at which specific scientific principles are introduced, and the extent to which the material is covered, is helpful in planning appropriate lessons. Some Hispanic ELLs may be able to explain their formal science education background, while others may not have the English language skills to adequately do so. Science content that ELLs have learned in the science classrooms in Latin America may be the foundation of science enrichment activities for the mainstream classroom (Cornell, 1995).

Science content in Latin American classrooms varies by country, but varies more by the social class, ethnicity, and home language of the student (Argueta, 2006). Mayan children in Guatemala often attend schools where the teacher speaks only Spanish and the children primarily speak a Mayan language. Such circumstances make education ineffective for these children, leading to high early dropout rates (Solorzano, 2008). This pattern is repeated in other countries with indigenous populations who speak Spanish as their second language. Some Latin American governments are addressing this issue by training indigenous speakers as teachers and providing bilingual education (Solorzano, 2008). These indigenous ELLs will come with science content knowledge learned through informal education in the areas of agriculture, hydrology, geological phenomena, and weather patterns, rather than in formal classrooms (Nicaragua Ministerio de Educación, Division General de Curricula y Desarrollo Tecnológico, 2008b; El Salvador Ministerio de Education, 2008, Costa Rica Ministerio de Educación Pública, 2008; Vazquez, 2007).

School curricula vary considerably by social class, and this affects the quality and quantity of science education. Prestigious private schools in Latin America may provide high levels of sophisticated science (Academia Maddox, 2008; Instituto Thomas Jefferson, 2008; Fregeau, 2006), while public schools serving low-income populations may teach little science other than that related to agriculture (Fregeau, 2006). Specialized schools exist in some places, such as The Cloud Forest School in Monte Verde, Costa Rica, offering intensive science across the K–12 curriculum (El Centro de Educación Creativa, 2008).

There is a lack of studies comparing science curricula in Latin American countries to those in the USA. We have researched science education and curricula in Latin American countries, therefore, and provide a variety of examples below from Mexico, Guatemala, Costa Rica, and El Salvador. The curricula are examples rather than comparisons among countries or all the possible curricula from any one country. Examples of primary and secondary science curricula from Mexico are presented first, followed by secondary science curricula from Guatemala, and finally, primary science curricula from Costa Rica and El Salvador. Since the majority of immigrant Hispanic ELLs are from Mexico, we present more in-depth information from that system.

The Mexican Education System and Sample Science Curricula

The Mexican Education system is divided into four levels: (1) Pre-escolar–Kindergarten, equivalent to USA Pre-school–Kindergarten; (2) Primaria, equivalent to USA 1st through 6th grade; (3) Secundaria, equivalent to the US 7th through 9th grades; and (4) Preparatoria or Technical (vocational), equivalent to the US high school, 10th through 12th grade (Center for Migrant Education, 2001). The two secondary curricula (vocational and college preparatory) are exclusive of one another. Mexican students may go through a Preparatoria (high school) program that articulates with Mexican post-secondary education or they may go through a technical (vocational) program that does not articulate with Mexican postsecondary education. A student who begins school in technical vocational curriculum may not switch to the high school curriculum (E. Salinas, personal communication, September 16, 2008). Indigenous students may go through a special K–9 program for indigenous and rural people (México Secretaria de Educación Pública, 2008).

A typical curriculum for Primaria, grades 1–6, has four hours of general sciences per week. Students in Secundaria (7th–9th grades) take from eight to nine hours of science classes per week, which include courses in physics, chemistry, and vocational-applied science.

Courses in Preparatoria (high schools) are semester-based and each class meets from four to five hours per week. According to the Center for Migrant Education (2001) there are courses many Mexican high school programs have in common. Science classes include biology, chemistry, ecology and environmental education, physics, integrated physics and chemistry (Center for Migrant Education, 2001). The integrated physics and chemistry courses include motion, waves, energy transformations, properties of matter, changes in matter, and solution chemistry (Texas Education Agency, 2008).

Mexican Preparatoria courses of study for science also vary between the general studies and specialized diplomas. They vary between Mexican states and even schools within a state for the general diploma track.

At Colegio Labastida, in Garza, for example, a student earning a General Diploma would take:

- 10th grade: Chemistry I and II, Biology I and II, Computer Science I and II, and Physics I
- 11th grade: Research Methods I, Physics II, Computer Science III and IV, Ecology and Environmental Education
- 12th grade: Psychology I and II, Sociology, and Research Methods

At Universidad Autónomo de Tamaulipas, Escuelas Preparatorias Dependientes e Incorporadas, for example, a student earning a General Diploma would take:

- 10th grade: Physics I and II, Environmental Problems, Biology I, Computer Science I, Research Seminar

- 11th grade: Physics III, Biology II, Inorganic Chemistry, Psychology, and Organic Chemistry
- 12th grade: data unavailable

Mexico also offers technical vocational secondary schools focusing on applied science and mathematics. These programs also are semester-based, with each class meeting four or five hours per week. A diploma can be earned in two to three years.

Such schools focus on agronomy, food production, technologies, and include high levels of chemistry, math, biology, and physics (depending on the curricular focus) (Fregeau, 2006).

A student earning a diploma in business technology (two year/four semester program) from the El Centro de Bachillerato Tecnológico, for example, would take:

- 10th grade: Chemistry I and II, Drafting, and Biology
- 11th grade: Research Methods I and II, and Physics I and II

In comparison, a student at El Centro de Bachillerato Tecnológico Agropecuario earning a diploma in Technical Administration and Rural Accounting (three-year/six semester program) would take:

- 10th grade: Chemistry I and II, and Biology I
- 11th grade: Physics I and II, and Research Methods II
- 12th grade: Psychology, Livestock Husbandry, and Veterinary Nursing

A good general overview of the Mexican educational system is "Schooling in Mexico: A Brief Guide for USA Educators," an ERIC Digest by H. James McLaughlin (2003).

The Guatemalan Secondary Science Curriculum

The Guatemalan secondary science curriculum varies by region and school, typically including physics, chemistry, environmental science, and biology. Science curricula are based on questions students try to solve about the world around them.

Students study and solve problems through applied scientific methods that help them to be better observers and develop critical thinking skills. Guatemala believes that "hands-on" education changes the mental schemes of students and creates curiosity and healthy skepticism (Solorzano, 2008; Barahona Vela, 2005).

The Guatemalan physics program covers vector mathematics with a focus in the resolution of physical problems involving synergy, dynamics, energy, work, and strength, as well as the study of the electrical properties of matter. The program develops capabilities in analyzing, reasoning, communicating, and interpreting ideas efficiently (Solorzano, 2008; Barahona Vela, 2005).

The chemistry course of study in Guatemala focuses on the study of matter, its properties, and how matter changes. Students analyze diverse theories, such as atomic theory and synergy theory; the organization of information through the study of quantum numbers, the periodic table, the law of gases, etcetera; and conservation of mass, energy laws, and all their applications in the fields of thermodynamics, stoichiometry, and calorimetry. An emphasis is placed on scientific vocabulary necessary to the fields. Students learn to create models that help explain the behavior of matter and environmental problems like acid rain, global warming, and the green house effect (Solorzano, 2008; Barahona Vela, 2005).

The Guatemalan biology course of study includes genetics, evolution (study of the origin of species that explains the diversity, creation, and extinction of the species), ecology, and the human body (the physical, social, and environmental factors that influence health). Students study living and non-living organisms and everything that is related to them, from the study of the cell as a basic unit of life to the study of ecosystems where the biotic and abiotic are in constant struggle for equilibrium (Solorzano, 2008;

Barahona Vela, 2005). Cell biology includes basic metabolic aspects of the cell, photosynthesis, cellular respiration, fermentation, synthesis and duplication, protein synthesis, mitosis, and meiosis. The theory of evolution is analyzed along with the principles that rule genetics. The human body is studied and analyzed in a systematic manner.

The Costa Rican Primary Science Curriculum

The Costa Rican primary grades' natural science curriculum is based on the concept of constructing a well-developed foundation that science teachers can use to effectively build upon (Costa Rica Ministerio de Educación Pública, 2008). Natural sciences include all sciences that deal with phenomena observable in nature, both biological and physical. The first- through third-grade science curriculum is divided into integrated themes, including two titled "knowing my body" and "energy and matter." These themes are taught through experimentation to increase student interest, generate ideas, and make discoveries related to facts of daily reality.

Knowing my Body

This theme concentrates on the knowledge of human anatomy and physiology that allows the student to conceptualize his/her body image and sexual identity. It also focuses on the relationships between major body systems, how to maintain a healthy body and prevent physical and sexual abuse, as well as diseases, addictions, and accidents, and the influences of advances in science and technology on medicine, health, and the environment.

Energy and Matter

This theme focuses on the knowledge of various sources and types of energy and matter, its advantages, and the risks of its uses in the betterment of quality of life. Four basic sub-themes are included: material objects, machines, work, and energy. The theme emphasizes human beings' ability to take advantage of managing and exploiting these sources to improve life. Students learn basic concepts of the compounds and elements that are essential for the organisms and matter that surround them.

The El Salvadoran Primary Science Curriculum

In contrast to the Costa Ricans' integrated thematic approach, El Salvador separates its natural science curriculum into discrete content and skill objectives and lessons. This incorporation can be seen in the first-grade curriculum, which we use here as an example. The first-grade science curriculum is divided into the areas of plant and animal anatomy and physiology; health, diet, and disease prevention; ecology and the environment; physics and chemistry; and geology and astronomy (Library of Congress, 1988).

Plant and Animal Anatomy and Physiology

Students learn external physical characteristics and vital functions of plants and animals, and they compare analogous physical characteristics and vital functions in humans and other living things. They also study human anatomy and physiology, specifically to understand external parts and functions of the human body, the senses, and some basic organ systems, and the symptoms of common illnesses associated with these systems.

Health, Diet, and Disease Prevention

Students learn the characteristics, origins, and types of foods and their nutritional value and identification, as well as the value of cleanliness in food preparation and sale. Students also learn basic first aid for common accidents; the importance of vaccination to reduce the incidence of childhood diseases, such as polio, measles, and whooping cough; and the prevention of some diseases with animal vectors, such as rabies and trypanosomiasis, and illnesses spread by other vectors, such as typhoid.

Ecology and the Environment

Students learn that humans are part of the environment and develop responsibility and respect for the environment. This part of the course of study also helps the student identify components and interrelationships of the environment as an open system; realize the value of natural resources and learn how to protect and conserve the planet; and identify high-risk environmental situations, such as earthquakes and hurricanes, and how to stay safe during such events.

Physics and Chemistry

Students learn basic concepts, including states of matter in the natural environment and how these states are important to living things, as well as forms of energy and their uses in the home and work environment. Students also develop positive attitudes towards energy conservation.

Geology and Astronomy

Students learn the basic systems of the Earth: atmosphere, hydrosphere, geosphere; the effects of the atmosphere on seasons (rainy and dry); and the effects of seasons on plants and animals. Students gain comprehension of basic astronomical components: the sun, the moon, and the stars, as well as the movements of the Earth and its part in the solar system (El Salvador Ministerio de Education, 2008).

CONCLUSIONS

This chapter addressed the cultural characteristics, socio-political issues, language acquisition theory, language content issues, science curricular content issues, pedagogical issues, and Latin American K–16 science curricular content that teachers need to consider when teaching science to Hispanic ELLs. It looked at the definition of a Hispanic ELL by identifying Latino and indigenous characteristics and by describing socio-political issues that Hispanic ELLs face, such as immigrant, migrant, and USA citizenship status. Understanding language acquisition theory, including the difference between social language and academic language, and language content issues, such as grade-appropriate language content standards, is paramount in being able to relate to the academic needs of the ELL. Maintenance of the Spanish or indigenous language through bilingual materials support is important to ELL student success. It should be noted that although race, social class, religion, and gender characteristics play an important role in the education of Hispanic ELLs, and all students, these issues were addressed in a limited context in this chapter.

Successful teaching of science to Hispanic ELLs involves teacher knowledge of the complexity of issues that these students bring to the classroom. Science teachers must employ a blended knowledge of Hispanic culture, student prior science knowledge, and second language acquisition theory to create a successful learning experience in the science classroom. An understanding of the science knowledge Hispanic ELLs bring with them will assist science teachers and schools in correctly placing these students in the appropriate and challenging science classes. This knowledge also can contribute to higher teacher expectations of Hispanic ELLs in science classes. Teacher knowledge of Hispanic culture can assist science teachers in incorporating culturally responsive material and approaches into their classroom as well as in better relating to their Hispanic students. Knowledge of second language acquisition prepares mainstream teachers for adapting their content lessons to reach all proficiency levels of ELLs.

Knowledge of science content issues, such as grade-appropriate science content standards and science academic language, will be valuable for the science teachers of Hispanic ELLs. Familiarity with the scientific accomplishments of Latin Americans and what students in Latin America learn in their science classes will help science teachers build on the schema that Hispanic ELLs bring with them. This also will promote a positive attitude about education, build pride through education, and thus promote academic success.

Science teachers should use pedagogical strategies that promote the teaching of science across the curriculum, not only in science classrooms, and employ teaching approaches and strategies that work for Hispanic ELLs. It is important that science teachers utilize authentic assessments for formative instruction and provide accommodations to ensure the success of their Hispanic ELL students on standardized tests. Finally, science teachers should pay particular attention to how the parents of Hispanic ELLs and their communities can work in conjunction with the schools to assist science teachers in teaching their children.

Effective teachers of Hispanic ELLs will not find a comprehensive, single source of information on cultural or political issues, second language, science content, or pedagogical approaches to address the needs of their students; research will be necessary to gather the information needed to best serve the specific populations in the classroom. The ideas presented in this chapter should be a guide to direct further research for science teachers to facilitate a successful school experience for Hispanic ELLs.

SUGGESTIONS FOR FURTHER RESEARCH

The comparative science curricula research completed for this chapter is limited. We suggest that further research is needed for an in-depth comparison of science curricula from Latin American countries to U.S. science curricula and standards. Useful research would include grade-by-grade content comparisons of science curricula from each Latin American country, with emphasis on countries most widely represented among current immigrant Hispanic ELLs (Mexico, Cuba, Colombia, the Dominican Republic, El Salvador, Guatemala, and Peru) (Migration Policy Institute, 2009; Department of Homeland Security, 2008). This research would aid science teachers to better serve Hispanic ELL students.

REFERENCES

Academia Maddox. (2008). *Currículo secundaria.* Retrieved October 25, 2008 from www.academiamaddox.com/secundaria.htm.

Argueta, B. (2006). *Informe final de consultoría para la coordinación de la elaboración de una propuesta curricular para el bachillerato en ciencias y letras.* UNESCO/Ministerio de Educación.

Asher, J. (1979). The total physical response approach to second language learning. *The Modern Language Journal, 53*(1), 3–17.

Asher, J. (2003). *Learning another language through actions, 6 th ed.* Los Gatos, CA: Sky Oaks Productions, Inc.

August, D., & Shanahan, T. (Eds.). (2006). *Developing literacy in second-language learners: Report of the national literacy panel on language-minority children and youth.* London: Lawrence Erlbaum Associates.

Barahona Vela, L. (2005). *Informe Primer de consultoría para la elaboración del currículo del área de ciencias naturales 1: Biología y medio ambiente del ciclo básico del nivel medio.* UNESCO/Ministerio de Educación.

Burger, H. (1968). *Ethno-Pedagogy: Cross cultural teaching techniques.* Albuquerque: Educational Cooperative.

Cagle, J. (April, 1998). *One teacher's perspective on the difference of academic expectations for Hispanic students: A case study.* Paper presented at the Annual Meeting of the American Educational Research Association San Diego, CA.

Calabrese, R., Goodvin, S. & Niles, R. (2005). Identifying the attitudes and traits of teachers with an at-risk student population in a multicultural urban high school. *International Journal of Educational Management, 19*(5), 437–449.

Center for Migrant Education. (2001). *Suggested guidelines for grade-level placement of migrant students for Mexico attending Texas schools for the first time.* San Marcos, Texas: Southwest Texas University.

Coe, M. D. (2002). *The Maya, 6 th ed.* Thames & Hudson.

Coley, R.J. (1999). *Opportunity offered—opportunity taken: Course-taking in American high schools.* Princeton, NJ: Educational Testing Service.

Cornell, C. (1995). Reducing failure of LEP students in the mainstream classroom and why it is important. *The Journal of Educational Issues of Language Minority Students, 15*, 123–146.

Costa Rica Ministerio de Educación Pública. (2008). Unidades de estudio por nivel. Retrieved October 25, 2008 from www.mep.go.cr/curricular/index.aspx.

Christensen, M. (1995). Definitions of hands-on and minds-on. North Central Regional Educational Laboratory. Retrieved October 20, 2008 from http://www.ncrel.org/sdrs/areas/issues/content/cntareas/science/sc500.htm.

Cruz, W. (2001). Differences in nonverbal communication styles between cultures: The Latino-Anglo perspectives. *Leadership and Management in Engineering, 1*(4), 51–53.

Cummins, J. (1979). *Cognitive/academic language proficiency, linguistic interdependence, the optimum age question and some other matters. Working Papers on Bilingualism, 19*, 121–129.

Deep Culture in *Teaching from a Hispanic perspective, A handbook for non-Hispanic adult educators.* Retrieved January 5, 2009 from http://www.literacynet.org/lp/hperspectives/deepcult.html.

Department of Homeland Security. (2008). Annual Flow Report. Retrieved October 20, 2008 from http://www.dhs.gov/ximgtn/statistics/publications/yearbook.shtm.

Dobb, F. (2004). *Essential elements of effective science instruction for English learners, 2nd ed.* California Science Project: Los Angeles, CA.

El Centro de Educación Creativa (The Cloud Forest School), Monte Verde, Costa Rica (2008). Retrieved October 20, 2008 from http://www.cloudforestschool.org/.

El Salvador Ministerio de Educación. (2008). *Programas de estudios.* Retrieved October 20, 2008 from www.mined.gob.sv/docentes/programas-estudios.asp.

Echevarria, J., Vogt, M. & Short, D. (2004). *Making content comprehensible for English learners: The SIOP model, 2nd ed.* Boston: Pearson, Allyn and Bacon.

Farmer, F. & Moon, Z. (2007, August 2). *Human capital endowment of Hispanic migrants: Rural and urban differences.* Paper presented at the annual meeting of the Rural Sociological Society, Marriott Santa Clara, Santa Clara, California. Retrieved October 20, 2008 from http://www.allacademic.com/meta/p187556_index.html.

Fathman, A.K. & Crowther, D.T. (Eds.). (2006). *Science for English language learners: K–12 classroom strategies.* National Science Teachers Association.

Feryok, A. (2008). The impact of TESOL on math and science teachers. *ELT Journal, 62*(2), 123–130.

Fregeau, L. & Leier, R.D. (2008). Assessing ELLs in ESL or mainstream classrooms: Quick fixes for busy teachers. *The Internet TESL Journal, XIV*(2). Retrieved October 20, 2008 from http://iteslj.org/Techniques/Fregeau-assessingELLs.html.

Fregeau, L. (2006, August 7–18). Onsite observation notes from Instituto Thomas Jefferson, Mexico City and Queretaro.

Garcia, E. (2002). *Student cultural diversity: Understanding and meeting the challenge.* Boston: Houghton Mifflin.

Geisinger, K. F. & Carlson, J. F. (1992). *Assessing language minority students*. ERIC Digest, No. ED356232.

Gottlieb, M., Oliver. A. & Cranley, M. (2007). *Understanding the WIDA English language proficiency standards, a resource guide*. Board of Regents of the University of Wisconsin System.

Hispanic PR Wire. (2007). Retrieved October 25, 2008 from http://www.hispanicprwire.com/news.php?l = in&id = 9720&cha = 9.

Instituto Nacional de Estadística, Geografía e Informática. (2000). *La diversidad cultural en México, Los pueblos indígenas y sus 62 idiomas*. Retrieved October 20, 2008 from http://www.inegi.gob.mx.

Instituto Thomas Jefferson. (2007). *Secundaria, preparatoria*. Retrieved October 20, 2008 from http://www.itj.edu.mx/.

Irujo, S. (1989). Do you know why they all talk at once? Thoughts on cultural differences between Hispanics and Anglos. *Equity and Choice, 5*(3), 14–18.

Klopfenstein, K. (2004). Advanced placement: Do minorities have equal opportunity? *Economics of Education Review, 23*(2), 115–131.

Kress, J. E. (2008). *The ESL/ELL teacher's book of lists (2nd ed.)*. 22267-6. (2008), Jossey-Bass.

Landon, C. (1993). *American Indian contributions to science and technology*. American Indian Baseline Essays. Portland Public Schools. Retrieved October 20, 2008 from http://www.pps.k12.or.us/depts-c/mc-me/be-ai-sc.pdf

Latino Coalition of Hillsboro County. (2000). *They are our kids: Findings from the Latino dropout study*. University of South Florida, Tampa Louis de la Parte Florida Mental Health Inst. ERIC Digest, No. ED461464.

Leontiev, A.N. (1977). *Activity and consciousness*. Moscow: Progress Publishers.

Library of Congress. (1988). *El Salvador country study*. Retrieved March 12, 2009 from http//countrystudies.us/el-salvador/31.htm.

Mantero, M. (2002). *The reasons we speak: Cognition and discourse in the second language classroom*. Westport, CT: Bergin and Garvey.

Marx, S. (2008). Popular white teachers of Latina/o kids: The strengths of personal and limitations of whiteness *Urban Education, 43*(1), 29–67.

Migration Policy Institute. (2009). Country and comparative data. Retrieved October 20, 2008 from http://www.migrationinformation.org/datahub/country data/data.cfm

McLaughlin, H. J. (2003). *Schooling in Mexico: A brief guide for U.S. educators*. ERIC Digest, No. ED470948.

México Secretaria de Educación Pública. (2008). *Programa binacional de educación emigrante*. Retrieved October 20, 2008 from www.sep.gob.mx/wb/sep1/sep1_Programa_Binacional_de_Educación_Migrante.

Molino, M. (2006). Autobiography. The Nobel Foundation. Retrieved March 25, 2009 from http://nobelprize.org/nobel_prizes/chemistry/laureates/1995/molina-autobio.html.

Moya, S.S. & O'Malley, J. M. (1994). A portfolio assessment model for ESL. *Educational Issues of Language Minority Students, 13*, 13–36.

National Women's History Project. (2008). Retrieved March 29, 2009 from http://www.nwhp.org/resourcecenter/honoredlatinas.php.

Nicaragua Ministerio de Educación, División General de Curricula y Desarrollo Tecnológico. (2008). *Matriz de contenidos educación secundaria.* Retrieved October 25, 2008 from www.mined.gob.ni.

O'Malley, J.M. & Valdez Pierce, L. (1996). *Authentic assessment for English language learners.* New York: Addison Wesley.

Ovando, C., Combs, C. C. & Collier, V. P. (2006). *Bilingual and ESL classroom: Teaching in multicultural contexts, 4 th ed.* Boston: McGraw-Hill.

Padrón, Y., Waxman, H.C. & Rivera, H.H. (2002, August). *Educating Hispanics students: Effective instructional practices.* Center for Research on Education, Diversity & Excellence, Practitioner Brief #5.

Peng, S., Wright, D. & Hill, S.T. (1995). *Understanding racial–ethnic differences in secondary school science and mathematics.* Washington, DC: National Center for Educational Statistics.

Riches, C., & Genesee, F. (2006). Literacy: Crosslinguistic & crossmodal issues. In F. Genesee, K. Lindholm-Leary, W. Saunders, & D. Christian (Eds.), *Educating English language learners: A synthesis of research evidence.* New York: Cambridge University Press.

Rodriguez, C.E., & Fordham Univ., B. (1992, April 1). Student voices: High school students' perspectives on the Latino dropout problem. Interim Report: Volume II. Student Research Project. ERIC Digest, No. ED359286.

Secretaria de Educación Pública. (2007). *Calendario Escolar.* Retrieved October 20, 2008 from www.sep.gob.mx.

Shuter, R. (1976). Nonverbal communication: Proxemics and tactility in Latin America. *Journal of Communication, 26*(3), 46–52.

Solorzano, L. (2008, February 12). *Education and teacher preparation in Guatemala.* Invited presentation to Seminar in Education Policy Studies.

Stonich, S.C. (2001). *Endangered peoples of Latin America: Struggles to survive and thrive.* Santa Barbara, CA: Greenwood Publishing Group.

Suro, R. & Passel, J.S. (2003). *The rise of the second generation: Changing patterns in Hispanic population growth.* Pew Hispanic Center.

Texas Education Agency. (2008). *Science Curriculum: Integrated Physics and Chemistry FAQS.* Retrieved October 20, 2008 from http://www.tea.state.tx.us/curriculum/science/ipcfaq.html.

Tharp, R.G. & Gallimore, R. (1988). *Rousing minds to life: Teaching, learning, and schooling in social context.* Cambridge: Cambridge University Press.

Thorson, A. (Ed.). (2002). Mathematics and science across the curriculum. *ENC Focus, 9*(2), 20.

Tienda, M. & Mitchell, F. (Eds.). (2006). *Hispanics and the future of America.* The National Academy Press.

Vazquez, A. (November 29–December 1, 2007). *Demystifying the Mexican education system.* Presented at Southeastern Teachers of English To Speakers of Other Languages Regional Conference, Louisville, KY.

Vygotsky, L.S. (1978). *Mind in society.* Cambridge, MA: Harvard University Press.

Wells, G. (1999). *Dialogic inquiry: Toward a sociocultural practice and theory of education.* Cambridge: Cambridge University Press.

World-class Instructional Design and Assessment Consortium (WIDA). (2007). *English language proficiency standards for English language learners in grade 6 through grade 12.* Board of Regents of the University of Wisconsin System. Retrieved March 20, 2008 from http://www.wida.us/standards/6-12%20Standards%20web.pdf.

Zainuddin, H., Yahya, N., Morales-Jones, C. & Ariza, E. (2002). *Fundamentals of teaching English to speakers of other languages in K–12 mainstream classrooms.* Dubuque: Kendall/Hunt Publishing.

Zuniga, K., Olson, J., & Winter, M. (2005). *Science* education for rural Latino/a students: Course placement and success. *Science Journal of Research in Science Teaching, 42*(4), 376–402.

APPENDIX A: Resources for Implementing Science Across the Curriculum

1. The National Center for Quality Afterschool of The Southwest Educational Development Laboratory offers an After School Training Toolkit on Integrating Science Across the Curriculum. The online kit includes an explanation integrating science across the curriculum, a section on planning science across the curriculum lessons, two sample lessons, links to nine resource websites with other sample lessons, and a list of three print resources. The kit is available online at: http://www.sedl.org/afterschool/toolkits/science/pr_integrating.html.
2. Another toolkit from the National Partnership for Quality Afterschool Learning under Integrating Science Across the Curriculum is the "Exploring Earthquakes: Earth Foldable." The illustrated lesson integrates vocabulary, art, geography and science. It is available online at: http://geography.uoregon.edu/edge/projects/Wakefield/wakefield_The_Earth_foldable.pdf
   Other toolkits are available at: www.sedl.org/afterschool/toolkits
3. A search for *science across the curriculum* on the National Science Foundation website yields resources such as *Beyond Penguins and Polar Bears,* a magazine produced by Ohio State University and the National Science Digital Library that integrates science and literacy (and social science). Available at: http://beyondpenguins.nsdl.org/.
4. Volume 9, number 2 of *ENC Focus* is titled "Mathematics and Science across the Curriculum." The issue is available as an Eric Document online and includes essays and a guide to instructional materials for integrating science across the curriculum in music, language arts, and art at the elementary and secondary levels.
   Thornson, A. (Ed.) (2002). ENC Focus, v9 n2 Eisenhower National Clearinghouse for Mathematics and Science Education, 1929 Kenny Road, Columbus, OH 43210-1079. Web site: http://www.enc.org. ED463169
   Available online at: http://eric.ed.gov/ERICDocs/data/ericdocs2sql/content_storage_01/0000019b/80/19/eb/c8.pdf.
5. The Teacher Resources for Instructional Planning (TRIP) website offers ESL suggestions for science lessons by grade level and standard (for Alabama). Available online at: http://www.tripforteachers.org/.

CHAPTER 4

PROMOTING SCIENCE UNDERSTANDING AND FLUENCY AMONG HISPANIC ENGLISH LANGUAGE LEARNERS

Strategies, Explorations, and New Directions

Ann M. L. Cavallo and Patricia A. Gomez

ABSTRACT

This chapter focuses on how education may be structured to meet the linguistic needs of Hispanic English language learners (ELLs) in learning science. The chapter addresses teacher change necessary for our schools to meet the challenges of a diverse and global society, and of preparing all students for the science, technology, engineering, and mathematics (STEM) careers our society so greatly needs. This chapter then presents exploratory research on fourth-grade Hispanic ELL and English-speaking students' understandings

Teaching Science with Hispanic ELLs in K–16 Classrooms, pages 65–99

of the nature of science (NOS), self-efficacy in learning science, enjoyment of science, and science learning preferences in terms of teachers' use of lecture, textbook reading, Internet, science picture books, and laboratory experience. Results revealed Hispanic ELLs held a more authoritative view of NOS compared to English-speaking students, and had lower self-efficacy toward learning science outside of the classroom setting. Differential patterns were observed in science learning preferences between Hispanic ELL and English-speaking students, particularly on use of the Internet and, marginally, reading the science textbook, though some preferences between the groups were similar. The chapter provides insights on effective practices in science teaching and learning of Hispanic ELL students.

INTRODUCTION

Upon entering U.S. schools, linguistically diverse students encounter unique challenges largely eclipsed by the need to learn the English language simultaneously with academic content. Learning science is especially demanding because the discipline itself is its own subculture within the larger cultural context, replete with unique scientific language and discourse, ways of thinking and processing knowledge, and experiences (Luria, 1976; Vygotsky, 1978). Consider this excerpt of an abstract published in professional science journal:

> Mutants of *Escherichia coli* defective in phosphatidylserine synthase (encoded bypss) and phosphatidylserine decarboxylase (encoded by psd) make cell membranes deficient in phosphatidylethanolamine. In this report we show that wild-type pss and psd genes are required for motility and chemotaxis. Null mutants or strains with temperature-sensitive pss or psd mutations grown at high temperature (35°C) were nonmotile. They lacked flagella and showed reduced rates of transcription of the flhD master operon (encoding FlhD and FlhC), the fli4 operon (encoding OF), and the fliC operon (encoding flagellin). (Shi, Bogdanov, Dowhan, & Zusmani, 1993, p. 7771)

Even the most articulate English-speaking students would have difficulty interpreting and becoming fluent with the scientific language represented in the passage. The language of the passage demonstrates that science is a distinct community or subculture with its own language and form of discourse. As educators, we must consider the added complexity the scientific subculture brings to Hispanic ELLs' meaningful understanding of science in an English-speaking society such as the USA.

The culture and NOS are endemically misunderstood by all students worldwide (Flick & Lederman, 2006). Such misunderstanding may be more pronounced among young ELLs who are, at the same time, learning a language and acclimating to the larger school and community cultures.

The expectation is that teachers will understand how to lead Hispanic ELLs through equitable and effective instructional practice, and at the same time promote their intellectual and emotional development, foster their understanding of newly learned scientific concepts, and help them find their "fit" in the larger school and society arenas.

We are now bridging into the middle of the 21st century and schools are faced with the critical call to focus on understanding how to meet the needs of a linguistically diverse population. What in the larger U.S. population is considered the minority population is, or is increasingly becoming, the majority population of students in many classrooms. From 2004–2005, the number of students identified with limited English proficiency in public school (K–12) grew 138 percent, while total enrollment increased only by 21 percent (National Clearinghouse for English Language Acquisition, 2006). These data mean that in many U.S. schools, the proportion of ELLs represents the majority of students entering as kindergartners and as new students. According to the report *Latinos in Education* (White House Initiative on Educational Excellence for Hispanic Americans, 1999), Hispanic students make up 75 percent of all students in ESL, bilingual, and other English language support programs. These changes in classroom demographics raise the call for teachers, teacher educators, and school administrators to change practices in ways that accommodate the learning needs of our growing Hispanic ELL population.

The purposes of bilingual education and English as a second language programs are to facilitate fluency and literacy in English. Second language learners produce English with a range of fluency that often deviates from Standard English in terms of accent, vocabulary, expression, and grammar. Facilitating such fluency, however, may be lost for this diverse population of students as educators succumb to pressures of teaching toward standardized testing at the expense of helping promote deeper-level understandings. Though Hispanic ELLs may be able to use the language of science, the extent to which these students understand the underlying concepts and processes represented by the language is of concern. Educators need to consider Hispanic ELLs' understanding of the dynamic and changing nature of science as these students grapple with learning English along with the larger culture of which they are now a part.

Teaching classrooms comprised of both Hispanic ELLs and English-speaking children involves rethinking views and changing traditional, monocultural ways we may practice teaching and learning. Such change engenders questions that must be addressed: "What aspects of science teaching and learning need to be adjusted or changed?"; "How can teachers make these changes in order to meet the challenges of helping increase Hispanic ELLs' scientific understandings and achievement and, ultimately, their participation in future science learning and careers?"; "How might teachers adjust their practices from what is currently in place to promote the learning of all students, inclusive of Hispanic ELLs?" In response to these questions, educators need to adopt the following strategies:

1. Understand and apply basic knowledge and understanding of second language learning theory and practice
2. Use pedagogical models and techniques that will help develop science content knowledge and processes in teaching Hispanic ELLs
3. Advocate language teaching in content area instruction for ELLs and long-term positive change
4. Implement programs, including science instruction, in dual language English/Spanish instruction

A disconnect exists currently between science teaching practices and the unique learning needs of Hispanic ELLs. This chapter focuses on building meaningful and resilient connections between science teaching and learning among Hispanic ELLs. With this focus in mind, we first discuss the aforementioned strategies for promoting needed change in science education. In an effort to better understand Hispanic ELLs' perceptions of science, we then explore fourth-grade students' views of NOS, self-efficacy to successfully learn science, and their enjoyment and preferences for learning science. We conclude by posing new directions for science education for teaching Hispanic ELLs in our schools.

PROMOTING SCIENCE UNDERSTANDING AND FLUENCY AMONG HISPANIC ELLS: STRATEGIES

Drawing from both theory and practice, four strategic areas emerge that are crucial for teaching and learning science in classrooms with Hispanic ELLs (Chamot & O'Malley, 1994; Freeman, Freeman, & Mercuri, 2005; Lee, Luykx, Buxton, & Shaver, 2007; Rodriguez & Kitchen, 2005). These areas are (1) understanding and applying ELL theory and practice, (2) using effective pedagogical models and techniques in teaching Hispanic ELLs, (3) becoming an advocate of change in current school and classroom structure in teaching ELLs, and (4) implementing effective programs such as dual language learning programs.

Understanding and Applying Hispanic ELL Theory and Practice

A two-year study by Lee et al. (2007) reported that most elementary teachers are not prepared to effectively teach science and are "lacking both science content knowledge and familiarity with inquiry-based science instruction" (p. 1269). The authors further contended that teachers are unaware of the need to teach toward cultural diversity and tend to overlook differences among children in their classes. Many believe that ELLs must learn English before they are able to learn subjects such as science (Garcia, 1999). Doing so, however, delays these second language learner's accomplishments in learning the subject matter, placing them even further behind English-speaking students. Teachers must recognize and work to incorporate students' home languages and cultures in instruction and articulate science with attention to student diversity (Lee et al., 2007). One of the first steps in this process is to understand language theory and practice.

Second language theory and practice play significant roles in how language professionals think about teaching content subjects to second language learners. The following five key hypotheses about second language acquisition and theory were identified by Krashen (1981),

Hypothesis One: Acquisition-Learning Distinction

This hypothesis contends that acquiring a language is different than learning a language (Krashen, 1981). Language acquirers are not consciously aware of the grammatical rules of the language, but rather develop a *feel* for correctness. Language learning, on the other hand, refers to the conscious knowledge of a second language, knowing the rules, being aware of them, and being able to apply them.

Hypothesis Two: Natural Order Hypothesis

This hypothesis implies the acquisition of grammatical structures that proceed in a predictable order. For a given language, some grammatical structures tend to be acquired. Language learning, however, necessitates conscious knowledge of a second language, knowing the rules of that language, being aware of them while speaking the language, and being able to talk about them, regardless of the first language of a speaker.

Hypothesis Three: Monitor Hypothesis

This hypothesis proposes that the language one has subconsciously acquired initiates one's utterances in a second language and is responsible for one's fluency. On the other hand, the language that we have consciously learned acts as an editor in situations where the learner has enough time to edit, is focused on form, and knows the rules, such as on a grammar test in a language classroom or when carefully writing a composition.

Hypothesis Four: Input Hypothesis

This hypothesis contends that whenever language acquirers try to produce language beyond what they have acquired, they tend to use the rules they have already acquired from their first language, thus allowing them to communicate but not really make progress in the second language.

Hypothesis Five: Affective Filter Hypothesis

This hypothesis states that motivation, self-confidence, and anxiety all affect language acquisition, which can impede or facilitate learning the second language (Krashen, 1981).

The challenge for educators lies with uncertainty on how to put these theoretical aspects of second language learning into practice. When teachers know how language is both learned and acquired, learn the rules of grammar to form meaning of text, and understand how to use the language in context to produce meaning, they may then be able to effectively facilitate comprehensible discourse in the classroom (Long, 1996; Mackey, 1999). Science teachers, then, must view themselves as language teachers as well as content area teachers, and would be well served to understand the effects specific strategies have on acquiring and learning a second language. In teaching ELLs, for example, learning a second language is facilitated when learners feel comfortable and thus have low anxiety about being "correct" in using the language. High stress and discomfort, in contrast, may develop among Hispanic ELLs in classrooms where they are expected to accurately use language, scientific or otherwise, and to do so out of context with actual experience and events. Learning new information using a non-threatening approach in which students can connect the language with experiences equates to better comprehension and use of the language in learning.

When students attempt to learn a second language, there is interference from the first language, leading students to overcorrect a word or set of words while trying to produce meaning. Classroom learning should allow discourse to take place, within whole class and small group settings, and with direct experiences, where ELLs will be able to hear the new language, connect it with objects and events, and form meanings. Hispanic ELLs need to be able to use their native language and practice their second language in an encouraging, secure environment as they learn relevant science concepts.

Pedagogical Models and Techniques in Learning Science among Hispanic English Language Learners

Pedagogical strategies for Hispanic ELLs include those that promote open discussion and discourse in which students engage in expressions of science ideas and vocabulary in conversational and written forms. The exchange of discourse should be done in small group conversations and as students are experiencing a common scientific exploration or activity. While students engage in such discourse, they also are experiencing the tentative, socio-cultural nature of science as they argue and negotiate meanings of scientific ideas in a laboratory setting. The exchange of dialogue, in verbal and written forms, supports the thinking and meanings that may be drawn from the scientific experiences. Inquiry teaching models, therefore, best support students' development of language, meanings, and understandings of not only science concepts, but also the nature of the discipline (Marek & Cavallo, 1997). It is posed that the use of inquiry discovery models in the classroom would also facilitate opportunities for ELLs to negotiate meanings about what they are learning with native English speakers.

Inquiry science teaching and learning using models such as the learning cycle is, by design, consistent with the nature of science, promotes scientific ways of thinking, and matches how children learn (Lawson, Abraham, & Renner, 1989; Marek & Cavallo, 1997). The model also promotes engaging the students in discourse, editing and adjusting words and language, using multiple forms of communicating science, and constructing meanings from experience. As an exemplar, the learning cycle discussed here is based upon the traditional three phases: exploration, concept invention (term introduction), and concept application (Lawson, 1995; Lawson, Abraham, & Renner, 1989; Marek & Cavallo, 1997). In exploration, students work in collaborative groups engaging in investigations, making observations, gathering data, discussing ideas, and attempting to make sense of observations. Students are not a priori given explanations of what is to occur or what will be the result of their exploration, which may be unknown in certain experiments (Cavallo & Laubach, 2001). Once observations are made and data has been collected by students, the teacher guides them through concept invention with a series of questions, and they display and share findings gathered with the class. Here, again, students discuss ideas, negotiate meanings, and engage in scientific logic and reasoning. Using the displayed data, student group members work together to construct a statement of the concept or main idea *in their own words*. After these statements are written, verbalized, displayed, and discussed, the teacher attaches the language of science, namely, the scientific terms or labels, to the concept (Marek & Cavallo, 1997). In the learning cycle's application phase, the teacher guides students to extend and expand conceptual understandings by using and describing the concept in various contexts. Through application activities, which may include reading, engaging in laboratory experiences, and using computers, writing, or other techniques, students use the concept in ways that connect it with prior knowledge and everyday experiences.

Several aspects of the learning cycle and like models are important to Hispanic ELLs in learning science. In the model, students are immersed in the culture and language of science as they interact with the materials of science in their investigations. Students have the experiences first (exploration phase), prior to introduction of scientific language. The exploration provides a concrete experience to which it would be valuable for Hispanic ELLs to attach scientific language. Students engage in the discourse of science and have opportunities to adjust and modify their words to make meaning. Students' ideas and concepts are written and stated in their own words (concept invention) prior to the introduction of scientific terms. Students work in collaborative groups where they discuss ideas, argue, modify, and formulate their concept statement. These processes help students form meanings from experiences, and do so with teacher guidance. Meanings may be expressed by all students in Spanish, as well as in English. Placing new ideas in multiple contexts, as done in concept applications, allows students to connect the newly learned science concept to their experiences and to practice the associated science vocabulary. The Hispanic/Latino culture has been described by Tyler et al. (2008) and Greenfield, Quiroz, and Raeff (2000) as ascribing to a collectivist ideal, where the members are interdependent and share responsibilities for the good of the family or society. The collaborative nature of constructivist or inquiry models, exemplified here by the learning cycle, is familiar to Hispanic ELLs, and is thus supportive their science learning.

Acquiring new knowledge and learning new skills in ways that enable them to be linked with students' prior knowledge and experience promotes meaningful learning for all students. Meaningful learning was described by Ausubel (1963) and Novak (1988) as the learner actively connecting information and ideas and concepts in ways that make sense in their minds. Such connections are best made when learners have relevant prior knowledge or experiences that allow for the connections of new knowledge to be made.

In *The Cognitive Academic Language Learning Approach Handbook* (CALLA), Chamot and O'Malley (1994) describe a design to meet the academic needs of students learning English as a second language in American schools that supports using a theme-based approach to teaching science. Such a theme-based approach facilitates selecting science content that is relevant to students' lives. As with the learning cycle model, theme-based instruction provides the context and concrete experiences that serve as an anchor for new knowledge. The theme-based approach means content and language being learned is integrated through all subjects in the curriculum, rather than compartmentalized within subjects. Theme-based instruction also facilitates drawing upon the diverse cultural experiences of students in the classroom. Relevant science concepts are learned within the context of culture and are connected with social studies, language arts, and mathematics. This approach is of particular importance for schools participating in dual language instructional programs, where the instruction is presented in two languages.

Science language can be made more understandable by recognizing that the terms frequently have a root word that stems from various languages, as well as the most common language used in science, Latin. There is value in having all students learn scientific terms by learning their root words and realizing that some of the language of science may be drawn from different cultures. Knowing the root words and meanings further helps students initially learn and retain the words and their meanings in memory. To promote meaning and interrelationship among concepts, the newly acquired terms that label these science concepts may be connected using tools such as semantic mapping or concept mapping (Novak & Gowin, 1984).

The language of science is a community-based way of speaking, meaning that it is unique to that community and occurs primarily, if not only, in a scientific context (Magnusson, Palinscar, & Templin, 2006). According to Werstch (1991) and Magnusson et al. (2006), it is the community that actively promotes specific ways of speaking or dialogue within that group of people. The form of dialogue used in science groups is clearly not the same used when one is casually speaking with friends or family. This form of dialogue may be awkward and unfamiliar to many. Classrooms that promote discourse and group collaborative leaning activities allow children to practice the language of science as well as the second language, and in doing so, dialogue with peers becomes a powerful language learning tool. The use of collaborative groups in science consisting of mixed ELL and English-speaking students enriches learning as students engage in discourse where they are paraphrasing, providing definitions, using drawings and pictures, and modeling concepts. The consistent engagement of Hispanic ELLs in dialogue that is characteristic of the scientific community will help them gain fluency and membership within this subculture.

While Hispanic ELLs experience unique language challenges, all students share a native or intuitive understanding of scientific phenomena. Students' natural curiosity about the world around them can serve as a springboard to scientific investigation designed to answer scientific questions (Chamot & O'Malley, 1994). The language of science can be naturally embedded to teach language skills as the teacher focuses on making connections between the English language and the students' first languages in the process of teaching science concepts. Students learn new concepts by building on prior knowledge, creating learning that "spirals" by connecting one concept to the next in continuous, overlapping cycles. Taking the students from the known to the unknown through interconnecting, spiraling learning experiences helps them build upon their understanding of each concept learned. Strategies that make new learning learnable help students build confidence in themselves as learners (Reiss, 2008).

Prospective teachers may choose not to teach for diversity and understanding for a variety of reasons, but among these may be a lack confidence and/or skills (Rodriguez & Kitchen, 2005). Teacher education programs typically have prepared teachers to teach to a single culture of students, even when ELL and English-speaking students are found often in the same classrooms. Increasingly, teacher education programs are providing an option for teachers in which they prepare for classrooms with students who are second language learners. Many classrooms today—not only bilingual classrooms, but also regular education classrooms—

have second language learners. Content-based English as a Second Language (ESL) instruction has not been sufficient to help all ELLs succeed academically. It is now generally understood that students can benefit from teachers adjusting pedagogical practice in content area instruction and in ways specific to the content (Echavarria, Vogt, & Short, 2008). The approach known as "sheltered instruction" can extend the time students have available to learn a second language while learning content specific skills. In sheltered instruction, students develop knowledge in specific subject areas through the medium of their second language. Teachers modify their use of English to teach core subjects in order to ensure that the material is comprehensible to learners and that it promotes their second language development. They adjust the language demands of the lesson in many ways, including by modifying speech rate and tone; simplifying vocabulary and grammar; repeating key words, phrases, or concepts; extensively using context clues and models; and relating instruction to students' background knowledge and experience (Cloud, Genesee, & Hamayan, 2000). Sheltered instruction draws from and compliments the instructional approaches of teaching by giving students more time to make connections between concepts and to negotiate meanings (Echavarria et al., 2008). According to Echavarria et al., sheltering of content instruction is an effective approach to teaching students who are not native English speakers because it provides more modeling of grammar and simplification of concepts.

Advocating Language Teaching in Content Instruction

Teachers must become advocates for change or "change agents" of educational approaches to ensure the achievement of Hispanic ELLs in science. According to Caldwell (2006), long-term change requires the concerted efforts and actions of all stakeholders in the educational arena. Transformation is change—especially under challenging circumstances—that is significant, systematic, and sustained, resulting in high levels of achievement for all students in all settings. The first step in producing transformation is preparing a teacher workforce that is open to learning new skills, techniques, and approaches that will help all students achieve academic success, including adapting to teaching students with different language needs in the classroom. There are teachers who truly believe that all students need to do is work hard enough to be successful in the science classroom (Rodriguez, 1998). Such teachers contend that students' language abilities, gender, ethnic backgrounds, or socioeconomic statuses are not important to student success, as long as the student works hard. This perspective toward learning disregards the child as an individual and as a member of a distinctive culture with unique concerns and struggles. Teachers may show resistance to change not necessarily because they disagree with the change, but because they lack the awareness, confidence, and/or knowledge and skills to implement a more culturally responsive and socially relevant curriculum (Rodriguez & Kitchen, 2005). Teacher educators and other stakeholders in the education of children must provide the edification and support needed for classroom teachers to be successful.

Preparing teachers who have confidence in their ability to teach content along with understanding language barriers must be a "team effort," if change is to be sustained over the long term. Fullan (2001) contends that understanding the change process is less about innovation and more about innovativeness. The most effective change and solutions are specific to the situation, and change should be viewed as a process. This process requires that educators utilize what is known or is familiar in their practices, yet, at the same time, strategically and perhaps gradually implement new research-based techniques in order to make progress. Practitioners, for example, already know the importance of adapting teaching to whole groups in ways that will address different learning styles. Teachers may alter instruction to incorporate a variety of approaches, including laboratory work, writing, using the computer, and reading to accommodate diverse learners. Teaching second language learners involves further adaptations to such teaching practices to accommodate ELLs' needs. In this case, teachers specifically use examples drawn from Hispanic culture in ways that have meaning to students so that they may connect learning to such experiences. The difficulty lies in teachers who may be unwilling to learn and make such adaptations to their current practices and/or implement new approaches in teaching Hispanic ELLs.

Teachers who take on the challenges described above are the true change agents in our schools. These teachers will make a lasting impact on teaching Hispanic ELLs if they are able to build the capacity necessary to maintain momentum toward implementing change (Fullan, 2007). Capacity building has been described by Fullan as "a policy, strategy, or action taken that increases the collective efficacy of a group to improve student learning through new knowledge, enhanced resources, and greater motivation on the part of people working individually and together" (p. 75). To make and sustain change requires team building among teacher colleagues, knowledge of the research, and support from all involved, including teachers, teacher educators, and administrators. The change happens when there is acceptance and support by all stakeholders in the educational arena toward implementing instruction that may differ from current practices in science classrooms. Teachers may need to rely less on direct or lecture-oriented instruction, for example, and more on hands-on, group instructional activities, as well as on using pictures and models to help students construct meanings of concepts. Hispanic ELLs will better understand what is to be learned in the face of language barriers when given the opportunity to see, touch, and experience in multiple ways the concept to be learned. Overreliance on expository instruction, for example, will quickly lose ELLs in the learning process and perhaps lose them from science as subject or career path altogether.

The following are cited by Fullan (2001) as critical to the process of change: (1) developing the capacity and commitment to solve complex problems; (2) recognizing the weaknesses as well as the strengths in their approaches; (3) maintaining confidence as one encounters an innovation that requires new skills and new understandings; (4) respecting resisters and, instead, working with them in situations of diversity, complexity, or the tackling of difficult problems; (5) changing current monocultural views and teaching practices to be culturally sensitive, and toward developing relationships, building knowledge, and striving for coherence in a non-linear world; and (6) realizing and welcoming the complexity of change. Consistent, deliberate awareness and implementation of these processes will promote positive changes in teaching Hispanic ELLs.

The greatest long-term effects on Hispanic ELLs occur when science classroom teaching is no longer focused only on skills and concept building for a general audience, but is cognizant of the needs of all students and inclusive of language teaching. Culturally aware educators aim to understand the cultures of all students in the classroom and use this information in implementing a variety of teaching techniques that facilitate learning among both linguistic minority students *and* English-speaking students (Quintanar-Sarallena, 1990). In effective, culturally aware subject area instruction teachers understand their students are from diverse backgrounds and know their students may have unique ways of thinking and talking about science. However, according to Tyler et al. (2008), classrooms in American culture typically reflect only mainstream European or Western worldviews. This cultural view may result in a discontinuity between the child's heritage and the ways of thinking to which they are accustomed in their everyday home lives. Obligating Hispanic ELLs to adopt the dominant European/Western culture in the classroom may explain reported difficulties among this group in terms of achievement and motivational factors such as self-efficacy. In teaching to accommodate only European or Western worldviews, the message communicated by schools is that, "to achieve, ethnic minority students must decide to cease exhibition of their specific cultural value-based behaviors" (Tyler et al., 2008, p. 282). In classrooms that hold only this European/Western world view, Hispanic ELLs must abandon their own cultures while in the classroom, changing to that of the extant monoculture, and at the same time learn the culture and language of the content area—in this case, science. Not only is abandoning one's cultural heritage and ways of thinking stressful to the individual, it is deleterious to learning in today's classrooms as it eliminates the new and valuable insights differing cultural perspectives bring to the table. Teachers must work to advocate change in current monocultural practices toward meeting the demands of an increasingly diverse population of students. These changes present an exciting opportunity for a classroom's diverse culture to enrich the teaching and learning of science for all students.

Supporting Instruction in Dual Language English/ Spanish Instruction

The term *dual language education* has been defined by Soltero (2004) as a long-term additive bilingual and bicultural program model that consistently uses two languages for instruction. The emphasis of dual language education is on learning and communication with a balanced number of students from two language groups integrated for instruction during at least half of the school day in the pursuit of bilingual, bi-literate, academic, and cross-cultural knowledge acquisitions and competence. The students who make up these classrooms often are both ELLs and Spanish language learners. Such classrooms are set up so that the English-speaking students are immersed in an all-Spanish instructional setting and the Spanish speaking students are immersed in an all-English instructional setting. Both groups at one point or another, depending on the design of the program, are using the two languages during instructional time. In such programs, students need to understand, develop, and master science skills in English as well as master the same skills in Spanish. In dual language education programs, it is eminently important that educators understand how to apply specific approaches to teaching when *all* learners in a classroom are second language learners.

There presently are two basic dual language program models: the 90/10 dual language program model and the 50/50 dual language program model. The difference between the two models is in the amount of time spent in each language of instruction. In the 90/10 model, the non-English language is used 90 percent of the time in the early grades. More English is gradually added until the sixth grade, when students have equal instructional time in both languages. In the 50/50 model, students learn in both English and the other language 50 percent of the time throughout the program (Freeman, Freeman, & Mercuri, 2005).

It is important to emphasize that in the 90/10 model the delivery of instruction is in the minority language 90 percent of the time. In many of the U.S. states, and particularly in southern states and urban areas, the minority language usually is Spanish. Students in a 90/10 dual language program model reach a 50/50 distribution in the language of instruction by the time they reach grade three. In teaching science, the thinking and processes of science must be maintained as students transition between languages, and while facilitating language transference in teaching concepts. Cultural experiences and understandings of the differing languages provide opportunities to enrich the learning of science concepts and experiences for both Hispanic ELLs and English-speaking students.

Supporting dual language instruction extends not only to classroom teachers, but also to the professoriate who prepare teachers. Science teacher educators must equip new teachers as well as experienced teachers with language and tools that will help pre-K through grade 12 students learn science concepts alongside English and Hispanic/Latino language skills. Since both dual language instruction and traditional bilingual education are found in schools, it is now critical that changes being espoused for schools must extend to the university. Science teacher educators need to examine their teacher preparation programs and ensure those responsible for pre-service and in-service teacher education have concomitant and relevant expertise in teaching Hispanic ELLs and use this expertise in instruction.

The various cultures of Hispanics/Latinos are the fastest growing populations presently in the U.S.A. It is exceedingly clear that teaching and learning science to accommodate a monoculture is insufficient and inadequate for educating our diverse student population and will not effectively prepare them for a globally connected society. Teacher educators, classroom teachers, administrators, and students need to understand various cultures, ways of thinking, and the language and discourse of others to be able realize the greatest achievements in our changing world.

Despite substantial gains over the past decade, minorities, including Hispanics, are largely underrepresented in science, technology, engineering, and mathematics (STEM) careers (National Science Foundation [NSF], 1996). According to data collected on the STEM workforce, minorities have remained underrepresented in STEM jobs over the past 50 years with the exception of Asian workers (Lowell & Regets, 2006). Underrepresented minorities constitute a growing population within the pool of students from which a highly skilled workforce in the needed areas of STEM will be drawn (Clark, 1999). The statistics reveal a clear and urgent need for educators to gain better understandings of teaching and learning among our nations' fastest growing population, Hispanics/Latinos, toward increasing participation in science and mathematics careers. The exploratory study presented here is a first step in the process of gaining such understanding of science learning among Hispanic ELLs.

PROMOTING SCIENCE UNDERSTANDING AND FLUENCY AMONG HISPANIC ELLS: EXPLORATIONS

The premise that science is unique as a discipline with its own culture or community, language, and ways of thinking about the world may pose difficulties among students new to this subculture of science and scientist. To become a member of the scientific community entails acquiring fluency with the scientific language, thinking processes, and "traditions" of science, including those put into place in the classroom. Becoming members of the community of science may challenge all students, but perhaps more so ELLs who may be experiencing discontinuity between home and school cultures while learning the culture and nature of science, and at the same time struggling to maintain their own cultural identities.

Children's views and beliefs about nature of science (NOS) often differ from the scientific view (Flick & Lederman, 2006). The literature reveals two opposing views among students about NOS. Students tend to either view NOS as an unchanging, authoritative body of knowledge or as a dynamic, tentative process (Cavallo, Blickenstaff, Rozman, & Walker, 2007). The view of science as fixed has been documented as a prevailing view among college students in a number of studies (Saunders, Cavallo & Abraham, 1999; Cavallo, Rozman, Blickenstaff, & Walker, 2003; Cavallo et al., 2007; Cavallo, Potter, & Rozman, 2004). Beliefs about NOS are less well understood among children at the elementary school level. Questions that need to be investigated by researchers include the following: "Do young children in the elementary grades also hold similar opposing views about NOS?" and "Might we find differences in NOS beliefs between students who are familiar with the language in the classroom and the culture of science, typically based in European or Western culture, and the Hispanic/Latino students who do not speak English as their first language?"

Confidence in the ability to be successful in science is important to achievement and understanding in science. Self-confidence, or self-efficacy, is students' beliefs in their own capability to be successful in a subject area or course (Bandura, 1986, 1995; Cavallo et al., 2003; Woolfolk, 2001). Self-efficacy may be used to describe students' perceptions of their ability to successfully perform an activity or reach a stated goal (Woolfolk, 2001). Task-specific self-efficacy has been described by Bandura (1997) as involving "judgments of capabilities specific to a particular task" (as cited in Woolfolk, 2001, p. 388). Self-efficacy is considered to be the competence individuals ascribe to themselves within a specific context (Marsh & Shavelson, 1985; Pajares, 1997) or content area (Marsh, 1992). Students' views on their ability in specific areas, including science and mathematics, are found to be related to their achievement in that subject, and not more general measures, including their overall school achievement (Marsh, 1992). In a study of college students, self-efficacy was found to be a significant predictor of physics achievement (Cavallo et al., 2003).

A number of studies have reported that students with high self-efficacy toward learning a particular subject, including science, have higher achievement in the subject, more positive attitudes, greater enjoyment of the subject, and view the subject as more useful, compared to those with low self-efficacy (Cavallo, Miller, & Saunders, 2002; Cavallo et al., 2003; Handley & Morse, 1984; Harty, Hamrick, & Samuel, 1985; Meece, Blumenfeld, & Hoyle, 1988; Schibeci & Riley, 1986). Little research, however, has been done to explore the salience of these same views and patterns among ELLs who may, at the same time, be struggling with self-efficacy toward the language and culture as a whole, in addition to the culture of science in their classroom. These studies lead us then to ask the question, "How confident are Hispanic ELLs about their ability to successfully learn science in their classroom and toward science as a subject in general as compared to English speakers?"

We are focused, as an educational community, on using the most effective teaching and learning procedures for our students. Our society and the students within our society, however, are changing at a rapid pace, particularly in terms of diversity and the influx of Hispanic ELLs. Educators, consequently, need to consider their teaching practices in science in regard to how adaptable such practices are to the students we serve in today's classrooms. In reference to second language learning, Ellis (2005) states that

> learning will be more successful when:
> 1. The instruction is matched to students' particular aptitude for learning.
> 2. The students are motivated. (p. 220)

Toward this end, educators need to better understand ELLs' learning preferences, as these preferences may differ from those we commonly ascribe to and use among English-speaking learners. Reading science in textbooks, for example, and being able to understand meanings from text is considered critical to achieving scientific literacy; as Kelly (2007) notes, "The significance of the textbook in classroom learning extends beyond its direct influence on student comprehension of the subject matter, as it typically serves as a guideline for instructional choices and for the sequencing of learning events" (p. 459). The science textbook may be a guide for instruction and at the same time communicate structure, design, and language usage in the classroom teaching of science. There is limited research, however, in regard to the extent to which Hispanic ELLs find the textbook a useful or preferred way of learning science.

Experiments are the hallmark of science and of the inquiry nature of the discipline. As such, inquiry teaching and learning is strongly advocated in our prevailing national documents and standards (American Association for the Advancement of Science [AAAS], 1989; National Research Council [NRC], 1996). Inquiry-based teaching is motivating to students (Hofstein & Lunetta, 1982; Lunetta, Hofstein, & Clough, 2007) and is appropriate, especially for young children, because of the concrete experiences that experiments provide (Piaget, 1964). In addition, Lunetta et al. (2007) contend that "learning, in part, depends on interactions with adults and peers" (p. 406). The inquiry-based laboratory also affords opportunity for productive dialogue to take place among the teacher and peers toward negotiating and achieving meaning (Lunetta et al., 2007). The discourse taking place in the laboratory can allow teachers to observe and monitor student learning of science, as well as learning and use of language in expressing ideas. Science learners, in general, often express positive attitudes toward conducting laboratory experiments in science. The extent to which laboratory experiments in science are favored among Hispanic ELLs, however, has not been well researched.

Other techniques in science teaching include using picture or trade books, direct instruction (lecture and notes), and the Internet (Ebenezer & Conner, 1999; Rennie, 2007). In using pictures books or trade books, the goal is to draw students' interest and attention to science, with the pictures and stories providing a context for learning concepts and vocabulary. This type of book format is characterized by large photos or pictures with minimal amount of text for children to read. Direct instruction or lecture is where science concepts are given or told to students. Direct instruction is characterized by a structured presentation of content, often using tools such as the board, overhead projector, or power point slides. Direct instruction—also called lecture, "teacher talk," or exposition—requires attention to oral and verbal information, and the processing of this information by students. Technology use is emphasized as a tool in learning science, with the Internet being available to students for content support. Students may use the computer as a tool to search the Internet for factual information and/or answers to their questions. The Internet provides students with opportunities to visualize objects and/or phenomena that would otherwise be unavailable to them. The Internet also affords an alternative way to communicate and use social learning tools and language norms beyond those typical in direct discourse.

Such practices—textbook use, laboratory, picture books, lecture, and Internet applications—may commonly be used in science classroom teaching to varying extents, yet it is not well understood how these techniques may impact Hispanic ELLs. A question that remains unanswered is, though these teaching techniques are commonly used in classrooms in the U.S., "To what extent do Hispanic ELLs in such classrooms have interest or prefer using picture books, listening to their teachers lecture or talk about science, or using the computer/Internet in learning science?"

Purpose

The overall goals of this study were to explore Hispanic ELLs' views on NOS, self-efficacy toward achieving in science in the classroom and in science in general, their science enjoyment and learning preferences, and how these may differ from their peer English-speaking students. This study's research questions are:

1. What are possible differences between English-speaking students' and Hispanic ELL students' views about NOS?
2. What are possible differences between English-speaking students and Hispanic ELL students regarding self-efficacy toward learning science in their classroom and toward learning science as a subject in general?
3. What are the descriptive patterns and the possible differences between Hispanic ELL students and English-speaking students relative to their enjoyment of science and their preferences for learning science in the classroom?

Methods

This exploratory survey study took place early in the school year, fall semester. The sample consisted of 176 fourth-grade students (87 females; 89 males) enrolled in four different schools in the same public school district. The school district is located in a large urban area in the southwestern United States and serves a diverse population of students, with 53.1% White, 21.8% Hispanic, 14.4% African American, 9.7% Asian/Pacific Islander, and 0.9% Native American. Forty-one percent of the students in this school district are economically disadvantaged. In this sample, 97 were native English-speaking students and 78 students were Hispanic ELLs.

Instrumentation

The students were given a Likert-type instrument with several sections or subscales. One section of the instrument measured students' beliefs about NOS. The questions used in this section were a combination of two NOS instruments, one developed by Abd-El-Khalick (2006) and a second used in prior research (Cavallo & McCall, 2008; Ryan & Aikenhead, 1992). The beliefs in NOS subscale consisted of 10 items. The items were on a four-point Likert-scale with choices of *Strongly Agree, Agree, Disagree,* and *Strongly Disagree.* Items on the NOS instrument indicating the view that science is authoritative and already known were reverse-scored such that a high score on this subscale indicated a belief or view of science as tentative, dynamic, and ever-changing, whereas low scores indicated students' beliefs in science as authoritatively known, unchanging, and fixed.

Sample items on the NOS subscale include:

- What we know about science does not change. (Authoritative View of NOS)
- What we know in science changes when solid, new observations are found. (Dynamic View of NOS)
- Science is an exact report of the truth about our universe. (Authoritative View of NOS)
- The science that we know today may change if new evidence is found. (Dynamic View of NOS)

A second section or subscale of the instrument measured students' self-efficacy in their ability to be successful in science. This subscale was divided into two parts: self-efficacy toward successfully learning science in their current classroom, and self-efficacy toward successfully learning science in general. The items on the self-efficacy subscale were adapted from a questionnaire used in earlier research (Cavallo, Miller, & Saunders, 2002; Miller, Behrens, Greene, & Newman, 1992). The self-efficacy toward science in their own classroom and toward learning science in general included items on a four point Likert-scale, and, as in the NOS section, ranged in choices from *Strongly Agree* to *Strongly Disagree.* Item examples for self-efficacy toward science in their classroom included, "I know I can do well on the science questions and activities given in this class," and "If I were to try a science activity on my own in this class, I am sure I would have trouble." A sample item on self-efficacy toward science in general is, "I am not very good at science."

A third section indicating overall enjoyment of science asked students, "How much do you enjoy learning science in school?" with four choices that ranged from "very much" to "not at all." A fourth section asked students to rank the activities they prefer to do in learning science from one to five, with a score of "1" indicating the learning activity is their most favorite, and "5" indicating the learning activity is their least favorite. In this section, students were to rank the following five activities: (1) reading textbooks about science, which refers to students learning science by reading expository text about the topics they are learning; (2) going on the Internet to learn about science, described as students engaging in searches, viewing pictures and animations, and reading about topics they are to learn in science; (3) doing experiments in science, indicating laboratory work where students engage in hands-on experiences in science (a distinction was not made in this study as to whether the labs were inquiry-based or verification, as the fourth graders may have had difficulty discerning the differences); (4) listening to my teacher tell us about science, described as teachers lecturing and giving notes on the overhead projector, chalkboard, or whiteboard for the students to record in their notebooks or journals; and (5) reading picture books about science, described as students viewing trade books with primarily large photos on each page and with few words. In the analyses, rankings of 1 and 2 for the science activity were combined together as one score, indicating more preferred activities, a score of 3 was mid-range, and rankings of 4 and 5 were scored as least preferred activities. The test was administered in English and was translated verbally to students in Spanish for students in all classrooms.

Results

Differences in Hispanic ELL and English-Speaking Students' Beliefs about NOS and Self-Efficacy

The first two questions of this research were to explore possible differences between Hispanic ELL students' and English-speaking students' beliefs about NOS and their self-efficacy to successfully learn in their science class and in science in general. The *t*-test analysis determines if there are significant differences between the means of two groups. Given the exploratory nature of this study, the *t*-test was selected as the analytical procedure to respond to the research questions. Results of *t*-test analyses on the means for each variable are shown in Table 4.1.

TABLE 4.1 Differences between ELL and English-Speaking Students' Views of NOS and Self-Efficacy

	English Speakers		ELLs				
Variable	*M*	*SD*	*M*	*SD*	*t*	*df*	*p*
Nature of Science beliefs	26.0	3.9	24.1	3.0	3.47	158	.001
Self-efficacy toward Science class	9.2	1.9	8.9	1.7	1.05	170	.294
Self-efficacy toward Science in general	3.3	0.98	2.92	1.00	2.16	163	.033

Note: Because of unequal groups, the assumption of equal variances was not assumed in these analyses and statistical results are reported accordingly.

As shown in Table 4.1, significant differences were found between Hispanic ELL students' and English-speaking students' beliefs about the nature of science ($p < .01$). The means in Table 4.1 reveal that English-speaking students scored higher on the NOS instrument, indicating more scientifically accurate views of science as changing and dynamic. Hispanic ELL students scored significantly lower on their beliefs in NOS, indicating a view of science as fixed and authoritative.

The data shown in Table 4.1 indicate no significant differences between ELL students' and English-speaking students' self-efficacy toward learning and solving problems in their school science class. There was, however, a significant difference found between the groups on their self-efficacy toward being successful in science in general. This difference indicated that the ELL students have lower self-confidence in their ability to successfully learn science in general as compared to the English-speaking students.

Descriptive Patterns and Possible Differences in Hispanic ELL Students' and English-Speaking Students' Enjoyment of Science and Preferences for Learning Classroom Science

The third question of this research was to examine, first on a descriptive level, the patterns between the two groups, ELL students and English-speaking students, in terms of their enjoyment of science and classroom science learning preferences, and then to examine these patterns for possible statistical differences. Because this study was exploratory, it was determined that the descriptive patterns may provide insight into the learning among both groups and also present information to serve as a springboard for future research. Thus, the descriptive patterns are presented first, followed by further analyses as to possible statistical differences between the groups

The first part of the question examined students' *enjoyment of science* and possible differences between the two groups. The descriptive information revealed that the Hispanic ELL students' mean score was 1.58 (SD = .91), and the mean score for the English-speaking students was 1.58 (SD = .79). Descriptively, the means appear similar, with somewhat greater variation around the means among the Hispanic ELL students as indicated by the standard deviations. The statistical procedure utilized was a t-test analysis to determine possible significant differences between the means of two groups. The t-test revealed no significant differences in enjoyment of science between Hispanic ELL and English-speaking students, with t (172) = .76, $p > .05$. In addition, the assumption of equal variance between the two groups was not violated in the analysis; however, because of unequal group sizes, and in order to be conservative, the results reported here are those generated under the assumption of unequal variance. Thus, enjoyment of science can be considered statistically equivalent between English-speaking and Hispanic ELL students.

The second part of this research question examined descriptive patterns in the students' *preferences for learning science.* In the data analyses, rank ordered scores for each teaching approach were tallied for frequency of responses within categories, where 1 = most preferred, 2 = midrange, and 3 = least preferred. To first explore *patterns* in preferences for learning between the two groups on a descriptive level, frequencies in students' responses were transformed to percentages. The descriptive data, in the form of percentages, is presented graphically in Figures 4.1 through 4.5.

Patterns in ELL and English-speaking students' preferences toward learning science by reading science textbooks are shown in Figure 4.1.

As the percentages shown in Figure 4.1 indicate, reading science textbooks was not a highly preferred activity for learning science among either English-speaking students or Hispanic ELL students. The percentage of Hispanic ELL students rating reading science textbooks as most preferred was somewhat higher than for English-speaking students.

Figure 4.2 presents patterns in ELL and English-speaking students' preferences toward learning science by searching, viewing, and reading science using the Internet.

Observation of the percentages shown in Figure 4.2 indicate that using the Internet to learn about science was considered to be a most favored activity by the majority of students in both groups. A higher percentage

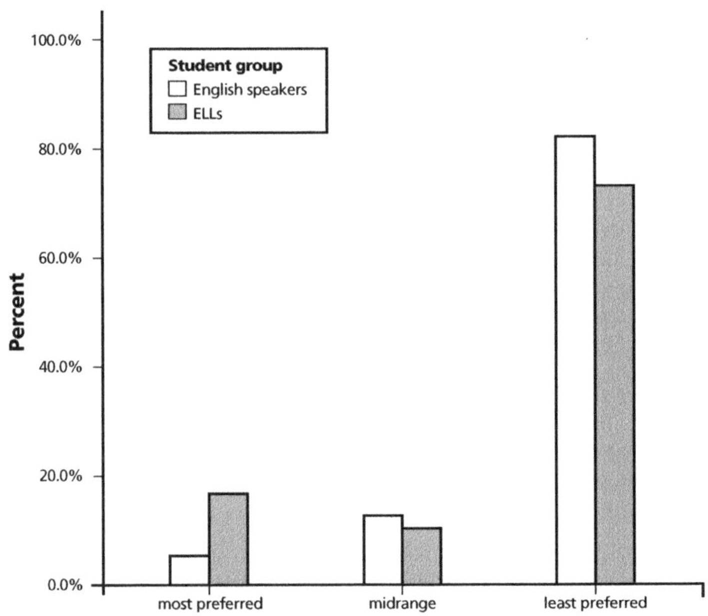

Figure 4.1 ELL and English-speaking students' preferences for learning science by reading science textbooks.

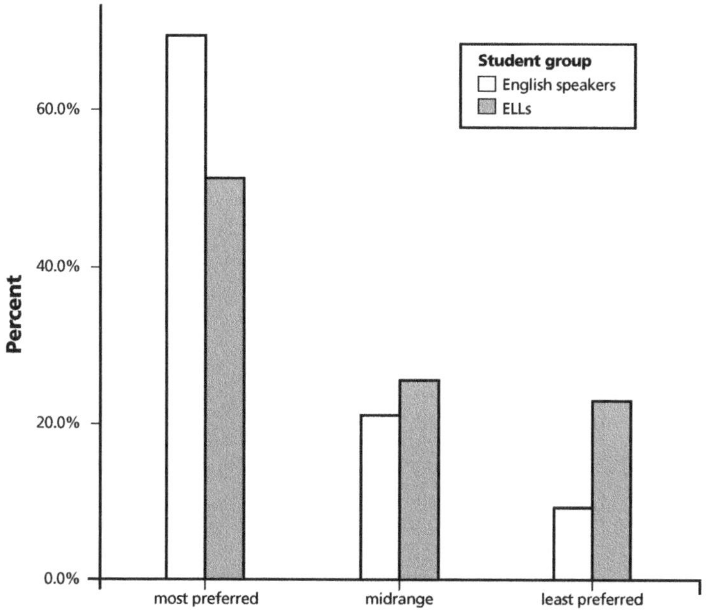

Figure 4.2 ELL and English-speaking students' preferences for learning science by using the Internet.

of Hispanic ELL students, however, ranked the Internet as their least preferred method of learning science compared to English-speaking students.

Figure 4.3 presents patterns in the percentages of ELL and English-speaking students' preferences toward learning science by doing hands-on laboratory activities.

The pattern revealed in Figure 4.3 is that both ELL and English-speaking students have strong preferences for laboratory experiments in science, as most students ranked experiments in the highest category.

Patterns in ELL and English-speaking students' preferences toward learning science through listening to the teacher lecture or talk about science are shown in Figure 4.4.

As Figure 4.4 indicates, the highest percentage of rankings for both groups indicated that listening to the teacher lecture about science is their least preferred activity. This trend showed fairly even responses between the two groups, in terms of lecture being midrange. A higher percentage of Hispanic ELL students, however, ranked talking about science or lecture as most preferred, while a higher percentage of English-speaking students ranked lecture as their least preferred when compared to ELL students.

Figure 4.5 presents patterns in Hispanic ELL and English-speaking students' preference toward learning science by looking at picture books.

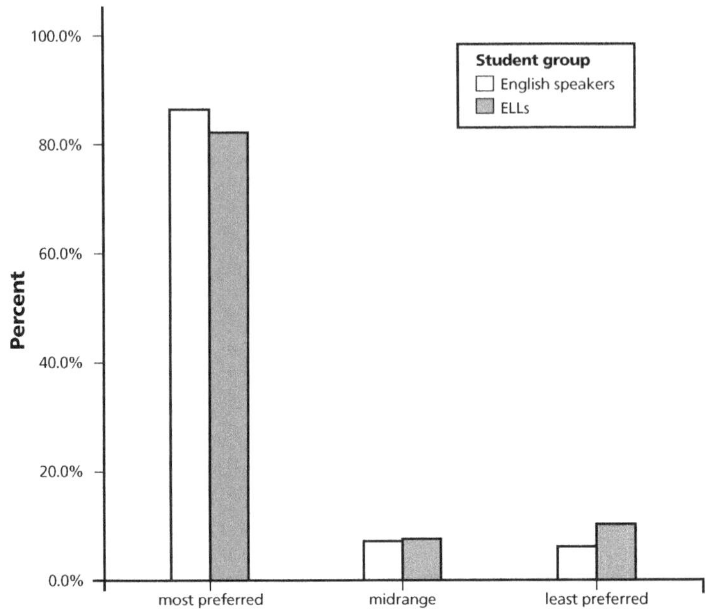

Figure 4.3 ELL and English-speaking students' preferences for learning science by doing experiments.

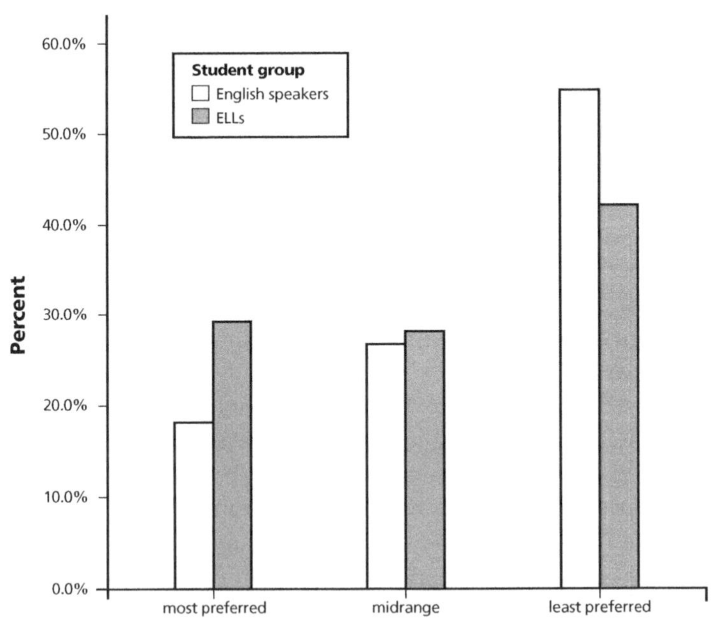

Figure 4.4 ELL and English-speaking students' preferences for learning science by listening to teachers talk about science (lecture).

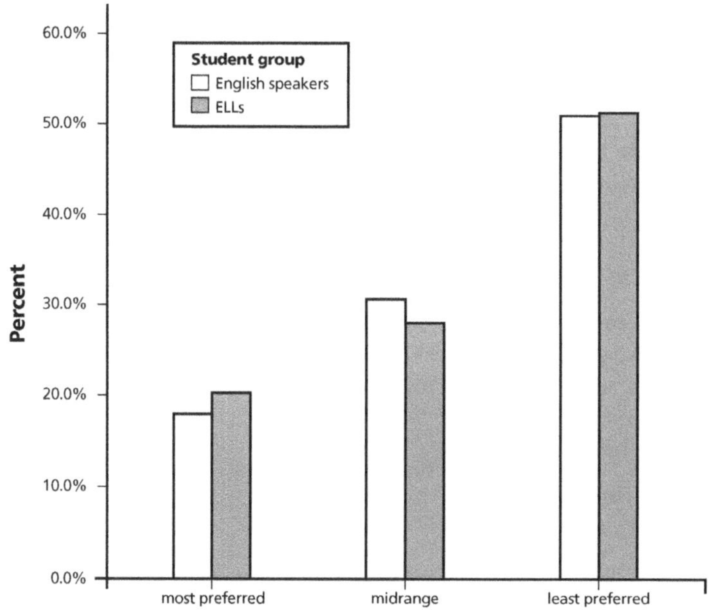

Figure 4.5 ELL and English-speaking students' preferences for learning science by looking at picture books.

The patterns illustrated in Figure 4.5 generally show that learning science by looking at science-themed picture books was least preferred by most students of both groups.

In addition to revealing and describing patterns, this research question aimed to determine if the observed descriptive patterns would indicate statistically significant differences in Hispanic ELL students' and English-speaking students' learning preferences. Chi-square analyses were conducted to determine if the observed descriptive patterns in learning preferences were significantly different between the Hispanic ELL students and English-speaking students in this study. It was determined that the chi-square responds best to the research question of this study in exploring differential patterns between Hispanic ELL and English-speaking students' learning preferences. These percent values and results of Chi-square analyses on ELL and English-speaking students' preferences for learning science are shown in Table 4.2.

The results of the analyses shown in Table 4.2 indicate that observed patterns in Hispanic ELL and English-speaking students who prefer learning science by reading the textbook could not be considered significantly different; however, the Chi-square value equaled that accepted in this study for significance ($p = .05$). In the Chi-square analysis on Hispanic ELL and

TABLE 4.2 Percent Values and Chi-Square Analyses on English-Speaking and ELL Students' Preferences in Learning Science

	Most Preferred	Midrange	Least Preferred
Reading Science Textbooks			
English-Speaking	5.3%	12.6%	82.1%
ELL	16.7%	10.3%	73.1%
Total Group	10.4%	11.6%	78.0%
$\chi^2 = 6.01$, $p = .05$			
Using the Internet			
English-Speaking	69.5%	21.1%	9.5%
ELL	51.3%	25.6%	23.1%
Total Group	61.3%	23.1%	15.6%
$\chi^2 = 7.78$, $p = .02$			
Doing Science Experiments			
English-Speaking	86.5%	7.3%	6.3%
ELL	82.1%	7.7%	10.3%
Total Group	84.5%	7.5%	8.0%
$\chi^2 = .97$, $p = .62$			
Listening to Teacher Talk			
English-Speaking	18.3%	26.9%	60.7%
ELL	29.5%	28.2%	39.3%
Total Group	23.4%	27.5%	49.1%
$\chi^2 = 3.66$, $p = .16$			
Looking at Picture Books			
English-Speaking	18.1%	30.9%	51.1%
ELL	20.5%	28.2%	51.3%
Total Group	19.2%	29.7%	51.2%
$\chi^2 = .23$, $p = .89$			

English-speaking students' preferences for learning science by conducting searches, viewing animations, and reading about science using the Internet, there was a significant difference found between the two groups (see Table 4.2). In reviewing the percentages reported in Table 4.2, it appears more English-speaking students prefer learning science via the Internet compared to Hispanic ELL students. Chi-square analyses revealed no significant differences between the two student groups regarding engaging in hands-on laboratory experiments, listening to the teacher lecture about science, or in reading/viewing picture books about science as learning preference categories ($p > .05$).

Discussion

ELLs certainly are faced with challenges related to becoming part of a new culture and community, and with learning the language and ways of knowing that may be unique to the new culture. In addition to these challenges, ELL students also are expected to adapt to the ways of thinking and knowing that are embedded within subcultures, including science. The significant difference in ELL students' and English-speaking students' views of NOS, with the higher mean being among English-speaking students, indicates that English-speaking students had a more tentative, dynamic view of science compared to ELL students. The findings mean, then, that ELL students have a comparatively more fixed view of NOS. Explanations for this finding may relate to difficulties in understanding the culture of science among Hispanic ELL students due to language or culture. According to Greenfield, Quiroz, and Raeff (2000), Hispanic students are of a culture that is very respectful of adults. Hispanic ELL parents, in fact, have been found in research to be more concerned about whether their children are respectful to the teacher than about their academic successes (Greenfield et al., 2000). A study by Tyler et al. (2008) noted that Hispanics/Latinos have high respect for adults and defer authority to older family members. Perhaps this respect for authority translates to the science arena, where it is part of the Hispanic/Latino culture not to question authority, but accept what is told to them as truth. This cultural aspect may relate to findings of this study, which indicated that Hispanic ELL students had a more fixed, authoritative view of science compared to their English-speaking counterparts. It is important for teachers to understand the aspect of the Hispanic/Latino culture where authority is unquestioned, so they may be prepared to help ELLs—and all students, for that matter—view science as dynamic and changing, rather than authoritatively known. It is also important to note the young age of the students and that many have been in the U.S. for only a short time. It is reasonable that these students may defer to authority more so than their U.S.-born, English-speaking peers. Holding fixed views of science, however, would make understanding the sub-culture and ways of thinking in the science discipline more difficult for students. A previous study found that students who viewed science as fixed also tended to learn science by rote, rather than by attempting to make sense of concepts and learn at a deeper, more meaningful level (Cavallo et al., 2007). Without an understanding of science as a dynamic discipline that is built on experimentation and evidence, students may be less likely to develop the thinking processes needed to succeed in this discipline, as they may tend toward accepting or memorizing science as a set of facts. It is posed that Hispanic ELL students, and all students, may be at risk of falling behind in science if they persist with a view of science as authoritative and fixed rather than dynamic and changeable. More research is needed to explore this issue.

Hispanic ELL students in this study were found to have levels of self-efficacy in learning science that were equivalent to their English-speaking peers in their current classrooms. Hispanic ELL students, however, had lower self-efficacy in regard to their ability to be successful in science in general compared to their English-speaking peers in the current study. Perhaps the prospect of science outside of their classroom was more of an unknown to the ELL students, thus self-efficacy was lower compared to English-speaking students. This finding is supported by MacIntyre, Clement, Dornyei, and Noels (1998), in that ELLs have more confidence in situations they have experienced before. However, new situations are detrimental to the confidence of ELLs because "the speaker will be uncertain of his or her ability to meet the communicative demands of the moment" (p. 549). Hispanic ELLs may feel more confident in their known classroom settings, where their teachers provide support for their learning, where they are familiar and used to communicating, and where teachers are able to communicate with them in their first language. Science beyond the classroom, however, may have uncertain levels of familiarity and support, perhaps attributable to perceived or real language and/or cultural barriers.

Science was rated similarly as a highly enjoyed subject among both ELLs and English speakers. Thus, science is a subject that appeals to the range of students in this study, regardless of the language and cultural barriers they may be experiencing.

Most students did not rate reading science textbooks as a highly preferred way to learn science. The finding of a Chi-square value that equaled the acceptable significance level in this study on students' preference for learning science by reading the textbook warrants further research. Examination of the percentages in Figure 4.1 regarding this finding reveals that comparatively more Hispanic ELL students tend to prefer the science textbook than English-speaking students. The textbook may provide a structure for these ELL students and afford them some level of support in case they become confused or lost in the instruction from their language and/or cultural difficulties. The textbook may be a road-map for these students, as they learn the culture and language of science. According to Kelly (2007) and Glynn and Muth (1994), the textbook provides students with a guideline for what is being taught and in what order. Teaching techniques such as lecture may not provide students with a detailed roadmap to follow because the content to be learned is verbally delivered, where the words being spoken are not seen (at least not verbatim) by students. ELLs have to contend with translation of words they cannot read along with new science vocabulary being introduced in attempting to make meaning of what is being said in a lecture. It may also be difficult for ELLs in particular to recall or recover what was misunderstood or missed in a lecture because they cannot retrace back through the words they may need to interpret or translate. The current study does not allow such claims since the Chi-square value approached, but did not reach, significance, but may point to the need for further research on the thinking processes used among ELLs toward making meaning during verbally delivered instruction.

Most Hispanic ELL and English-speaking students in this study prefer learning science through the Internet. However, use of the Internet as a preference for learning science was differently distributed between Hispanic ELL students and English-speaking students in the study. An analysis of Figure 4.2 on students' preference for learning science via the Internet indicates that fewer Hispanic ELL students ranked this activity as most preferred compared to English-speaking students, and more Hispanic ELL students ranked the Internet in the midrange or least preferred category than did English speakers, indicating the source of the significant difference found in this analysis. These findings may reveal a lack of comfort with the Internet among Hispanic ELL students. In research on mathematics learning, Freeman and Crawford (2008) found that learning via computer was less effective for ELLs compared to English language speakers in their study. In reviewing the collaborating school's demographic information and comparatively lower socio-economic status of the Hispanic ELLs, it may be that fewer of the Hispanic ELLs have access to computers at home and therefore do not have access to the Internet. Hispanic ELL students likely have less experience using the Internet, therefore, when compared to the English-speaking students of this study. The study by Freeman and Crawford indicates that "other cultural (nonlinguistic) group risk factors (e.g., low socioeconomic status, limited parental education, low-achieving schools, etc.) contribute to difficulties faced by new language learners" (p. 17). Students, particularly Hispanic ELLs, who lack computer and Internet access would be at a tremendous disadvantage in learning, as they are unable to achieve equitable educational opportunities as their English-speaking and/or more affluent peers. Teachers must address this potential lack of access to computers among all students, regardless of circumstances, by providing technology experiences for children in their classes. More research needs to examine the extent to which Hispanic ELLs have access to Internet and technology to support science learning. It also needs to be determined to what extent technological support is available or provided in their home language, as Snow, Burns, and Griffin (1998) emphasize that one basis for success in learning English is furthering the students' language and literacy development in the their native language.

Consistent with research results reported elsewhere, most students in this study prefer learning science through laboratory exploration. Inquiry lab investigations provide concrete experiences students need to connect new knowledge, meanings, and scientific language (Lawson, Abraham, & Renner, 1989; Marek & Cavallo, 1997). As previously discussed, Ausubel (1963) and Novak (1988) contended that students need "anchor concepts" upon which to connect new understandings in order for new, related concepts to be meaningfully learned. The laboratory exploration provides the anchor—a common experience—so that concepts based on this experience will make sense to the students. Research is limited, and hence more is needed, to investigate the levels and quality of learning that takes place among Hispanic ELLs during inquiry-based laboratory investigations.

Surprisingly, the majority of both Hispanic ELL and English-speaking students ranked picture books as a least preferred activity. Picture books provide visual images of science related phenomena, objects, and organisms, typically with limited textual explanation. This finding was unexpected, but it is speculated that picture books may not have received the highest preference among students because they would rather use the computer (Internet) or engage in hands-on laboratory activities, both of which were given "most preferred" rankings among students in this study. Future research should examine the science understandings acquired by English-speaking and Hispanic ELL students using these two science learning formats.

PROMOTING SCIENCE UNDERSTANDING AND FLUENCY AMONG HISPANIC ELLS: NEW DIRECTIONS

This chapter was aimed at teacher educators and researchers, as well as at teachers, administrators, and other practitioners in U.S. schools who now must change standard, monocultural science teaching practices to address a new classroom of diverse learners with unique and rich cultural backgrounds that will enhance the learning experiences of Hispanic ELLs, while at the same time enrich learning for all. This chapter poses a challenge to educators to develop the capacity and commitment to solve complex educational issues and effectively meet the educational needs of a growing number of second language learners in their schools. Educators need to answer this call and continue to seek ways to improve science teaching and learning among Hispanic ELLs, allowing them to become well prepared to pursue STEM learning and careers in their futures.

Change, according to Fullan (2007), is "a policy, strategy, or action taken that increases the collective efficacy of a group to improve student learning through new knowledge, enhanced resources, and greater motivation on the part of people working individually and together" (p. 85). It is time to focus on how we look at and apply new approaches to current structures in the classroom. The National Clearinghouse for English Language Acquisition and Language Instruction Educational Programs reports that "between 1989–1990 and 2004–2005, enrollment of ELL students in US schools increased 150 percent, from roughly 2 million to well over 5 million" (Waters, 2007, p. 1). In response, education needs to emphasize furthering the knowledge base of teacher educators, teachers, and administrators toward enabling them to effectively direct, structure, and support dual language programs, particularly in learning science, mathematics, and technology needed for today's world. Such measures may positively impact all students' success in science and in school and lay a strong foundation for their futures.

In fourth grade, our schools have students who are energized and enthusiastic about learning science. So what happens to them between fourth grade and high school and onto college, such that a crippling majority of them turn away from science? Science education must continue to focus its efforts on finding answers to this question, improving the status quo toward accommodating differences in the many ways there *are* differences among students, and, importantly, envisioning and pursuing ever-changing new directions and emphases in science education for the future.

REFERENCES

Abd-El-Khalick, F. (2006). Over and over again: College students' views of nature of science. In L. B. Flick and N.G. Lederman (eds.), *Scientific Inquiry and Nature of Science: Implications for Teaching, Learning, and Teacher Education* (pp. 389–426). Dordrecht, The Netherlands: Springer.

American Association for the Advancement of Science (1989). Science for All Americans: A Project 2061 Report on Literacy Goals in Science, Mathematics, and Technology. Washington, DC.

Ausubel, D. P. (1963). *The Psychology of Meaningful Verbal Learning.* New York: Grune and Stratton.

Bandura, A. (1986). *Social Foundations of Thought and Action: A Social-Cognitive Theory.* Englewood Cliffs, NJ: Prentice Hall.

Bandura, A. (1995). Exercise of personal and collective efficacy in changing societies. In A. Bandura (Ed.), *Self-Efficacy in Changing Societies* (pp. 1–45). New York: Cambridge University Press.

Bandura, A. (1997). *Self-Efficacy: The Exercise of Control.* New York: Freeman.

Cavallo, A. M. L. & Laubach, T. A. (2001). Students' science perceptions and enrollment decisions in differing learning cycle classrooms. *Journal of Research in Science Teaching, 38,* 1029–1062.

Cavallo, A.M.L. & McCall, D. (2008). Seeing may not mean believing: Examining students' understandings and beliefs in evolution. *The American Biology Teacher, 70,* 522–530.

Cavallo, A. M. L., Miller, R. B., & Saunders, G. (2002). Motivation and affect toward learning science among preservice elementary school teachers: Implications for classroom teaching. *Journal of Elementary Science Education, 14,* 25–38.

Cavallo, A. M. L., Potter W. H. & Rozman, M. (2004). Gender Differences in Learning Constructs, Shifts in Learning Constructs and Their Relationship to Course Achievement in a Structured Inquiry, Yearlong College Physics Course for Life Science Majors. *School Science and Mathematics, 104,* 1–12.

Cavallo, A. M. L., Rozman, M., Blickenstaff, J., & Walker, N. (2003). Students' learning approaches, reasoning abilities, motivational goals, and epistemological beliefs in differing college science courses. *Journal of College Science Teaching, 33,* 18–24.

Cavallo, A. M. L., Blickenstaff, J., Rozman, M., & Walker, N. (2007). Learning college science: An investigation of biology and physics students' learning approaches, scientific reasoning, motivational goals, and beliefs about the nature of science. In L. Cooke (Ed.), *Frontiers in Higher Education.* Hauppauge, NY: Nova Science Publishers, pp. 207–228.

Caldwell, B. J. (2006). *Re-Imagining Educational Leadership.* Thousand Oaks, CA: SAGE Publications.

Chamot, A. U., & O'Malley, J. M. (1994) *The Cognitive Academic Language Learning Approach Handbook: Implementing the Cognitive Academic Language Learning Approach.* Reading, MA: Addison-Wesley.

Clark, J. (1999). *Minorities in Science and Math.* (ERIC Document Reproduction Service No. ED 433 216). Darling-Hammond, L. (1992). Teaching and knowledge: Policy issues posed by alternative certification for teachers. *Peabody Journal of Education, 67*(3) 123–154.

Cloud, N., Genesee F., & Hamayan, E. (2000). *Dual language instruction: A handbook for enriched education.* Boston, MA: Thomson-Heinle.

Echavarria, J., Vogt, M. E. & Short, D. (2008). *Making Content Comprehensible for English Language Learners: The SIOP Model, Third Edition.* Boston: Allyn & Bacon.

Ebenezer, J., & Connor, S. (1999). *Learning to Teach Science.* Scarborough, ON, Canada: Prentice-Hall.

Ellis, R. (2005). Principles of instructed language learning. *System, 33,* 209–224.

Flick, L.B., & Lederman, N.G. (2006). *Scientific Inquiry and Nature of Science: Implications for Teaching, Learning, and Teacher Education.* Dordrecht, The Netherlands: Springer.

Freeman, B, & Crawford, L. (2008). Creating a middle school mathematics curriculum for English language learners. *Remedial and Special Education, 29,* 9–19.

Freeman S., Freeman E., & Mercuri S. P. (2005). *Dual Language Essentials for Teachers and Administrators.* Portsmouth, NH: Heinemann

Fullan, M. (2001). *Leading in a Culture of Change.* San Fransisco, CA: John Wiley & Sons.

Fullan, M. (2007). *The New Meaning of Educational Change.* New York: Teachers College Press.

Garcia, E. E. (1999). *Student cultural diversity: Understanding and meeting the challenge* (2nd ed.). Boston, MA: Houghton Mifflin.

Glynn, S. & Muth, K. (1994). Reading and writing to learn science: Achieving scientific literacy. *Journal of Research in Science Teaching, 31,* 1057–1073.

Greenfield, P. M., Quiroz, B., & Raeff, C. (2000). Cross-cultural conflict and harmony in the social construction of the child. In S. Harkness, C. Raeff, & C. M.

Super (Eds.), *Variability in the Social Construction of the Child* (pp. 93–108). San Francisco: Jossey-Bass.

Handley, H. M., & Morse, L. W. (1984). Two-year study relating adolescents' self-concept and gender role perceptions to achievement and attitudes toward science. *Journal of Research in Science Teaching, 21*, 599–607.

Harty, H., Hamrick, L., & Samuel, K. V. (1985). Relationships between middle school students' science concept structure interrelatedness competence and selected cognitive and affective tendencies. *Journal of Research in Science Teaching, 22*, 179–191.

Kelly, G. J. (2007). Discourse in science classrooms. In S. K. Abell and N. Lederman (Eds.) *Handbook of Research on Science Education.* Mahwah, NJ: Lawrence Earlbaum.

Krashen, S. D. (1981) *Principles and Practice in Second Language Acquisition.* English Language Teaching series. London: Prentice-Hall International.

Lawson, A. E. (1995). *Science Teaching and the Development of Thinking.* Belmont, CA: Wadsworth.

Lawson, A. E., Abraham, M. R., & Renner, J. W. (1989). *A Theory of Instruction: Using the Learning Cycle to Teach Science Concepts and Thinking Skill.* National Association of Research in Science Teaching, Monograph No. 1.

Lee, O., Luykx, A., Buxton, C., & Shaver, A. (2007). The challenge of altering elementary school teachers' beliefs and practices regarding linguistic and cultural diversity in science instruction. *Journal of Research in Science Teaching, 44*, 1269–1291.

Long, M. H. (1996). The role of the linguistic environment in second language acquisition. In W. D. Riechie & T. T. Bhatia (Eds.), *Handbook of research on language acquisition: Vol. 2. Second language acquisition* (pp. 413–468). New York: Academic Press.

Lowell, B. L., & Regets, M. (2006). *A half-century snapshot of the STEM workforce, 1950 to 2000.* Washington, DC: Commission on Professionals in Science and Technology.

Lunetta, V, N., Hofstein, A., & Clough, M.P. (2007). Learning and Teaching in the school science laboratory: An analysis of research, theory, and practice. In S. Abell & N. Lederman (EdS), *Handbook of Research in Science Education*, (pp 393–342). Lawrence Erlbaum.

Luria, A. R. (1976). *Cognitive development: Its cultural and social foundations.* Cambridge, MA: Harvard University Press.

MacIntyre, P.D., Clement, R., Dornyei, Z., & Noels, K. (1998). Conceptualizing willingness to communicate in a L2: A situational model of L2 confidence and affiliation. *The Modern Language Journal, 82*, 545–562.

Mackey, A. (1999). Input, interaction, and second language development: An empirical study of question formation in ESL. *Studies in Second Language Acquisition, 21*, 557–587.

Magnusson, S. J., Palincsar, A. S. & Templin, M. (2006). Community, culture, and conversation in inquiry based science instruction. In L. B. Flick and N.G. Lederman (Eds.), *Scientific Inquiry and Nature of Science: Implications for Teaching, Learning and Teacher Education* (pp. 131–155). Dordrecht, The Netherlands: Springer.

Marek, E. A., & Cavallo, A. M. L. (1997). *The Learning Cycle: Elementary School Science and Beyond* (rev. ed.). Portsmouth, NH: Heinemann.

Marsh, H. W. (1992). Content specificity of relations between academic achievement and academic self-concept. *Journal of Educational Psychology, 84,* 35–42.

Marsh, H. W., & Shavelson, R. J. (1985). Self-concept: Its multifaceted, hierarchical structure. *Educational Psychologist, 20,* 107–125.

Meece, J. L., Blumenfeld, P. C., & Hoyle, R. H. (1988). Students' goal orientations and cognitive engagement in classroom activities. *Journal of Educational Psychology, 80,* 514–523.

Miller, R. B., Behrens, J. T., Greene, B. A., & Newman, D. (1992, April). Motivation to learn statistics. Paper presented at the Annual Meeting of the American Educational Research Association, San Francisco.

National Clearing house of English Language Acquisition (NCELA). (2006). NCELA frequently asked questions. Retrieved July 27, 2009 from www.ncela.gwu.edu/expert/faq.

National Research Council. (1996). *National Science Education Standards.* Washington, D.C.: National Academy Press.

National Science Foundation. (1994). *Women, Minorities, and Persons with Disabilities in Science and Engineering.* Washington, DC.

National Science Foundation. (1996). *Women, Minorities, and Persons with Disabilities in Science and Engineering.* Washington, DC.

Novak, J.D. (1988). Learning science and the science of learning. *Studies in Science Education, 15,* 77–101.

Novak, J. D., & Gowin, D. B. (1984). *Learning How to Learn.* New York: Cambridge University Press.

Pajares, F. (1997). Current directions in self-efficacy research. In M. L. Maehr & P. R. Pintrich (Eds.), *Advances in Motivation and Achievement,* Volume 10 (pp. 1–49). Greenwich, CT: JAI Press.

Piaget, J. (1964). Cognitive development in children: Piaget, development and learning. *Journal of Research in Science Teaching, 2*(3), 176–180.

Quintanar-Sarellana, R. (1990). *Teachers' perceptions of the language and culture of linguistic minority students.* Unpublished doctoral dissertation, Stanford University, Stanford.

Reiss, J. (2008). *102 content strategies for English language learners: Teaching for academic success in grades 3–12.* Upper Saddle River, New Jersey: Prentice Hall.

Rennie, L. J. (2007). Learning science outside of school. In S.K. Abell and N. Lederman (Eds.) *Handbook of Research on Science Education.* Mahwah, NJ: Lawrence Earlbaum.

Rodriguez, A. J. (1998). Strategies for counter-resistance: Toward socio-transformative constructivism and learning to teach science for diversity and for understanding. *Journal of Research in Science Teaching, 36*(6), 589–622.

Rodriguez, A. J. & Kitchen R. S. (2005). *Preparing Mathematics and Science Teachers for Diverse Classrooms: Promising Strategies for Transformative Pedagogy.* Mahwah, NJ: Lawrence Erlbaum.

Ryan, A. G., & Aikenhead, G. S. (1992). Students' preconceptions about the epistemology of science. *Science Education, 76,* 559–580.

Saunders, G. L., Cavallo, A. M. L., & Abraham, M. R. (1999, March). Relationships among epistemological beliefs, gender, approaches to learning, and implementation of instruction in chemistry laboratory. Paper presented at the Annual Conference of the National Association for Research in Science Teaching, Boston, MA.

Shi, W., Bogdanov, M., Dowhan, W. & Zusmani, D. R. (1993). The pss and psd genes are required for motility and chemotaxis in escherichia coli. *Journal of Bacteriology, 175,* 7711–7714.

Schibeci, R. A., & Riley, J. P., Jr. (1986). Influence of students' background and perceptions of science attitudes and achievement. *Journal of Research in Science Teaching, 23,* 177–187.

Snow, C., E., Burns, M. S., & Griffin, P. (Eds.). (1998). *Preventing reading difficulties in young children.* Washington, DC: National Academies Press.

Soltero, S. W. (2004). *Dual Language Teaching and Learning in Two Languages.* Boston: Pearson.

Tyler, K. M., Uqdah, A. L., Dillihunt, M. L. Beatty-Hazelbaker, R., Conner, T., Gadson, N., Henchy, A., Hughes, T., Mulder, S., Owens, E., Roan-Belle, C., Smith, L., & Stevens, R. (2008). Cultural discontinuity: Toward a quantitative investigation of a major hypothesis in education. *Educational Researcher, 37,* 280–297.

Vygotsky, L. S. (1978). *Mind in Society: The Development of Higher Psychological Processes.* Cambridge, MA: Harvard University Press.

Waters, John K. (2007). The Universal Language. *T H E Journal, 34,* 1–2.

Wertsch, J. V. (1991). *Voices of the Mind: A Sociocultural Approach to Mediated Action.* Cambridge, MA: Harvard University Press.

White House Initiative on Educational Excellence for Hispanic Americans. (1999). *Latinos In Education: Early Childhood, Elementary, Undergraduate, Graduate.* Washington, DC: Author. Retrieved June 30, 2009 from www.ed.gov/offices/OIIA/Hispanic/rr/ech.html.

Woolfolk, A. (2001). *Educational Psychology* (8th ed.), Needham Heights, MA: Allyn & Bacon.

CHAPTER 5

SYNERGISTIC TEACHING OF SCIENCE TO ENGLISH LANGUAGE LEARNERS

Common Components of Model English Language Learner and Science Instruction

Daniel J. Bergman

ABSTRACT

Science teachers frequently encounter time constraints when planning and teaching lessons, experiments, and activities. This pressure increases in the context of meeting and addressing needs of English language learners (ELLs). This chapter features an in-depth document analysis of the standards and strategies promoted by these two disciplines: English language learning and science education. Results indicate that the National Science Education Standards (NSES) and English Language Proficiency (ELP) Standards compliment one another, each offering guidance and benchmarks not completely addressed by its counterpart. Strategies for ELL instruction and science inquiry (including the learning cycle) contain common research-based elements.

Teaching Science with Hispanic ELLs in K–16 Classrooms, pages 101–133

Both feature the following teacher actions: using small groups and student interactions, encouraging students to learn by doing, effectively selecting and using materials, connecting content to student experiences, providing opportunities for application, managing students and time, and playing a key role in classroom interactions. Discrepancies between ELL and science strategies occur with respect to lesson objectives and the extent to which students know expected outcomes. The chapter concludes with a discussion of how teachers can negotiate between these seemingly different pedagogical approaches, taking advantage of synergetic strategies and standards to effectively teach ELL and all science students.

INTRODUCTION

A common lament in the teaching profession is a chronic lack of time. Consider the following comments from science teachers when asked about their work experiences (Bergman, 2008):

> I really wish I had more time to do the lessons... how to convert the lessons into more meaningful activities. (p. 169)
>
> With time limited, I don't spend as much time as I'd like to on planning good lessons. (p. 218)
>
> I've audio-taped [myself teaching] a number of times and I've video-taped just once. But that's something that I think I'd like to do more of. I feel like, unfortunately, I'm just trying to stay above water right now. (p. 316)

Research repeatedly finds that teachers face a "time crunch" in several aspects of education: planning and implementing innovative curricula (Hoover & Achilles, 1996), meeting state academic standards (Meek, 2002), and having enough instructional time and correcting student work (Fadiman & Howard, 1980). Time has also been found to be a crucial ingredient necessary for enduring school improvement (Corbett, Dawson, & Firestone, 1984).

Presently, the instruction of ELLs is a prominent topic in science classrooms (Fathman & Crowther, 2005). Past publications have highlighted "best practices from different but complementary fields of science education and English language teaching" (National Science Teachers Association [NSTA], 2008, p. 1). Science teachers are encouraged to "easily dip in and out of the topics you want" (p. 1). Unfortunately, teachers frequently find such promotions to be one more task added to their already daunting duties.

Many educators, however, have developed methods of incorporating science instruction for ELL students. Through a combination of necessity and ingenuity, strategies for joint English language learning–science lessons have proliferated in recent years (Armon & Morris, 2008; Hansen, 2006; Medina-Jerez, Clark, Medina, & Ramirez-Marin, 2007; Watson, 2004). When viewed in this manner, such efforts may appear isolated. If these methods are indeed applicable to their particular classroom situations, teachers must take time to locate and learn the necessary techniques. Another option is spending in-service time learning district initiatives to meet the needs of ELL students in all disciplines. Historically, professional development in schools lacked a consistent focus, promoted isolation among educators, and produced short-lived results (Bainer & Wright, 1998; Lanier & Little, 1986; Meister, 2000; Moore & Hyde, 1981; Schlechty & Crowell, 1983). Teachers can grow weary of these various attempts and prefer the ease and comfort of the status quo.

Despite fragmented presentation, recent publications about teaching science to ELL students would suggest that successful learning depends primarily on the quality of instruction, not the quantity (Hansen, 2006; Medina-Jerez et al., 2007). If teachers are to genuinely support these efforts, they must be shown how the synergetic benefits of ELL and science instruction are, indeed, worth their time.

FOCUS AND RESEARCH QUESTIONS

This chapter delves deeper into the standards and strategies of two apparently different areas—science and English language learning. For over a decade, the National Science Education Standards (NSES) have outlined the benchmarks of success for student science learning in grades K–12 (NRC, 1996). More recently, two separate organizations—the World-Class Instructional Design and Assessment Consortium (WIDA) and Teachers of English to Speakers of Other Languages, Inc. (TESOL)—have produced matching English language proficiency (ELP) standards for ELL students from pre-Kindergarten through grade 12 (TESOL, 2006; WIDA, 2007b). In addition to standards, strategies—or models of instruction—have outlined methods for effectively reaching students and teaching content. One primary model for planning for and teaching ELL students is sheltered instruction (SI), predominantly portrayed in published form through the Sheltered Instruction Observation Protocol (SIOP) (Echevarria, Vogt, & Short, 2008). Science education, meanwhile, has largely emphasized an inquiry-based approach to learning both content and process skills (NSTA, 2004).

The present study seeks to examine the above exemplars in English language learning and science education and determine the degree of alignment among these key documents. This study does not evaluate nor make any claims on the effectiveness of these documents on students' science or English language achievement. Rather, the focus is to compare components of these texts and determine how much they agree or disagree on educating ELL students. The following questions guide this research:

1. In what ways are ELL (ELP) and science education (NSES) standards similar? Different?
2. In what ways are ELL (SIOP) and science education (Science Inquiry) strategies similar? Different?
3. How might potential disparities among ELP/SIOP and NSES/Science Inquiry strategies be reconciled?

If true compatibility exists among these documents, how might teachers and other educators make the most of such cohesion—and make the most of their time—to produce synergetic instruction and successful ELL student learning?

METHODOLOGY

This chapter features a content analysis (Esterberg, 2002) of salient publications in English language learning (ELL) and science instruction. The primary data sources are found in Table 5.1.

Research methods included an open analysis to distinguish the dominant messages and subject matter within the text (McKeone, 1995). In-depth surveys of the documents in Table 5.1 resulted in a considerable collection of exemplars. These acquired outcroppings were chosen to demonstrate the principal tenets and traits of each document type. In content analysis, "the

TABLE 5.1 Documents Featured in Content Analysis

English Language Learner (ELL) Standards	Science Education Standards
English Language Proficiency (ELP) Standards Published by World-Class Instructional Design and Assessment (WIDA) Consortium, 2007[a]	National Science Education Standards (NSES) Published by National Research Council (NRC), 1996
ELL Instructional Strategies	**Science Education Strategies**
Eight "Model Components" of the Sheltered Instruction Observation Protocol (SIOP) (Echevarria, Vogt, & Short, 2008; Pearson Education, Inc., 2008)	Inquiry-Based Science Instruction (including the Learning Cycle, a prominent rendition) (Colburn, 2000b; Colburn & Clough, 1997; NSTA, 2004)

[a] Since WIDA has published the most recent ELP Standards, this updated version will be featured over the nearly identical TESOL product. Both documents, however, provide materials for analysis.

general assumption is that intention and meaning are discoverable in the frequency with which words, phrases, idioms or ideas occur in a text and the meaning can be captured in a set of predefined content variables" (Truex, 1996, p. 1). Data—content variables—were categorized and classified for the presentation of results and for the subsequent comparative analysis.

In addition to analyzing primary and supplementary documents, the author completed a formal three-day, interactive training session on the SIOP Model presented by representatives of the SIOP Institute. The author also acted as a science advisor for the Iowa Department of Education's Enhancing ELL Science Instruction Institute. These experiences—combined with the author's background in teaching science and science teachers—assisted to equip a more comprehensive review of English language learning and science education paradigms.

RESULTS OF DOCUMENT CONTENT ANALYSIS

Each document was first reviewed to provide a sufficient frame of reference. These data assisted with the subsequent comparative analysis. The following descriptions summarize historical developments and key features of the studied documents.

WIDA English Language Proficiency (ELP) Standards

Comprised of representatives from sixteen states and housed at the Wisconsin Center for Education Research, the World-Class Instructional Design and Assessment Consortium (WIDA) is a non-profit group dedicated to help ELL students achieve the requirements of the No Child Left Behind Act (WIDA, 2007a). WIDA first created the English Language Proficiency (ELP) Standards in 2004 and revised them in 2007. The standards have been described as "a curriculum and instruction planning tool. They help educators determine children's ELP levels and how to appropriately challenge them to reach higher levels" (WIDA, 2007b, p. 1). For the purposes of this study, ELP Standard 4 (the language of science) is the featured standard, as it is the only one that directly refers to science education. It states, "English language learners *communicate* information, ideas and concepts necessary for academic success in the content area of *Science*" (Gottlieb, Cranley, & Oliver, 2007, p. 10, emphases in original).

As noted earlier, ELP standards have been published both by WIDA and TESOL. Although separate entities, these two organizations have leapfrogged each other in publishing versions of the ELP standards, each one building off of the other in their revision efforts. Since WIDA produced the most recent

ELP standards, this edition received the most attention. In producing the 2007 ELP standards, WIDA provided further clarification of the various levels of language proficiency, including examples with science concepts.

The ELP standards feature the four domains of language: listening, speaking, reading, and writing. In most social or academic situations, all four domains are interdependent and used in any number of combinations. However, the ELP standards documents parse them out into fragments to ease their use in instruction, curriculum, and assessment (TESOL, 2005). With respect to ELL students' language development, the four domains are described in Table 5.2.

ELP standards also classify levels of language development: (1) Entering, (2) Beginning, (3) Developing, (4) Expanding, (5) Bridging, and (6) Reaching. These levels "reflect the complexity of second language development. They provide specific descriptions of language acquisition stages for use in instruction and assessment" (TESOL, 2005, p. 25). As students acquire a second language, they advance along these levels of proficiency and gradually move to more challenging facets of communication. Although second-language acquisition follows the general progression from (1) Entering to (6) Reaching, each student will proceed along a unique path, due to individual differences (Gottlieb, Cranley, & Oliver, 2007). Disparate factors include age and grade level, linguistic and cultural backgrounds, experiences in life and education, and possible diagnoses, such as learning disabilities.

The ELP grade level clusters (Pre-K–K, grades 1–2, grades 3–5, grades 6–8, and grades 9–12) and language domains (listening, speaking, reading, and writing) are used to further explicate each standard. For science, a certain topic is specified for a particular grade level cluster, identified by the language domain, and then delineated according to the language proficiency indicators. The ELP Science Standard does not elaborate on par-

TABLE 5.2 Definitions of Four Language Domains with Respect to ELP Standards

Listening	Process, understand, interpret, and evaluate spoken language in a variety of situations.
Speaking	Engage in oral communication in a variety of situations for a variety of purposes and audiences.
Reading	Process, understand, interpret, and evaluate written language, symbols, and text with understanding and fluency.
Writing	Engage in written communication in a variety of situations for a variety of purposes and audiences.

Source: "Understanding the WIDA English Language Proficiency Standards: A Resource Guide," by M. Gottlieb, M. E. Cranley, and A. R. Oliver, 2003. Retrieved July 22, 2008, from http://www.wida.us/standards/Resource_Guide_web.pdf.

ticular disciplines of science (biology, chemistry, etc.) or specific concepts (form and function, systems, mechanics, etc.). WIDA does provide model performance indicators (MPIs) with examples of science topics for each grade cluster and language domain within that age group. Specific MPIs are limited to only one language domain. Yet, student skills or behaviors are described according to different levels of proficiency in the English language. Table 5.3 shares an example of MPIs from the WIDA 2007 *ELP Resource Guide*.

Note that WIDA does not provide any description for students at the highest level of language proficiency—Level 6, "Reaching." This absence may be due to the creators assuming that students attaining this level of proficiency do not need a specialized standard for English language learning. The highest level described—Level 5, "Bridging"—is distinct from Levels 1–4 in that the first four levels of proficiency include a form of support (visual aids such as pictures, examples, diagrams; small groupings; oral or written prompts; etc.) for the language learner.

The MPIs of the ELP standards help science teachers acquire an understanding of what students can do to exhibit proficiency. However, these specific examples are limited both in science content and in language domain. In order to ensure successful ELL science instruction, teachers also need strategies from which they can choose and apply in the appropriate classroom setting.

SIOP Model ELL Instructional Strategies

Although not explicitly stated in prominent WIDA or TESOL standards documents, a fundamental tenet of the ELP standards is SI:

> Sheltered instruction is an approach that can extend the time students have for getting language support services while giving them a jump-start on the content subjects they will need for graduation. SI is not simply a set of additional or replacement instructional techniques that teachers implement in their classrooms. Instead, it draws from and complements methods advocated for both second language and mainstream classrooms. (Echevarria et al., 2008, p. 13)

The SI approach is typically used for ELL students who have enough proficiency in the English language to be able to function in a mainstream classroom, as opposed to ELL students with little or no English ability.

The Sheltered Instruction Observation Protocol, or SIOP, was originally developed as a classroom observation tool to gauge teachers' incorporation of techniques in SI. Specifically, the creators of SIOP sought a model that promoted and assessed consistent, effective SI (Echevarria et al., 2008). After

TABLE 5.3 Sample Model Performance Indicators (MPIs) for WIDA ELP Standard 4: The Language of Science

Level 1—Entering	Level 2—Beginning	Level 3—Developing	Level 4—Expanding	Level 5—Bridging
Grade Cluster: 3–5 Language Domain: *Writing* Example Topic: *Solar System*				
Copy names of astronomical objects associated with the solar system from labeled diagrams (e.g., "planets," "stars")	Describe features of astronomical objects from labeled diagrams	Compare/contrast astronomical objects from diagrams or graphs (e.g., size, distance from sun)	Discuss relationships between astronomical objects from diagrams or graphs	Evaluate potential usefulness of astronomical objects (e.g., life on the moon, solar power)
Grade Cluster: 6–8 Language Domain: *Listening* Example Topic: *Scientific tools or instruments*				
Match scientific tools or instruments with pictures from oral statements (e.g., sundial)	Classify scientific tools or instruments with pictures and labels from oral directions (e.g., "Telescopes and sundials go with the sky.")	Identify examples of scientific tools or instruments and their uses from pictures and oral descriptions	Compare/contrast examples of scientific tools or instruments and their uses from oral descriptions (e.g., differences between telescopes and microscopes)	Infer uses of scientific tools or instruments from oral reading of grade level material
Grade Cluster: 9–12 Language Domain: *Writing* Example Topic: *Taxonomic systems*				
Label examples from different taxonomies using illustrations and word/phrase banks (e.g., one-cell plants and animals)	Describe in sentences features of taxonomies depicted in illustrations or graphic organizers	Summarize in a series of related sentences features of taxonomies depicted in illustrations or graphic organizers	Compare and contrast in paragraph form features of taxonomies depicted in illustrations or graphic organizers	Integrate information about taxonomic systems into essays or reports

Source: From "English Language Proficiency Standards for English Language Learners in Grade 6 through Grade 12," by World-Class Instructional Design and Assessment, 2007. Retrieved July 25, 2008 from http://www.wida.us/standards/6-12%20Standards%20web.pdf and "English Language Proficiency Standards for English Language Learners in Pre-Kindergarten through Grade 5," by World-Class Instructional Design and Assessment, 2007. Retrieved July 25, 2008 from http://www.wida.us/standards/PreK-5%20Standards%20web.pdf.

its testing and refinement as a teacher observation protocol in the 1990s, the team of developers recommended that SIOP be used not only as an observation tool (SIOP protocol), but also as a model for planning and delivering lessons (SIOP Model). "The theoretical underpinning of the model is that language acquisition is enhanced through meaningful use and interaction" (Echevarria et al., 2008, p. 16). In an ideal SIOP lesson, ELL students study relevant subject content (science, for example), while simultaneously developing their English language skills—listening, speaking, reading, and writing.

Organization of the SIOP Model for instruction is based on the SIOP observation protocol. The protocol breaks lessons down into eight *components*, each containing between three and six features. A total of 30 *features* are assessed in the SIOP protocol during classroom observation. Through SIOP's transition from observation instrument into model, its eight components and subsequent features remain intact, but serve a different function. Rather than being criteria for scoring a teacher's lesson, the components have become tools for teachers in planning and delivering lessons for systematic, successful SI. The components are not in any particular order or hierarchy, except for the first, Lesson Preparation, which is essential for beginning effective SI lessons (Echevarria et al., 2008).

SIOP Model Component 1: Lesson Preparation

"If properly prepared, a lesson will include most of the SIOP features in advance. It is then up to the teachers and class to accomplish them as the lesson unfolds" (Echevarria et al., 2008, p. 24). Depending on the content and context, SIOP lessons may last for one class period or may require several days to complete. Features in the lesson plan component include the following, as organized by the creators (Echevarria et al., 2008, emphases in original):

1. Content objectives clearly defined, displayed, and reviewed with students.
2. Language objectives clearly defined, displayed, and reviewed with students.
3. Content concepts appropriate for age and educational background.
4. Supplementary materials used to a high degree.
5. Adaptation of content to all levels of student proficiency.
6. Meaningful activities that integrate lesson concepts with language practice opportunities.

SIOP Model Component 2: Building Background

SIOP reaffirms research and theory promoting the need for connecting students' backgrounds and experiences to their learning (Brooks & Brooks, 1999). Lessons without such a link breed confused or disinterested students. The following summarizes the three features of SIOP Model Component 2: Building Background.

1. Concepts explicitly linked to students' background experiences.
2. Links explicitly made between past learning and new concepts.
3. Key vocabulary emphasized (e.g., introduced, written, repeated, and highlighted for students to see).

SIOP Model Component 3: Comprehensible Input

The SIOP creators cite this component to be one area making their model different than other effective instructional strategies for mainstream students. Intentional efforts for comprehensible input (Krashen, 1985) can vary from using visual clues during a lesson to the teacher using slower, clearly enunciated speech with repetition. Decisions regarding comprehensible input rely on the particular needs of the students.

1. Speech appropriate for students' proficiency levels (e.g., slower rate, enunciation, and simple sentence structure for beginners).
2. Clear explanation of academic tasks.
3. A variety of techniques used to make content concepts clear (e.g., modeling, visuals, hands-on activities, demonstrations, gestures, body language).

SIOP Model Component 4: Strategies

This component addresses specific classroom *teaching strategies* for helping students make connections between past and new learning, solve problems, and retain information. It also includes teaching *learning strategies* to students so they can improve their abilities to process content. Learning strategies are defined as "the special thoughts or behaviors that individuals use to help them comprehend, learn, or retain new information" (O'Malley & Chamot, 1990, p. 1). The features of Component 4 highlight the use of both teaching and learning strategies in the classroom.

1. Ample opportunities provided for students to use learning strategies.
2. Scaffolding techniques consistently used, assisting and supporting student understanding.
3. A variety of questions or tasks that promote higher-order thinking skills.

SIOP Model Component 5: Interaction

In order to learn language, ELL students must have the chance to practice using it (Day, 1986). SIOP highlights the need for students to employ academic language in addition to social language.

1. Frequent opportunities for interaction and discussion between teacher/student and among students, which encourage elaborated responses about lesson concepts.
2. Grouping configurations support language and content objectives of the lesson.
3. Sufficient wait time for student responses consistently provided.
4. Ample opportunities for students to clarify key concepts in L1 (first language) as needed with aide, peer, or L1 text.

SIOP Model Component 6: Practice and Application

Teachers must plan opportunities for students to practice and apply both content and language. In this context, ELL students work toward mastery of academic language in all four domains: reading, writing, listening, and speaking. To assist this development, the SIOP Model promotes the following instructional features for practice and application:

1. Hands-on materials and/or manipulatives provided for students to practice using new content knowledge.
2. Activities provided for students to apply content and language knowledge in the classroom.
3. Activities integrate all language skills (i.e., reading, writing, listening, and speaking).

SIOP Model Component 7: Lesson Delivery

If preparing lessons (Component 1) is a key element of SI, lesson delivery is equally as important, both in following through and adapting the lesson to ensure successful learning. Features of this component measure the extent to which the teaching of a lesson matches the original intent and appropriate modifications for ELL students.

1. Content objectives clearly supported by lesson delivery.
2. Language objectives clearly supported by lesson delivery.
3. Students engaged approximately 90% to 100% of the period.
4. Pacing of the lesson appropriate to students' ability levels.

SIOP Model Component 8: Review and Assessment

Although it is last in order of components, reviewing and assessing student learning should occur throughout the entire lesson. Frequent review is essential for ELL students.

> Unless the teacher takes the time to highlight and review key information and explicitly indicate what students should focus on and learn, English learners may not know what is important.... [ELL students] are much less able to

> evaluate which pieces of information among all the input they receive are important to remember. That is why the teacher must take the time to review and summarize throughout a lesson and particularly as a wrap-up at the end. (Echevarria et al., 2008, p. 168)

Like instruction, assessment requires multiple modes to completely reach and teach students. Assessment also involves gathering data about students to determine the extent to which objectives have been met.

1. Comprehensive review of key vocabulary.
2. Comprehensive review of key content concepts.
3. Regular feedback provided to students on their output (e.g., language, content, work).
4. Assessment of student comprehension and learning of all lesson objectives (e.g., spot checking, group response) throughout the lesson.

Supplementing SIOP Model Components

SIOP's creators provide multiple scenarios, examples, and activities for teachers to elaborate on their SI model components and features. These include their "favorites" for classroom preparation and instruction. Some of the supplemental materials are specific toward grade level or subject, although many are applicable or adjustable for teachers' particular classrooms and students' needs. Additional details of the SIOP components and features will be examined in the comparative analysis with science education strategies.

National Science Education Standards

The National Research Council published the National Science Education Standards (NSES) in 1996. The NSES were developed to facilitate the goal of scientific literacy for all students.

> An understanding of science makes it possible for everyone to share in the richness and excitement of comprehending the natural world. Scientific literacy enables people to use scientific principles and processes in making personal decisions and to participate in discussions of scientific issues that affect society. (NRC, 1996, p. ix)

The NSES state a variety of standards appropriate for various aspects of science education: science teaching, professional development, assessment, education programs, and systems. The focus of the present analysis will be the "Standards for science content" (NRC, 1996, p. 3), as these standards are the essential benchmarks for instructing and measuring students' scientific literacy—"what students should know, understand, and be able to do in natural science" (p. 103).

The NSES divide content standards into eight categories. Each category contains content or skills to be learned within that topic. With the exception of the first NSES category—Unifying Concepts and Processes in Science—each category features content unique to the particular grade level cluster: K–4, 5–8, or 9–12 (NRC, 1996).

The publishers of NSES point out that the presentations of each standard and subsequent development of student understanding "are illustrative, not proscriptive. Similarly, the discussion of each standard concludes with a guide to the fundamental ideas that underlie that standard, but these ideas are designed to be illustrative of the standard, not part of the standard itself" (NRC, 1996, p. 6). What follows is a brief summary of each NSES content standard.

NSES Content Standard: Unifying Concepts and Processes in Science

When scientists examine and describe the complex mechanisms of the natural universe, they look to identify patterns, constancy, and other simplified relationships. Likewise, students can learn science concepts and processes through recognition and application of similar models. This first content standard is identical across all grade level clusters, "because the understandings and abilities associated with these concepts need to be developed throughout a student's educational experiences" (NRC, 1996, p. 6).

NSES Content Standard A: Science as Inquiry

Interestingly, this standard and components are also identical across grade levels. The promotion of these qualities, however, will be different according to the developmental capabilities of the age group. According to the NSES, students in all grades must learn science through the inquiry process—asking questions, doing investigations, using observations, constructing explanations, and communicating ideas among peers. Science as inquiry is both a concept and a process. Inquiry-based science as a teaching method will be explored in greater detail when discussing science education strategies.

NSES Content Standards B, C, D: Physical, Life, Earth, and Space Sciences

These three standards follow the consensus of dividing science into domains. This is the "subject matter" commonly recognized and remembered as school science curriculum. "Science subject matter focuses on the science facts, concepts, principles, theories, and models that are important for all students to know, understand, and use" (NRC, 1996, p. 106). As students progress in grade level, they experience and learn concepts of a more abstract nature with more meticulous vocabulary.

NSES Content Standard E: Science and Technology

This standard is not simply technology education. Instead, it addresses the link between the two fields and the actions necessary for designing and doing science. Like science as inquiry, this standard promotes the dual development of conceptual understanding and applied ability.

NSES Content Standard F: Science in Personal and Social Perspective

A complete science education requires meaningful connections to personal and social issues. "Central ideas related to health, populations, resources, and environments provide the foundations of students' eventual understandings and actions as citizens" (NRC, 1996, p. 139). This standard is also identified to be essential in promoting students' decision-making skills.

NSES Content Standard G: History and Nature of Science

A historical perspective enables students to learn about the interactions between culture and science. The "nature of science" (NOS) refers to the central values and assumptions found in the development of scientific knowledge (Lederman & Zeidler, 1987). It includes what science is, how it works, the foundations of science, the social interactions of scientists, and the reciprocal role between science and society (Clough, 2003). To different degrees, students at all grade levels can learn how society and individuals impact the development of science and vice versa. Understanding historical contexts and NOS enhances and equips students' learning of science concepts and skills.

Assessment and Teaching Standards

In addition to science content standards, the NSES do present standards for assessment and teaching (NRC, 1996). However, neither is featured in this study, due to space considerations and the peripheral connection of these particular standards to the present research. The function of NSES assessment standards is to evaluate the quality of assessment, including consistency, fairness, data collection, and application. The NSES teaching standards, too, are more programmatic in nature and outline teacher qualities, as opposed to student expectations. Teaching standards include planning inquiry-based programs, facilitating learning, preparing assessments, and developing environments conducive to learning. These issues are important and are more appropriate in the upcoming discussion of science education strategies, namely, inquiry-based instruction.

Science Inquiry Instructional Strategies

In their introduction of the NSES, the leading creators note that the standards emphasize "inquiry as a way of achieving knowledge and understanding about the world" (NRC, 1996, p. ix). According to the NRC (1996, p. 105), inquiry-based science helps students develop

- Understanding of scientific concepts.
- An appreciation of "how we know" what we know in science.
- Understanding of the nature of science (NOS).
- Skills necessary to become independent inquirers about the natural world.
- The dispositions to use the skills, abilities, and attitudes associated with science.

As a way to learn science, inquiry frames itself around the processes scientists use in their studies and investigations. The NSTA's Position Statement on Scientific Inquiry describes the origins of this approach to teaching science:

> Scientific inquiry reflects how scientists come to understand the natural world, and it is at the heart of how students learn. From a very early age, children interact with their environment, ask questions, and seek ways to answer those questions. Understanding science content is significantly enhanced when ideas are anchored to inquiry experiences. (NSTA, 2004, p. 1)

For the purposes of this discussion, "inquiry" will refer to its use as a strategy for teaching and learning science content and skills.

Inquiry is not a new concept to science teaching (Victor, 1974), but it can be a topic of confusion to science educators, partly due to its multiple meanings. Colburn (2000b) defines inquiry-based instruction as "the creation of a classroom where students are engaged in essentially open-ended, student-centered, hands-on activities" (p. 42). Even as a teaching technique, inquiry can come in several different forms, often depending on the degree of student involvement and decision-making. The three most common approaches to inquiry are summarized in Table 5.4. All three are inquiry in that the teacher does not inform students of the potential outcomes or "answers" of the investigation.

Another common approach to inquiry-based instruction is the learning cycle, a product of the K–6 Science Curriculum Improvement Study (SCIS) program in the early 1970s (Atkin & Karplus, 1962; Coe, 2001; Karplus, 1977, 1979). The learning cycle typically contains three main phases: (1) Exploration: students have concrete experiences with content in an inquiry

TABLE 5.4 Common Forms of Inquiry-Based Science Instruction

Structured Inquiry	The teacher provides the question or problem to investigate, as well as the procedures and materials to use for investigation.
Guided Inquiry	The teacher provides the question/problem and materials. Students develop the procedures to complete the investigation.
Open Inquiry	Students devise the question/problem they want to investigate, as well as the procedures and materials for their investigation.

Source: From "An Inquiry Primer," by A. Colburn, 2000. *Science Scope, 23*(6), 42–44.

setting, during which they make their personal knowledge public and relate their prior ideas to the new learning; (2) Concept Introduction or Invention: with the teacher's guidance, students discuss, develop, and practice their ideas and skills, connecting experiences to relevant science concepts; and (3) Application or Expansion: students apply what they have learned to a new situation (Colburn & Clough, 1997; Lawson, Abraham, & Renner, 1989). These features of the learning cycle are not static steps to passively follow. Rather, the learning cycle offers teachers a framework with important pieces of effective science lessons.

As with science inquiry (Fradd, Lee, Sutman, & Saxton, 2001), the learning cycle has been modified into various forms and divisions (Barman, 1989; Chessin & Moore, 2004; Osborne & Wittrock, 1983; Ramsey, 1993), including the popular 5-E model (Bybee, 2002; Trowbridge & Bybee, 1990). These variations all support the notion that an inquiry-based approach is context-sensitive and adjustable according to teachers' and students' needs. Regardless of specific deviations, classroom inquiry can be identified by five "essential features," outlined by the National Research Council (2000, p. 29) in its text relating inquiry to the NSES:

1. Learner engages in scientifically oriented questions.
2. Learner gives priority to evidence in responding to questions.
3. Learner formulates explanations from evidence.
4. Learner connects explanations to scientific knowledge.
5. Learner communicates and justifies explanations.

Table 5.5 highlights key components of inquiry-based science education strategies, as found in multiple publications from the NSTA, including *Pathways to the Science Standards* (Lowery, 2000; Rakow, 2000; Texley & Wild, 2004). Used together, these science inquiry components emphasize student-focused investigation and initiation of learning. A popular idiom promoted by science educators is to "decookbook" traditional cookbook-type activities, modifying the latter by inserting features like those in Table 5.5 (Clark, Clough, & Berg, 2000; Shiland, 1997).

TABLE 5.5 Key Components of Inquiry-Based Science Instruction

Early Experiences	Unlike traditional lessons with a pedantic lecture-lab-practice cycle, an inquiry-based lesson typically begins with students directly experiencing natural phenomena (Colburn & Clough, 1997). Such introductions can range from open exploration to focused investigation guided by challenge questions or prompts.
Hands-On Materials	To ensure meaningful learning, students' early experiences must involve appropriate hands-on materials, equipment, and tools (NSTA, 2004). Concrete encounters provide a foundation for students in developing understandings of more abstract concepts.
Student Ownership	In order for students to directly experience inquiry, they must be active and responsible for their efforts. Common practices for increasing student ownership are having them decide which observations and data are significant (Clark, Clough, & Berg, 2000), how they will record and organize these findings, what experimental procedures they will use, or what questions they will investigate (Colburn, 2000b).
Student–Student Interactions	Students develop responsibility when they hold each other accountable. As scientists must collaborate in their research endeavors, so too must students work in groups during inquiry-based lessons. Discussions in small group and whole class settings facilitate sharing of ideas and provide students the opportunity to communicate and defend their ideas (NSTA, 2004).
Teacher's Role	The teacher plays a key role in the inquiry-based science classroom. "Effective teaching is a highly interactive activity, and curricula alone cannot create this interaction" (Clark, Clough, & Berg, 2000, p. 43). Open-ended questioning, sufficient wait-time, student-centered responses without answering, and management are all essential for fostering engaged student thinking and productive classroom interactions (Colburn, 2000b; Penick, Crow, & Bonnstetter, 1996).
Assessment	Questions and discussions provide opportunities for the teacher to gauge students' levels of comprehension. Through interactions, assessment occurs continually during an inquiry-based lesson. Other formal assessments may include having students present their findings and conclusions or having students review and evaluate additional investigations (Shiland, 1997).
Sense of Wonder	Modeled after means by which scientists question and study the surrounding universe, inquiry-based science instruction seeks to instill that same sense of curiosity in all students. Teachers are encouraged to "implement approaches to teaching science that cause students to question and explore and to use those experiences to raise and answer questions about the natural world" (NSTA, 2004, p. 1).

RESULTS: COMPARATIVE ANALYSIS

Research Question 1: Complementary Standards

Both the NSES and ELP Standards allude to meeting the needs of all students' complete education. Understandably, the NSES elaborate in much more detail on science, and ELP on English language learning. The science standards claim to "apply to all students, regardless of age, gender, cultural or ethnic background, disabilities, aspirations, or interest and motivation in science" (NRC, 1996, p. 2). Such an encompassing spirit is admirable, but the NSES are mostly silent as to the particular needs of ELL students. The ELP standards go further with respect to the specific, significant group of ELLs. In their description of the ELP, TESOL cites standards of the NSTA as the "anchor" for their science language proficiency standard: "Communication of information, ideas, and concepts of science" (TESOL, 2005, p. 10).

Besides highlighting communication, the ELP standards themselves are considerably bare with respect to science education for ELL students. The MPI do, indeed, provide further information, although these serve less as standards and more as exemplars for teachers when planning assessments and activities. Nevertheless, these examples can provide material for comparison with the content promoted by the NSES. As stated in the discussion of ELP standards, the MPIs provide support strategies for students in Levels 1–4 of language acquisition. At Level 5 (Bridging), the MPIs do not include explicit language supports. Indicators at this level are identical to a standard science lesson objective written in alignment with the NSES, with no particular consideration for students' language abilities. Teachers with ELL students at the Bridging level could, in theory, write lesson objectives common for the entire class. It is essential, therefore, for teachers to be aware of the particular language proficiency level (1–6) of their ELL students.

Table 5.6 itemizes the MPIs for the ELP science standard with respect to the NSES content standards categories. The MPIs are divided according to ELP grade cluster and include both summative and formative frameworks, since the NSES does not delineate on measuring at the process or outcome level. Each MPI featured in the WIDA 2007 ELP *Resource Guide* was included in the analysis. Matching to the closest NSES standard was based on the most prominent topic provided in the individual MPI. For visual representation, the symbol "×" is used to designate each match.

The three NSES standards mentioned most often in the ELP MPIs were all traditional science content: physical (B), life (C), and earth and space science (D). Other categories with substantial attention in the ELP exemplars are unifying concepts/processes and science in personal and social perspectives (NSES F). The inclusion of the first NSES category may be due to the underlying presence of unifying concepts and processes among the

TABLE 5.6 Comparison of Content in NSES and ELP Standards

NSES Standard	Number of Appearances in Model Performance Indicators (MPIs) of ELP Standard for Science (WIDA 2007 Resource Guide)					
	PreK–K	1–2	3–5	6–8	9–12	Total Matches
Unifying Concepts and Processes	×	×		××	×	5
A. Science as Inquiry	×				×	2
B. Physical Science	×	××	×	×××	×××	10
C. Life Science	××××	×	××		××	9
D. Earth and Space Science	×	××	×××	×		7
E. Science and Technology						0
F. Science in Personal and Social Perspectives		××	××	×	×	6
G. History and Nature of Science				×		1

science subjects. Students enrolled in life, physical, and earth and space science courses will learn specific content along with fundamental principles such as form and function; systems, order, and organization; evolution and equilibrium; and more. The other non-subject-specific NSES category with considerable attention—science in personal and social perspectives (F)—may have this level of attention due to SI's emphasis on language development "through meaningful use and interaction" (Echevarria et al., 2008, p. 16). Students learning about the personal and social implications of science content will most likely find content meaningful and have several opportunities for interaction.

Interestingly, the two NSES standards with the fewest appearances in the ELP standards' MPIs are history and nature of science (G, one match) and science and technology (E, zero matches). Similar to Standard F, these two categories would create occasions for students to experience meaningful content and interact with peers. Likewise, science as inquiry (NSES A) lends itself to multiple opportunities for development and assessment of ELL students' language in all four domains—writing, reading, speaking, and listening. Unfortunately, the inquiry category receives little mention in the ELP standards' model indicators (two matches). The relative absence of this category is significant due to its level of prominence both in the NSES and in its primary strategic role in science instruction.

Individually, the standards for science and ELL instruction are not sufficient to address the needs of ELL students in science. Despite noble intentions, the NSES fail to identify explicit expectations or modifications for students at various levels of language development. This topic is thoroughly developed and described by the ELP standards. Although the ELP Standards provide only a general allusion to language use and learning in a science setting, the NSES outline particular science content and processes for grade levels. Therefore, the two standards documents are interdependent and complementary, both necessary for the successful science education of ELL students.

Research Question 2: Common Strategies

English language learning and science teaching strategies also complement one another. As opposed to the standards—in which one field supports the counterpart's weaknesses—the strategies are nearly identical. The term "strategies" is featured as one component (#4) in the SIOP model for sheltered ELL instruction. Nevertheless, all eight create a framework for curriculum and instruction decisions. Similarly, science inquiry is a framework of doing and learning science, including strategies for planning and delivering lessons. Both SIOP and science inquiry follow a theoretical foundation on research into how students learn. The following paragraphs discuss these common themes, featuring several elements essential to the strategies of science inquiry and SI.

Connecting Content to Students' Experiences

Whether the link is to students' cultural backgrounds or to popular media, both inquiry and SIOP emphasize connections. Students' lives and experiences are a foundation from which to scaffold learning (Brooks & Brooks, 1999). In order to produce meaningful learning, teachers should seek to create vivid experiences for students and avoid heavy emphasis on early text reading. ELL students often struggle with reading comprehension as well as making connections between a school text and their current schemata. Regardless of cultural background or developmental level, learners who read without any concrete encounter often struggle to make sense of the abstract concepts. Early, tangible encounters enhance students' development of language abilities and science understandings. This approach—advocated by both SIOP and science inquiry—is also supported by the Piagetian perspective of developmental learning (Caine, 1990; Joyce & Weil, 2008; Piaget, 1964).

Two of the three features of the SIOP Model Component 2 (Building Background) focus on helping students connect present learning to their past experiences and education. Science inquiry also emphasizes early experiences from which students can build understanding, including the motto "do the lab first" (Colburn & Clough, 1997, p. 31) before extensive lecture or oral presentation. "Giving students direct experience with a concept before providing verbal instruction is critical in helping them relate the verbal abstractions to more meaningful concrete experiences" (Colburn & Clough 1997, p. 30). Such an approach teaches students science content as well as the science investigation processes. Students develop "the skills of observation, controlling variables, and asking operational questions" (Shiland, 1997, p. 18).

Meaningful and Memorable Materials

Meaningful connections to content often come through students' classroom experiences with concrete materials. According to Colburn (2000b), "the more familiar the activity, materials, and context of the investigation, the easier it is for students to learn through inquiry" (p. 43). The use of visual and hands-on materials is addressed in three of the eight components in the SIOP Model: Lesson Preparation (1), Comprehensible Input (3), and Practice and Application (6). As supported by both constructivist (Colburn, 2000a) and developmental learning theories (Piaget, 1964), content becomes more accessible when students encounter concrete examples and applications. "Supplementary materials provide a real-life context and enable students to bridge prior experience with new learning" (Echevarria et al., 2008, p. 33). Classroom materials promoted by both SIOP and science inquiry include the following:

- Hands-on manipulatives
- Realia (real-life objects)
- Pictures and illustrations from a variety of sources
- Visual aides such as models, transparencies, graphs, charts, maps, props
- Multimedia resources: videos, DVDs, interactive CD-ROMs, internet resources
- Demonstrations

Learning by Doing

Science inquiry promotes physical and mental activity in students. When modifying science labs to promote inquiry, teachers are encouraged to involve students in several key actions: choosing a question to investigate, defending relevant data, applying math in their analysis, setting goals, making decisions, and assessing progress (Clark, Clough, & Berg, 2000). Student involvement is central to several steps toward implementing inquiry-based science investigations (Colburn & Clough, 1997, pp. 31–33):

- Require students to decide how lab findings will be communicated.
- Have students invent the procedures to answer a lab question.
- Be sure students apply what they learned in the content phase.

The creators of SIOP cite learning to ride a bicycle as an illustration for students' need to use language in order to practice and demonstrate ability (Echevarria et al., 2008). The Practice and Application Component (#6) of SIOP contains two features emphasizing the habit of learning by doing. This includes learning four domains of language, even if only one is in primary use at a time. "Although the relationships among the [domains] are complex, practice in any one promotes development in the others" (Echevarria et al., 2008, p. 143). This student-centered process is "mutually supportive" for the language domains as well as science content. Through learning by doing, both SIOP and inquiry-based strategies employ the behavioral principles of guided and open practice and application (Champagne, & Hornig, 1987).

Opportunities for Application

An extension of the learning-by-doing strategy is creating opportunities for students to apply their learning. From a science inquiry perspective, "Application means using or recognizing previous ideas in a new situation" (Colburn & Clough, 1997, p. 33). SIOP's creators elaborate on the value of application in language learning: "For students acquiring a new language, the need to apply new information is critically important because discussing and 'doing' make abstract concepts concrete" (Echevarria et al., 2008, p. 141). This critical lesson component can feature a variety of activities in which students use knowledge, including those actions listed in Table 5.7.

Application activities involve additional practice to solidify student mastery (Champagne, & Hornig, 1987). Furthermore, as students apply their learning, they have opportunities to socialize and learn from each other (Vygotsky, 1978). As seen in Table 5.7, several occasions for application require interactions among students, another commonality between inquiry and SI strategies.

Student Groups and Interactions

Component 5 of the SIOP Model is Interaction. Two of its features are "frequent opportunities for interactions and discussion between teacher/student and among students" (Echevarria et al., 2008, p. 16) and "grouping configurations support language and content objectives of the lesson" (Echevarria et al., 2008, p. 17). Students' practice and application of language necessitate interaction not only with the teacher, but also their more numerous peers. SIOP promotes all kinds of interactions among students, emphasizing the need for students to use their academic language in such settings. The elements of inquiry-based science instruction afford and often require students to collaborate and communicate with each other.

Both English language learning and science inquiry strategies employ a variety of grouping structures, ranging from partners to whole group. De-

TABLE 5.7 Activities Requiring Student Application of Knowledge

Science Inquiry Applications (Colburn & Clough, 1997)	SIOP Applications (Echevarria, Vogt, & Short, 2008)
1. Design new or extended research projects	1. Cluster activities
2. Research literature on applications of concepts	2. Use graphic organizers
3. Employ skills in a field trip	3. Write a journal
4. Role play	4. Discussion circles
5. Debate	5. Work in groups, partners
6. Work with younger students on related topics	6. Oral and written reports

pending on the context, cooperative groups may be homogenous or heterogeneous via assorted categories such as gender, language background, language proficiency, conceptual understanding, developmental level, and more. Intentional social interactions enhance student learning through common experiences and language, creating opportunities for students to use and review their learning (Vygotsky, 1978).

The Critical Role of the Teacher

Frequent exchanges among students also allow the teacher to continually observe and assess individuals' development of content understanding and language ability. Both SIOP and science inquiry refute the teacher-centered, lecture-dominated classroom. In a student-centered classroom, however, the teacher is not a passive observer. SIOP and inquiry definitely stress the indispensable actions a teacher must take to ensure successful student learning. Six of the eight SIOP components include features that specifically refer to the teacher's actions and behaviors in the classroom. Likewise, science inquiry documents repeatedly include explicit attention to the teacher's role. Attention to the teacher reflects recurring research on the teacher's primary influence in student learning and educational change (Berliner, 1989; Cremin, 1961; Fullan, 1991; Good & Brophy, 1994; Langer & Applebee, 1987; Penick, Yager, & Bonnstetter, 1986; Shymansky & Penick, 1981). The following are ideal teacher behaviors common to both SIOP and science inquiry:

- Clearly enunciated speech
- Simple, straightforward language without confusing vocabulary
- Warm, welcoming gestures, body language, facial expressions, and eye contact
- Movement around classroom and repeated proximity with all students
- Frequent, individualized interaction with all students
- Modeling of expected behaviors and thinking processes

- A variety of open-ended, higher-order thinking questions (e.g., "Tell me about what you are thinking," "What do you think might happen if...?" or "How does that compare with...?")
- Sufficient wait-time I (after teacher question) and wait-time II (after student response) so all have time to think
- Symmetrical responses repeating or paraphrasing what students say (without criticism or excess praise) and encouraging more student contributions and critical thinking (e.g., "What do you mean by...?" or "Tell me more about...")

By itself, an activity will not ensure meaningful and correct learning. Teachers must impart explicit attention to connections among concepts and previous learning. Such remarks can often occur through the skillful use of questions. Open-ended, or divergent, questions are not reserved for proficient English language speakers. The creators of SIOP are quick to point out that it is possible for teachers "to reduce the linguistic demands of responses while still promoting higher levels of thinking" (Echevarria et al., 2008, p. 103). At issue is research supporting the need for teachers to draw out students' ideas and bring them to the forefront of discussion and thought. As students construct understanding, they each develop conceptions that may or may not be correct (Appleton, 1997; Posner, Strike, Hewson, & Gertzog, 1982). Drawing out students' ideas through interactions engages students in thinking and assists the teacher in measuring comprehension. "Asking questions that force students to show their thinking allows teachers to diagnose students' conceptions and then structure future lessons accordingly" (Colburn & Clough, 1997, p. 33). Truly effective questioning requires a combination of the behaviors previously listed. These actions are critical for creating positive interactions with students (Pintrich, Marx, & Boyle, 1993), so that productive learning occurs. Such behaviors are also key in ensuring effective time and student management.

Time and Student Management

The instructional approach promoted by both SIOP and science inquiry is a paradigm shift from the traditional classroom. Top-down, teacher-dominated lecture will not result in meaningful learning of science nor a mastery of language. The student-centered design common to both strategies still relies heavily on a proactive, purposeful teacher. Sheltered and inquiry-based instruction, therefore, require teachers to effectively manage the classroom. A teacher who wishes to create a classroom of student groups developing and implementing science investigations must maintain discipline (Colburn, 2000b). Student management often occurs through student engagement. According to SIOP, effective teachers "plan to use the entire class period efficiently, teach in ways that engage students, and make sure students are engaged in activities that specifically relate to the material on which they will be assessed" (Echevarria et al., 2008, p. 156).

Time management does not necessarily mean a rapid pace. Both SIOP and science inquiry strategies readily agree that lessons may require several days. As seen in the common strategies above, students need suitable time to explore, examine, discuss, process, practice, and apply their learning. Similarly, students need appropriate content that is not too easy or too difficult. "Maximum learning... occurs when the activities are 'just right'—cognitively challenging, but still doable" (Colburn, 2000b, p. 43). Teachers must be attentive to the specific student needs and classroom contexts. Successful learning of language and science forms when students have appropriate challenges and suitable time to master them.

Research Question 3: Strategy Discrepancies and Overview

A review of the common strategies presented indicates high agreement between sheltered ELL language instruction and inquiry-based science education. However, discrepancies do exist between the two models. The following paragraphs discuss the differences and possible reconciliation. This section finishes with an overview of the strategies that overlap and distinguish SIOP and science inquiry.

Lesson Objectives

Perhaps the most noticeable difference between SIOP and science inquiry is in lesson objectives. Both promote lesson preparation with planned objectives providing a framework for instruction. Objectives pertain to a specific lesson and "identify what students should know and be able to do" as a result of the lesson (Echevarria et al., 2008, p. 24). The SIOP model goes further and separates objectives into two categories: content and language. Whereas science typically focuses on content, SIOP adds objectives that focus on students' language development as a result of the lesson. Language objectives may involve one or a combination of language domains: writing, reading, listening, and speaking.

The addition of language objectives, however, is not meant to double teachers' work of planning lessons. Rather, SIOP encourages teachers to think of their subject with respect to how it promotes and practices students' language ability. Based on their ELL students' development, teachers can "write an objective that all students should attain based on content concepts in the lesson but adjust the intended outcomes to match the students' [language] ability levels" (Echevarria et al., 2008, p. 28). Language objectives may simply be an elaboration of how students will communicate their content understanding. Language objectives can also serve as an extension or application of the students' content learning. The following are two pairs of content/language objectives from the creators of SIOP (p. 30).

> *Content Objective*: Students will solve word problems using a two-step process.
>
> *Language Objective*: Students will write a word problem for a classmate to solve requiring a two-step process.
>
> *Content Objective*: Students will be able to identify specific landforms on a map of South America.
>
> *Language Objective*: Students will be able to present an oral report about one landform and its influence on economic development.

These two examples illustrate the variety of ways teachers can plan for both content and language development. Two kinds of objectives are not competitive, but rather complimentary. One technique for planning lessons is to consider content objectives as the "what" students will learn and language objectives as the "how" students will develop and display understanding. With some intentional thought, science teachers can quickly identify opportunities to incorporate language development with learning content.

Overt Outcomes and Objectives

Throughout its discussion of content and language objectives, the SIOP Model encourages teachers to be explicit with the class about expected lesson outcomes. This typically plays out through a routine in which the teacher writes daily lesson objectives somewhere in the classroom (e.g., the chalkboard or overhead screen), points at the objectives at the lesson's beginning, and has the class recite the agenda for that day. Feature 11 of SIOP Model Component 3 (Comprehensible Input) advocates "clear explanation of academic tasks." According to SIOP's creators, "It is critical for [ELL students] to have instructions presented in a step-by-step manner, preferably modeled or demonstrated for them" (Echevarria et al., 2008; p. 81).

Teachers using an inquiry-based approach to science instruction will clearly see the conflict between inquiry's support for exploration and SIOP's promotion of upfront exposition. SIOP advances a straightforwardness that—from an inquiry standpoint—may ruin the curious, open-ended nature of scientific investigation. The NSTA position statement (2004, p.2) on scientific inquiry recommends that teachers help students understand the following about inquiry:

- That science involves asking questions about the world and then developing scientific investigations to answer their questions.

- That there is no fixed sequence of steps that all scientific investigations follow. Different kinds of questions suggest different kinds of scientific investigations.
- That scientific inquiry is central to the learning of science and reflects how science is done.

Guided and open inquiry—along with several learning cycle variations—all create situations in which students should *not* know the expected outcomes or even procedures to use in learning content. Therefore, how do science teachers of ELL students reconcile these differences?

Perhaps the best insight is a restatement of the second trait of scientific inquiry shared by the NSTA above: Different kinds of lessons suggest different kinds of instructional approaches. Learning is context-sensitive. Depending on the situation, each lesson will lend itself to a different level of explicit explanation. Teachers may be able to clearly state expected *lesson* outcomes (e.g., complete the investigation; determine what affects reaction rate; observe as many changes as possible) without giving away the content outcomes of the inquiry activity itself. Furthermore, even in moments needing direct instruction, teachers can use appropriate behaviors to communicate the sense of wonder that perpetuates science inquiry. Another special consideration given by SIOP is for teachers to accompany any oral directions with written ones, so that ELL students "can refer back to them at a later point in time as they complete the assignment or task" (Echevarria et al., 2008; p. 81). According to the SIOP creators, such clarity of instructions is helpful for all students. This practice reaffirms research into the supportive role of visual and concrete representations on students' learning.

Overview of SIOP and Science Inquiry Strategies

Even with much overlap, SIOP and science inquiry do not promote identical instructional strategies. Nevertheless, potential conflict may be negotiated to create meaningful science and language learning experiences for all students. Figure 5.1 illustrates the common strategies of these two models as well as differences. Educators are encouraged to create constructive compromise in moments of disagreement, as discussed above.

CONCLUSIONS AND IMPLICATIONS: TIME TO DECIDE, THINK, AND ACT

The education profession can place strenuous demands on teachers' energy and time (Bergman, 2008; Bransford, Brown, & Cocking, 2000). As the population of ELL students increases, science teachers may assume additional responsibilities with accompanying pressures. The seminal standards

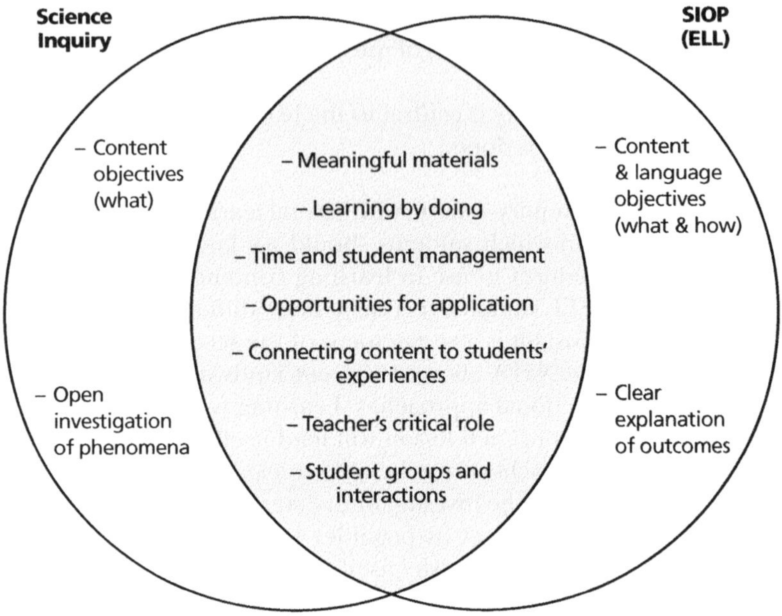

Figure 5.1 Strategies of model ELL and science instruction.

and model strategies for these two fields, however, actually synergize to alleviate such stress. They do not compete for time and attention, but rather compliment and corroborate with one another. Moreover, they equip teachers with habits necessary for truly exceptional instruction, regardless of subject matter or student.

Both English language learning and science education emphasize the critical role of teacher as decision-maker. This caste of considerate, judicious educator is ideal. "Teachers translate curricula by making decisions about the depth and breadth of content, how meaning will be conveyed, and why learning is important" (Clark, Clough, & Berg, 2000, p. 43). Both science inquiry and sheltered instruction are flexible to the needs of students and teachers, avoiding the "cookie-cutter" approach (Echevarria et al., 2008, p. 20).

The actual impact of standards and strategies on students' language and science achievement remains a question for future studies. Researchers must explore the long-term effects of sheltered science and language instruction. How do such streamlined efforts compare to those of traditional science classrooms with pull-out language learning lessons? To what extent must science teachers be explicit with language development as they promote English along with science learning? One must also investigate the most effective methods of preparing teachers (pre-service and in-service) to successfully teach and model sheltered science lessons. If traditional professional development does not produce lasting results (Bainer & Wright, 1998), educators must develop new approaches to help science teachers realize the value of sheltered ELL science instruction and then master these strategies.

In December of 2009, the NSTA approved a new position statement regarding science education for ELLs, recognizing the critical role of teachers and school systems in reaching and teaching all students (NSTA, 2009). Nevertheless, science teachers using inquiry-based instruction already possess most of the pedagogical skills required for reaching this significant group of students. A common maxim (or variations thereof) circulating schools is "What's good for English learning students is good for all students" (Hansen, 2006, p. 23). Certainly, a well-designed and well-delivered lesson for ELL students will produce successful learning in nearly every student. However, the opposite is not necessarily true. A "good" content lesson may not be sufficient to successfully meet the needs of ELL students.

As apparent in this study of strategies and standards, ELL students often require additional support and consideration. Such efforts are not especially extraneous, and they are ultimately relevant for the entire class. English language learning and science instruction should not occur in isolation from each other, as in past attempts. Instead, classroom teachers can create an embedded method that employs the traits and techniques of both fields. Implementation of this synergetic approach serves teachers as well as the students. Sheltered, inquiry-based science instruction promotes reflective and intentional teaching—pedagogical habits that advance the success of all students.

REFERENCES

Appleton, K. (1997). Analysis and description of students' learning during science classes using a constructivist-based model. *Journal of Research in Science Teaching, 34*(3), 303–318.

Armon, J., & Morris, L.J. (2008). Integrated assessments for ELL. *Science and Children, 45*(8), 49–53.

Atkin, J. M., & Karplus, R. (1962). Discovery or invention? *The Science Teacher, 29*(2), 45–51.

Bainer, D.L., & Wright, D. (1998, April 13–17). *Teacher choices about their own professional development in science teaching and learning.* Paper presented at the Annual Meeting of the American Educational Research Association. San Diego, CA.

Barman, C. (1989). Making it work. *Science Scope, 12*(5), 28–31.

Bergman, D.J. (2008, January 10–13). *The effects of two secondary science teacher education program structures on teachers' habits of mind and action.* Paper presented at the Annual Meeting of the Association for Science Teacher Education. St. Louis, MO.

Berliner, D.C. (1989). Effective schools: Teachers make the difference. *Instructor, 99*, 14–15.

Bransford, J., Brown, A., & Cocking, R. (2000). *How people learn: Brain, mind, experience, and school.* Washington, DC: National Academy Press.

Brooks, J., & Brooks, M. (1999). *In search of understanding: The case for constructivist classrooms (2nd ed.).* Alexandria, VA: Association for Supervision and Curriculum Development.

Bybee, R. W. (Ed.) (2002). *Learning science and the science of learning.* Arlington, VA: NSTA Press.

Caine, R., & Caine, G. (1990). Understanding a brain-based approach to learning and teaching. *Educational Leadership, 48*(2), 66–70.

Champagne, A., & Hornig, L. (Eds.) (1987). *Students and science learning: Papers from the 1987 National Forum for School Science.* Washington, DC: American Association for the Advancement of Science.

Chessin, D.A., & Moore, V.J. (2004). The 6-E learning model. *Science and Children, 42*(3), 47–49.

Clark, R. L., Clough, M. P., & Berg, C. A. (2000). Modifying cookbook labs. *The Science Teacher, 67*(7), 40–43.

Clough, M.P. (2003, July 30–August 3). *Learners' responses to the demands of conceptual change: Additional considerations for effective NOS instruction.* Proceedings of the Seventh International History, Philosophy and Science Teaching Conference. Winnipeg, Manitoba, Canada.

Coe, M.A. (2001). Inquiry approach: The 5E learning cycle model. Retrieved July 30, 2008, from http://faculty.mwsu.edu/west/maryann.coe/coe/inquire/inquiry.htm.

Colburn, A. (2000a). Constructivism: Science education's "grand unifying theory." *The Clearing House, 74*(1), 9–12.

Colburn, A. (2000b). An inquiry primer. *Science Scope, 23*(6), 42–44.

Colburn, A., & Clough, M.P. (1997). Implementing the learning cycle: A gradual transition to a new teaching approach. *The Science Teacher, 64*(5), 30–33.

Corbett, H.D., Dawson, J.A., & Firestone, W.A. (1984). *School context and school change: Implications for effective planning.* New York: Teachers College Press.

Cremin, L.A. (1961). *The transformation of the school: Progressivism in American education, 1876–1957.* New York: Vintage.

Day, R. (Ed.). (1986). *Talking to learn: Conversation in second language acquisition.* Cambridge, MA: Newbury House Publishers.

Echevarria, J., Vogt, M., & Short, D.J. (2008). *Making content comprehensible for English learners: The SIOP Model, 3rd Edition.* Boston, MA: Allyn and Bacon.

Esterberg, K. (2002). *Qualitative methods in social research.* New York: McGraw-Hill.

Fadiman, C., & Howard, J. (1980). The conditions of teaching. *American Educator, 4*(1), 6–11.

Fathman, A.K., & Crowther, D.T. (Eds.) (2005). *Science for English language learners: K–12 classroom strategies.* Arlington, VA: National Science Teacher Association.

Fradd, S.H., Lee, O., Sutman, F.X., Saxton, M.K. (2001). Promoting science literacy with English language learners through instructional materials development: A case study. *Bilingual Research Journal, 25*(4), 479–501.

Fullan, M.G. (1991). *The new meaning of educational change.* New York: Teachers College Press.

Good, T. L., & Brophy, J. E. (1994). *Looking in classrooms* (6th ed.). New York: Harper Collins.

Gottlieb, M., Cranley, M.E., & Oliver, A.R. (2007). Understanding the WIDA English Language Proficiency Standards: A resource guide. Retrieved July 22, 2008, from http://www.wida.us/standards/Resource_Guide_web.pdf.

Hansen, L. (2006). Strategies for ELL success. *Science and Children, 43*(4), 22–25.

Hoover, S.P., & Achilles, C.M. (1996, March). *The problem is only part of the problem.* Paper presented at the Annual Meeting of the American Association of School Administrators. San Diego, CA.

Joyce, B., & Weil, M. (2008). *Models of teaching* (8th ed.). Boston, MA: Allyn and Bacon.

Karplus, R. (1977). Science teaching and the development of reasoning. *Journal of Research in Science Teaching, 14*(2), 169–175.

Karplus, R. (1979). Teaching for the development of reasoning. In A.E. Lawson (Ed.), *1980 AETS yearbook: The psychology of teaching for thinking and creativity.* Columbus, Ohio: ERIC/SMEAC.

Krashen, S. (1985). *The input hypothesis: Issues and implications.* New York: Longman.

Langer, J. A., & Applebee, A. N. (1987). How writing shapes thinking: A study of teaching and learning. *National Council of Teachers of English Research Report No. 22.* NCTE: Urbana, IL. 87, 137.

Lanier, J., & Little, J. (1986). Research on teacher education. In M. Wittrock (Ed.), *Handbook of research on teaching, 3rd Edition* (pp. 527–569). New York: Macmillan Publishing.

Lawson, A. E., Abraham, M. R., & Renner, J. W. (1989). *A theory of instruction: Using the learning cycle in teaching science concepts and thinking skills.* Monograph of the National Association for Research in Science Teaching, No. 1. Cincinnati, OH: NARST.

Lederman, N., & Zeidler, D. (1987). Science teachers' conceptions of the nature of science: Do they really influence teacher behavior? *Science Education, 71*(5), 721–734.

Lowery, L. F. (Ed.) (2000). *NSTA elementary school pathways to the science standards: Guidelines for moving the vision into practice (2nd edition).* Arlington, VA: NSTA Press.

McKeone, D. (1995). *Measuring your media profile.* Aldershot, Hampshire, England: Gower Publishing Company.

Medina-Jerez, W., Clark, D.B., Medina, A., & Ramirez-Marin, F. (2007). Science for ELLs: Rethinking our approach. *The Science Teacher, 74*(3), 52–56.

Meek, C. (2002). Classroom crisis: It's about time. *Phi Delta Kappan, 84*(8), 592–595.

Meister, D.G. (2000, April 24–28). *Teachers and change: Examining the literature.* Paper presented at the Annual Meeting of the American Educational Research Association. New Orleans, LA.

Moore, D.B., & Hyde, A.A. (1981). *Making sense of staff development: An analysis of staff development programs and their costs in three urban school districts.* Chicago: Designs for Change.

National Research Council. (1996). *National science education standards.* Washington, DC: National Academy Press.

National Research Council. (2000). *Inquiry and the national science education standards.* Washington, DC: National Academy Press. Retrieved November 18, 2008, from http://www.nap.edu/openbook.php?isbn=0309064767.

National Science Teachers Association. (2004) NSTA position statement: Scientific inquiry. Retrieved July 23, 2008, from http://www.nsta.org/about/positions/inquiry.aspx.

National Science Teachers Association. (2008). Product Detail, *Science for English Language Learners: K–12 Classroom Strategies.* Edited by Ann K. Fathman and David T. Crowther. Retrieved July 22, 2008, from http://www.nsta.org/store/product_detail.aspx?id=10.2505/9780873552530.

National Science Teachers Association. (2009). *NSTA position statement: Science for English language learners.* Retrieved March 25, 2010, from http://www.nsta.org/about/positions/ell.aspx.

O'Malley, J.J., & Chamot, A.U. (1990). *Learning strategies in second language acquisition.* Cambridge: Cambridge University Press.

Osborne, R. & Wittrock, M. (1983). Learning science: A generative process. *Science Education, 67*(4), 489–508.

Pearson Education, Inc. (2008). *SIOP training for teachers: Participant workbook.* Upper Saddle River, NJ: Author.

Penick, J. E., Crow, L. W., & Bonnstetter, R. J. (1996). Questions are the answer: A logical questioning strategy for any topic. *The Science Teacher, 63*(1), 27–29.

Penick, J. E., Yager, R. E., & Bonnstetter, R. J. (1986). Teachers make exemplary programs. *Educational Leadership, 44*(2), 14–20.

Piaget, J. (1964). Development and learning. *Journal of Research in Science Teaching, 2*(3), 176–186.

Pintrich, P., Marx, R., & Boyle, R. (1993). Beyond cold conceptual change: The role of motivational beliefs and classroom contextual factors in the process of conceptual change. *Review of Educational Research, 63*(2), 167–199.

Posner, G., Strike, K., Hewson, P., & Gertzog, W. (1982). Accommodation of a scientific conception: Toward a theory of conceptual change. *Science Education, 66*(2), 211–227.

Rakow, S.J. (Ed.) (2000). *NSTA pathways to the science standards: Guidelines for moving the vision into practice, middle school edition (2nd edition).* Arlington, VA: NSTA Press.

Ramsey, J. (1993). Developing conceptual storylines with the learning cycle. *Journal of Elementary Science Education, 5*(2), 1–20.

Schlechty, P.C., & Crowell, D. (1983). *Understanding and managing staff development in an urban school system.* Chapel Hill: University of North Carolina.

Shiland, T.W. (1997). Decookbook it! *Science and Children, 35*(3), 14–18.

Shymansky, J.A., & Penick, J.E. (1981). Teacher behavior does make a difference in hands-on science classrooms. *School Science and Mathematics, 81,* 412–422.

Teachers of English to Speakers of Other Languages. (2005). *PreK–12 English Language Proficiency Standards in the core content areas: Preview for review and comment.* Retrieved July 25, 2008, from http://www.tesol.org/s_tesol/bin.asp?CID=95&DID=3461&DOC=FILE.PDF.

Teachers of English to Speakers of Other Languages. (2006). TESOL revises preK–12 English Language Proficiency Standards. Retrieved July 25, 2008, from http://www.tesol.org/s_tesol/sec_document.asp?CID=1186&DID=5349.

Texley, J., & Wild, A. (Eds.) (2004). *NSTA pathways to the science standards: Guidelines for moving the vision into practice, high school edition.* Arlington, VA: NSTA Press.

Trowbridge, L. W., & Bybee, R. W. (1990). *Becoming a secondary science teacher.* Columbus, OH: Merrill.

Truex, D. (1996, December 19). *Text-based analysis: A brief introduction.* Panel presentation for the Georgia State University College of Business Administration. Atlanta, GA.

Victor, E. (1974). The inquiry approach to teaching and learning: A primer for teachers. *Science and Children, 12*(2), 23–26.

Vygotsky, L. (1978). *Mind and society: The development of higher psychological processes* (M. Cole, V. John-Steiner, S. Scribner, & E. Souberman, Eds. and trans.). Cambridge, MA: Harvard University Press.

Watson, S. (2004). Opening the science doorway: Strategies and suggestions for incorporating English language learners in the science classroom. *The Science Teacher, 71*(2), 32–35.

World-Class Instructional Design and Assessment. (2007a). Background. Retrieved July 23, 2008, from http://www.wida.us/aboutus/background.aspx.

World-Class Instructional Design and Assessment. (2007b). English language proficiency (ELP) standards. Retrieved July 22, 2008, from http://www.wida.us/standards/elp.aspx.

World-Class Instructional Design and Assessment. (2007c). English language proficiency standards for English language learners in grade 6 through grade 12. Retrieved July 25, 2008, from http://www.wida.us/standards/6-12%20Standards%20web.pdf.

World-Class Instructional Design and Assessment. (2007d). English language proficiency standards for English language learners in pre-kindergarten through grade 5. Retrieved July 25, 2008, from http://www.wida.us/standards/PreK-5%20Standards%20web.pdf.

CHAPTER 6

ENHANCING CONTENT INSTRUCTION FOR ENGLISH LANGUAGE LEARNERS

Learning about Language in Science

Luciana C. de Oliveira

ABSTRACT

English language learners (ELLs) attend mainstream classes most of their time in school. When working with an increasing number of ELLs in K–12 mainstream classes, science teachers have the dual responsibility of facilitating ELLs' content learning while also supporting their ongoing English language development. Learning science is highly dependent on reading texts, especially at the intermediate elementary grades and beyond. Because of this, science teachers should know what is challenging in the texts they assign to students. This chapter provides a closer look at some potential linguistic challenges of the discourse of science for ELLs. Using fourth-grade examples from two science textbook series, it is argued that language-based science instruction is more than just technical terms and should draw on other aspects of school science discourse. In addition, the chapter shows that key linguistic features found in science at the secondary level are already present at the elementary school level.

Teaching Science with Hispanic ELLs in K–16 Classrooms, pages 135–150

INTRODUCTION

Over the last half decade, the number of ELLs in the United States has increased dramatically. Typically, ELLs take English as a second language (ESL) classes or participate in programs where both their native language and English are used to develop their language proficiency before they enter the mainstream classroom. Yet, the number of ESL specialists in K–12 schools is limited. Many school districts do not serve the full number of ELLs (Walker, Ranney, & Fortune, 2005). There is approximately one bilingual or ESL teacher for every 350 ELLs (Faltis & Hudelson, 1994). Most ELLs spend only a portion of their day with bilingual or ESL teachers. These students, then, attend mainstream classes most of their time in school. When working with an increasing number of ELLs in K–12 mainstream classes, content-area teachers have the dual responsibility of facilitating ELLs' content learning while also supporting their ongoing English language development. This is especially relevant in science. Learning science is highly dependent on reading texts, especially at the intermediate elementary grades and beyond.

Compared to their English-speaking peers, ELLs have an even more challenging task in reading textbooks (de Oliveira, 2007). My work in K–12 classrooms has demonstrated that, as ELLs progress at school, they need to understand how textbook authors construct the technical discourse of school science. Drawing on linguistic features of academic registers, the language of textbooks often becomes distanced from the everyday language of many students (Schleppegrell, 2004; Schleppegrell & de Oliveira, 2006).

General strategies, such as creating collaborative groups, using visuals, and building on students' background knowledge, often are cited as strategies that work well for ELLs in science (Keenan, 2004; Hansen, 2006). To make textbook content accessible to ELLs, many content area teachers of ELLs draw on a variety of strategies and techniques to simplify the language of textbooks and to dilute textbook content. While these strategies and techniques may be helpful for ELLs at the beginning levels of language proficiency, they are not appropriate for ELLs at intermediate to advanced levels, especially as they progress through the elementary grades. Under a watered-down curriculum, ELLs may not be taught the language of textbooks and they may never learn to read textbooks without modifications or adaptations (Gibbons, 2006). At intermediate–advanced levels, ELLs commonly must read content area texts and so need to have additional strategies and be able to access difficult content (de Oliveira & Dodds, in press).

My view on content knowledge development emerges from a functional linguistics perspective that sees content knowledge as constructed in language and through language. This perspective sees grammar as the means through which content knowledge is realized and presented (Halliday, 1993; Schleppegrell, 2004).

In this chapter, I provide a closer look at some potential linguistic challenges of the discourse of science for ELLs in addition to technical terms, using fourth-grade examples from two science textbook series. I argue that language-based science instruction is more than just technical terms and should draw on other aspects of school science discourse. In addition, I show that key linguistic features found in science at the secondary level are already present at the fourth grade level.

CONTENT-SPECIFIC LANGUAGE IN SCIENCE TEXTBOOKS AND ELLs

A close look at science textbooks reveals that textbook authors present disciplinary knowledge very differently from the ways in which meanings are constructed in students' everyday language (Schleppegrell & Colombi, 2002; Fang, 2006; Fang & Schleppegrell, 2008). To learn about science, ELLs need to be able to understand the language of science. The difficulty found among ELLs in comprehending textbooks has been connected to their lack of vocabulary knowledge in the content areas (Scarcella, 2002). But the language of science is much more than vocabulary, specifically technical terms in science, as Halliday and Martin (1993) describe:

> Of course, technical terms are an essential part of scientific language; it would be impossible to create a discourse of organized knowledge without them. But they are not the whole story. The distinctive quality of scientific language lies in the lexicogrammar (the "wording") as a whole, and any response it engenders in the reader is a response to the total patterns of the discourse. (p. 4)

Technical terms, then, are only part of the whole story of science and the challenges it presents to English language learners.

In K–12 science education curricula, textbooks have been shown to be a dominant instructional device (Tyson & Woodward, 1989; Good, 1993; Weiss, Banilower, McMahon, & Smith, 2001). Yet, more descriptions of the content-specific language in science textbooks are needed to better understand students' literacy needs (Unsworth, 1997a; Saul, 2004). The complexity of language in science teaching and learning for diverse populations has deserved special attention (e.g., Stevens, Jefferies, Brisk, & Kaczmarek, 2009).

Concerned about the complexity of language in science textbooks which students encounter, a number of researchers have analyzed science textbooks and highlighted their distinctive linguistic features at the secondary level (Unsworth, 1997b; Veel, 1997; Fang, 2006). Science textbook authors organize and condense science knowledge and information through complex clause structures, very different from the ways meanings are constructed in students' everyday language (Schleppegrell & Colombi, 2002). The gap between the language of science textbooks and students' everyday language presents obstacles for students' comprehension.

With a specific focus on science education for ELLs, Lee (2005) conducted a research synthesis of science learning outcomes, science assessment, and science teacher education. In regard to ELLs' equity of science educational opportunity, Lee points out that science researchers and teachers need to better understand that ELLs' home languages and cultures are one of the key elements constituting their knowledge base of school science learning. Similarly, Bruna and Gomez (2009), in the introduction to their edited volume, call for "valuing and centralizing" multilingual students' everyday science talk and ways of knowing (p. 6). It is important, then, to build on ELLs' experiences in helping them develop new ways of meaning in science.

POTENTIAL LINGUISTIC CHALLENGES FOR ENGLISH LANGUAGE LEARNERS IN READING SCIENCE TEXTBOOKS

The content that ELLs need to access in science is specialized. Science is highly dependent on language and uses language in particular ways. The language of science has changed over time and become more distanced from everyday language. Advances in science have led to new ways of using language to express scientific arguments and methods (Halliday & Martin, 1993; Martin, 1993). These new ways of using language have evolved for functional reasons and present particular challenges for all students, but even more for ELLs.

Drawing on the research done by applied linguists in science at the secondary level, this section describes and highlights the potential linguistic challenges of the language of science for ELLs at the intermediate elementary grades. The work presented here is informed by classroom-based research done in a school district with a high number of Hispanic ELLs (30% of the student population) in Indiana. As a consultant with this district since May 2007, my work has included summer ELL institutes and several workshops addressing content instruction for these learners. This work draws from other studies conducted in a fourth-grade classroom in which a teacher engaged in a process for identifying some language challenges of science and engaging her ELLs in talk about science (de Oliveira & Dodds, in press). My discussion is based on classroom observations of science teaching as well as my work in helping elementary teachers address the language and content needs of ELLs in mainstream classes.

The two textbook series analyzed in this study were *Harcourt Science* (Harcourt, 2005) and *Scott Foresman Science* (Pearson/Scott Foresman, 2006), two widely used textbooks throughout the United States. I selected fourth grade because, at this grade level, there is a decline in the literacy achievement of students, known as the "fourth-grade slump" (Chall & Jacobs, 1983; Chall, Jacobs, & Baldwin, 1990). Fourth grade is when students are required to read more advanced academic texts, placing much higher demands on all students.

The categories of linguistic challenges found in the textbooks identified are drawn from recent scholarship on the language demands of schooling (e.g., Schleppegrell, 2004; Fang, 2006; de Oliveira, 2007; Fang & Schleppegrell, 2008; de Oliveira, 2010; de Oliveira & Dodds, in press). I exemplify the challenges by using two text passages that are representative of other texts analyzed in this study. The first text is from *Harcourt Science* (Harcourt, 2005) and is part of the chapter entitled "Plant Growth and Adaptations," and appears as part of Lesson 3, "How Do Plants Reproduce?" This section is called "Plant Life Cycles" and describes the ways in which plants reproduce and how seeds are spread. The second text is from *Scott Foresman Science* (Pearson/Scott Foresman, 2006) and is part of the chapter entitled "Energy from Plants." The sections analyzed, "Parts of flowers" and "Pollen on the move," appear as Lesson 2, "How do plants reproduce?" The two passages have similar content and vocabulary and are representative of other texts selected from this textbook. A total of 34 text passages were analyzed for this study from both textbook series. The two texts are presented in Figures 6.1 and 6.2.

From Words to Sentences: Technical Terms and Definitions

Technical terms are a common feature of the language of science and constitute the first challenge for ELLs. Technical terms are essential for constructing the distinctive nature of scientific language. This analysis revealed that these terms often are defined, explained, or paraphrased and appear typically in bold, as in the sample passages (italics here represent bold in the original). In science textbooks, definitions are used often when technical terms need to be introduced and described.

In Figure 6.1, the technical terms *germinates* and *pollination* are both defined and appear in bold (italics here represent bold in the original). Some other terms that students must be familiar with and understand in this ex-

Plants from Seeds

Have you ever planted seeds in a garden? Seeds form in the cones of conifers. Seeds also form in the flowers of flowering plants. When a flower dries up and falls away, fruit forms around the young seeds. The fruit protects the seeds. Inside each seed is a tiny plant and the food it needs to start growing.

Seeds are the first part in a flowering plant's life cycle. To begin to grow, seeds need warmth, water, and air. Most seeds don't get what they need, so they don't grow. However, when a seed has its needs met, it *germinates*, or sprouts. Seedlings that sprout in soil may keep growing. They grow to become adult, or mature, plants. The mature plants form flowers.

Flower Parts and Seeds

Flowers have parts that work together to make seeds. The *stamen* makes pollen, a kind of powder. The top of the *pistil* collects pollen. The bottom of the pistil is the ovary where seeds form. An apple core is an apple tree ovary.

Seeds form after a flower is pollinated. *Pollination* happens when pollen is carried from a stamen to a pistil by wind or animals. For example, bees, birds, and bats feed on nectar, a sweet liquid in some flowers. Pollen sticks to the animals' bodies. When they leave or visit another flower, the pollen is transferred to the pistil.

Figure 6.1 Text 1: Text excerpt from *Harcourt Science* (Harcourt, 2005, pp. A84–85)

cerpt are *seeds, cones, conifers, flowers, stamen, pollen, pistil, ovary*, and *nectar*, but not all of these terms are in bold face.

The technical term *pollen* is defined as "a kind of powder." But this explanation occurs right after the technical term and is set out by only a comma. Identifying "a kind of powder" as a definition could be a special challenge for an ELL unfamiliar with this feature of the language of science. Because this word is not in bold face, it may not be easily recognized as an important term. But pollen is a very relevant term for ELLs if they are to understand the process of pollination. Another key technical term in this passage is *seedlings*. This technical term is never defined or explained, although it is also important for ELLs' understanding of *adult plants*, which appears in the next sentence.

In Figure 6.2, *sepals, stamens, ovary*, and *fertilization* are defined and appear in bold (italics here represent bold in the original). However, other terms appear in bold (italics) but are not defined within the passage. The definition of *pistil* and its role in plant reproduction are part of the photos

Parts of Flowers

One way that scientists classify plants is by how they make new plants, or reproduce. When plants are classified in this way, they are put into one of two groups. Plants in one group make seeds. Flowering plants and conifers are in this group.

Most flowers have four main parts. The part that you can see easily is the petals. Petals are often colorful. They are different shapes and sizes. They protect the parts of the flower that make seeds. They attract bees, butterflies, birds, and other living things.

The small, green leaves below the petals are sepals. They cover and protect the flower as it grows inside the bud. As the bud opens and the flower spreads its petals, the sepals are pushed apart.

If you look at the center of the flower in the photos, you can see small, knoblike structures. These structures are part of the pistil. The smaller stalks surrounding the pistil are stamens. The anthers are at the tips of the stamens. The anthers make tiny grains of pollen. Sperm cells in pollen combine with the flower's eggs to make seeds.

Pollen on the Move

In order for a seed to form, pollen has to get from a stamen to a pistil. Animals play a big part in making this happen.

Flowers make a sweet liquid called nectar. This is a tasty food for bats, bees, butterflies, and birds. These and other animals move from flower to flower in search of nectar. The colors of the petals and the flowers' scent attract animals and guide them to the nectar deep inside the flower. As the animal feeds, pollen from the stamens rubs off onto its body. The pollen may then rub off onto the pistil of the next flower the animal visits. This movement of pollen from stamen to pistil is called pollination.

Once a pollen grain lands on a pistil, a thin tube grows from the pollen down through the pistil. This pollen tube reaches the thick bottom part of the pistil called the *ovary*. Inside the ovary are egg cells. Sperm cells from the pollen travel down the pollen tube into the ovary. There the sperm cell and egg cell combine in a process called *fertilization*.

Figure 6.2 Text 2: Text excerpt from *Scott Foresman Science* (Pearson/Scott Foresman, 2006, pp. 54–56)

that appear on the textbook pages from which this excerpt was taken. The technical term *nectar* is defined but not in bold. Another very important technical term is *pollination*, which also is defined but not in bold. ELLs then need to be able to recognize that not all terms in bold may be defined within the passage, and sometimes definitions appear with terms that may not be in bold. These sample passages also reveal other challenges for ELLs in addition to just the technical terms themselves, in the ways that these terms are used with other words in the passage.

From Common Usage to Science Usage: Conjunctions with Specific Roles

The conjunction *or* in science has two very specific and distinct roles. The first role is introducing explanations or paraphrases, rather than indicating alternatives—more commonly used in everyday language (Fang, 2006)—as is the case in the clause "When they leave or visit another flower" in Figure 6.1.The technical term *germinates*, for instance, is followed by a conjunction, *or*, and a paraphrase, *sprouts*, in an attempt to paraphrase the more technical term. In this case, however, the word used as a paraphrase is itself a technical term and may be of little help to an ELL.

The second role for the conjunction *or* is the introduction of more abstract or technical terms. One example found in Figure 6.1 appears in the sentence "They grow to become adult, or mature, plants." Here, *or* is used to introduce the more technical term *mature*. In Figure 6.2, a similar example is found. In "they make new plants, or reproduce," the technical term *reproduce* follows the conjunction *or*. This dual role of the conjunction *or* in science constitutes another challenge for ELLs who may understand the conjunction as indicating alternatives or choices, which is common in more everyday language.

From Everyday to Academic: Everyday Meanings and Words with Specialized Meanings

Text 1 (Figure 6.1) starts out with a question: "Have you ever planted seeds in a garden?" Questions like this are a common feature of textbooks, especially in the lower elementary grades. They become less common in the upper elementary and secondary grades. This feature can be seen as an attempt to make the content more accessible to students, to connect to students' everyday language, and to spark students' interest in the topic. The use of the pronoun *you* also exemplifies this attempt to connect to students. Right after this introductory question, however, a sentence with several technical terms is already introduced and, from then on, the language of science is dense and packed, as will be described. Text 2 (Figure 6.2) also has an example of connection to students through the use of the pronoun *you* in "If you look at the center of the flower in the photos, you can see small, knoblike structures." In this case, there is an attempt to call students' attention to the photos, as they are an important part of the science content that appears in these textbook pages.

Everyday words with specialized meanings in science also can be a potential challenge for ELLs. In the first passage, for instance, the meaning of the verb *form* is very important for the understanding of this content. But ELLs may be more familiar with the noun *form*, as a printed document with lines and blank spaces one needs to fill out. The word *form* actually has very different and specialized meanings in each one of the content areas. In mathematics, for instance, *form* typically occurs as the shape or structure of something. These multiple meanings of the same word may present challenges for ELLs, who may be unfamiliar with additional meanings of everyday words.

From Simple to Complex: Noun Groups and Zig-Zag Structuring

Both simple and complex noun groups appear in science writing. Expanded and complex nominal groups—those that have pre-modifiers and post-modifiers—are typically present in science texts. Expanded nominal groups function to "accumulate less specialized meanings in a single lexical item" (Martin, 1993, p. 229). In the sentence "Temperature is a measure of the average energy of motion of the particles in matter" (Harcourt, 2005, p. E43), an expanded and complex nominal group is given as the definition for temperature: "a measure of the average energy of motion of the particles in matter." This nominal group has as its head the noun *measure*; *a* is a pre-modifier and *of the average energy of motion of the particles in matter* are post-modifiers. Such meaning accumulation allows science texts to pack more lexical content per sentence. This high lexical density of science texts establishes a language barrier for ELLs.

Both sample texts contain many nouns, most of which are the technical terms referred to earlier. But these nouns have specific roles in text structuring and must be understood in the context of the entire passage, not as isolated terms. Figure 6.3 shows the zig-zag movement found from one sentence to the next in Text 1; the same is done in Figure 6.4 for Text 2. The arrows show the connections between the nouns and previous discourse.

The noun groups are underlined and include simple nouns (noun with head only), pronouns, nouns with pre-modifiers, and nouns with post-modifiers. Table 6.1 shows these different noun group structures.

All of these noun groups are important for the understanding of this science topic and must be understood in the context of this passage. The

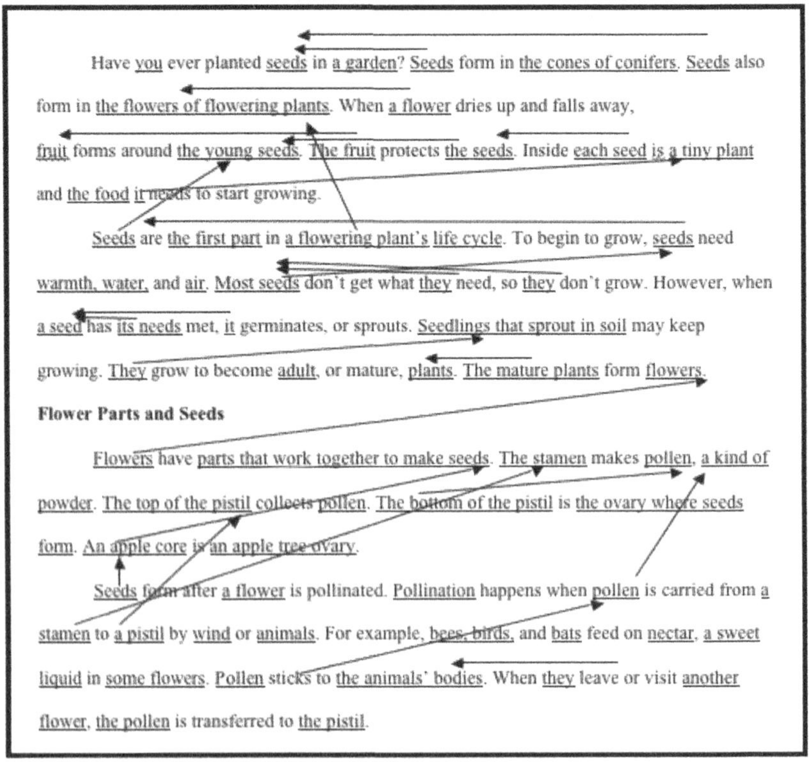

Have you ever planted seeds in a garden? Seeds form in the cones of conifers. Seeds also form in the flowers of flowering plants. When a flower dries up and falls away, fruit forms around the young seeds. The fruit protects the seeds. Inside each seed is a tiny plant and the food it needs to start growing.

Seeds are the first part in a flowering plant's life cycle. To begin to grow, seeds need warmth, water, and air. Most seeds don't get what they need, so they don't grow. However, when a seed has its needs met, it germinates, or sprouts. Seedlings that sprout in soil may keep growing. They grow to become adult, or mature, plants. The mature plants form flowers.

Flower Parts and Seeds

Flowers have parts that work together to make seeds. The stamen makes pollen, a kind of powder. The top of the pistil collects pollen. The bottom of the pistil is the ovary where seeds form. An apple core is an apple tree ovary.

Seeds form after a flower is pollinated. Pollination happens when pollen is carried from a stamen to a pistil by wind or animals. For example, bees, birds, and bats feed on nectar, a sweet liquid in some flowers. Pollen sticks to the animals' bodies. When they leave or visit another flower, the pollen is transferred to the pistil.

Figure 6.3 Nouns, noun groups, and zig-zag structuring in Text 1.

majority of nouns is introduced and referenced back within the text itself, with the exception of the pronoun *you* in both texts, which is a reference outside of the text. These referential relations need to be understood by ELLs, who must be able to track and connect these referrers so they can fully comprehend this text. For instance, in the sentence from Text 1, "Inside each seed is a tiny plant and the food it needs to start growing," the referent for the pronoun *it* may be very difficult for ELLs to understand. In this case, *it* refers back to "a tiny plant." This sentence also has an unusual pattern, since it starts with a prepositional phrase indicating place, "inside each plant." Teachers would need to deconstruct a sentence like this with ELLs to help them understand the important information presented.

The noun *pollination* is an example of a nominalization, a grammatical resource for the construction of nouns. It refers to the expression as a noun or nominal group of what would in everyday language be presented as a verb, an adjective, or a conjunction. The noun *pollen* is introduced in the first paragraph of the section. Then, *pollinated* is used as a verb in the passive voice construction, "is pollinated," in the second paragraph. *Pollina-*

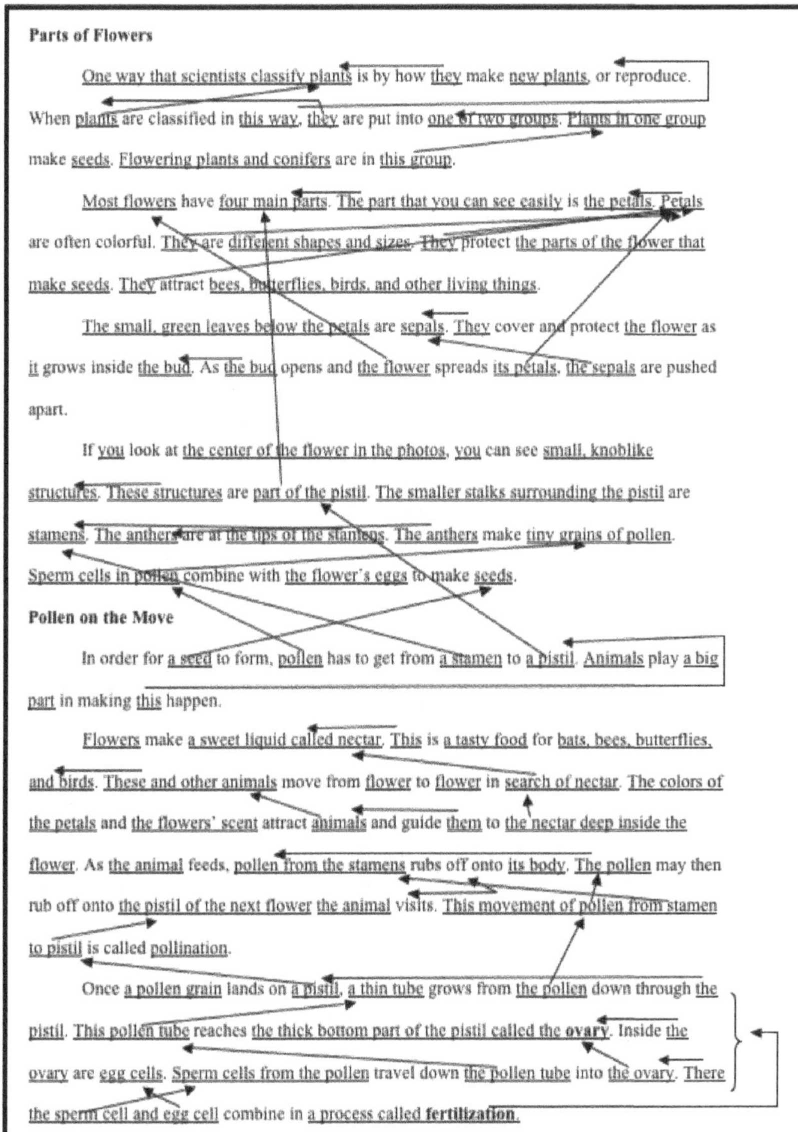

Parts of Flowers

One way that scientists classify plants is by how they make new plants, or reproduce. When plants are classified in this way, they are put into one of two groups. Plants in one group make seeds. Flowering plants and conifers are in this group.

Most flowers have four main parts. The part that you can see easily is the petals. Petals are often colorful. They are different shapes and sizes. They protect the parts of the flower that make seeds. They attract bees, butterflies, birds, and other living things.

The small, green leaves below the petals are sepals. They cover and protect the flower as it grows inside the bud. As the bud opens and the flower spreads its petals, the sepals are pushed apart.

If you look at the center of the flower in the photos, you can see small, knoblike structures. These structures are part of the pistil. The smaller stalks surrounding the pistil are stamens. The anthers are at the tips of the stamens. The anthers make tiny grains of pollen. Sperm cells in pollen combine with the flower's eggs to make seeds.

Pollen on the Move

In order for a seed to form, pollen has to get from a stamen to a pistil. Animals play a big part in making this happen.

Flowers make a sweet liquid called nectar. This is a tasty food for bats, bees, butterflies, and birds. These and other animals move from flower to flower in search of nectar. The colors of the petals and the flowers' scent attract animals and guide them to the nectar deep inside the flower. As the animal feeds, pollen from the stamens rubs off onto its body. The pollen may then rub off onto the pistil of the next flower the animal visits. This movement of pollen from stamen to pistil is called pollination.

Once a pollen grain lands on a pistil, a thin tube grows from the pollen down through the pistil. This pollen tube reaches the thick bottom part of the pistil called the **ovary**. Inside the ovary are egg cells. Sperm cells from the pollen travel down the pollen tube into the ovary. There the sperm cell and egg cell combine in a process called **fertilization**.

Figure 6.4 Nouns, noun groups, and zig-zag structuring in Text 2.

tion is used as a nominalization to refer to the process being described in the two paragraphs. Nominalization is a resource used in many academic and scientific genres (Halliday & Martin, 1993; Martin, 1993; Unsworth, 1999; Schleppegrell, 2004; de Oliveira, 2010), allowing writers to package information into a single unit and then use this in subsequent paragraphs

TABLE 6.1 Noun Group Structure

Type of Noun Group	Sample Text Examples
Noun (with head only)	*seeds* *fruit* *pollination*
Pronoun	*you* *they* *it* *this* *these*
Noun with pre-modifiers	***a*** *garden* ***the young*** *seeds* ***this pollen*** *tube*
Noun with post-modifiers	*seedlings* ***that sprout in the soil*** (*that sprout in the soil* is a relative clause that adds information about the head *seedlings*) *parts* ***that work together to make seeds*** (*that work together to make seeds* is a relative clause that adds information about the head *parts*)
Noun with pre- and post-modifiers	***the thick bottom*** part ***of the pistil called the ovary***

to expand the explanation. Understanding what nominalizations mean is challenging for all students (Schleppegrell, 2004), but may be particularly challenging for ELLs.

The zig-zag patterning appears as words are first introduced and then either repeated or referred to within the passage. For instance, *seeds* is first introduced in the question that appears at the beginning of the paragraph. This word is repeated again in the second sentence and throughout the passage. Pronouns are used as referrers and may present challenges for ELLs, so linking pronouns to their specific referents can be a reading strategy to work on with ELLs. These internal references are important for the information flow of the paragraph and the text as a whole, but ELLs may have difficulty understanding these patterns.

In Text 2, a similar zig-zag patterning occurs. Text 2 uses more references to build its text structure. For instance, the nominal group *this way*, used in the sentence "When plants are classified in this way, they are put into one of two groups," refers back to the sentence that came just before, "One way that scientists classify plants is by how they make new plants, or reproduce." The nominal group *this way* is used to pick up information from the previous sentence so that more information can be added. The same occurs with the pronoun *this* in "This is a tasty food for bats, bees, butterflies, and birds." *This* refers back to the nominal group that appeared in the sentence right before, "Flowers make a sweet liquid called nectar." *This* is referring to the nominal group "a sweet liquid called nectar." If ELLs do not under-

TABLE 6.2 Key Challenges of Science for ELLs at the Elementary Level

Challenge for ELLs	Reason for Challenge
Technical Terms and Definitions	• Technical terms occur throughout science textbooks; some are in bold and defined. • Technical terms may appear without a definition. • Definitions may be difficult to find; definition itself may contain complex language that ELLs may not understand.
Conjunctions with Specific Roles	• Conjunctions (e.g., *or*) may have specific roles in science. • All roles may occur within a few paragraphs.
Everyday Questions and Words with Specialized Meanings	• Everyday questions may occur at the beginning of paragraphs with highly technical language following. • Words with specialized meanings in science can occur throughout and be confusing as ELLs may know their everyday meaning but not how they occur in science.
Noun Groups and Zig-Zag Structuring	• Several noun group structures appear in science: head only, pronouns, noun with pre-modifiers, noun with post-modifiers, and noun with pre- and post-modifiers. • Nouns are introduced and referenced throughout a passage; zig-zag structuring complex to follow. • Lexical content is accumulated in complex and expanded noun groups, creating high lexical density.

stand the connection between the pronoun *this* and the nominal group to which it refers, they may not understand what is *the tasty food* that the text discusses. In addition, the sentence that comes right after "This is a tasty food for bats, bees, butterflies, and birds" also uses a zig-zag structuring: "These and other animals move from flower to flower in search of nectar." The pronoun *these* is referring back to the animals ("bats, bees, butterflies, and birds") from the previous sentence. It is clear in Text 2 how this zig-zag structuring is utilized.

This section highlighted some challenges of the language of science for ELLs in the intermediate grades at the elementary school level. Table 6.2 summarizes these key challenges.

LEARNING ABOUT LANGUAGE IN SCIENCE: SUGGESTIONS FOR THE ELEMENTARY METHODS COURSE

Building language and literacy in the context of content-area instruction is a key component in the education of ELLs (August & Shanahan, 2006; Short & Thier, 2006). Language and literacy can only be built in science if we more fully understand the demands of the language of science for ELLs. ELLs need to have experience with grade-level content in science. To provide such an experience, the language challenges of science of the intermediate grades at the elementary school level clearly need to be considered.

Reading textbooks may be a hurdle for students who are unfamiliar with the complexities embedded in science discourse. Complex and lexically dense clause structures present obstacles for ELLs as they try to grasp the full meaning of science textbooks. This chapter showed that vocabulary—and technical words in science, in particular—are only one of the challenges of the language of science for ELLs, who must know more than technical terms in order to comprehend complex texts (Unsworth, 1998; de Oliveira, 2007).

This chapter has demonstrated some of the linguistic demands of science discourse for ELLs. These demands should be better understood and addressed in elementary science methods courses. The findings and conclusions presented here suggest the need for an explicit attention to language in elementary science methods courses. These courses could integrate a close look at language from the perspective of identifying some potential language challenges for ELLs. For example, selecting a passage from a textbook and asking preservice teachers to identify some of the features described in this chapter may be a helpful activity that could be incorporated in such courses. Pre-service teachers can then plan instruction to not only teach particular science concepts but also work on the language of science with students.

Both teachers and students of school science can develop a certain linguistic awareness of some typical discourse features of science. To be able to read science textbooks effectively, ELLs need to be able to engage with the meanings presented in science textbooks. Discourse structure is seldom attended to in science classrooms. Content is never separate from the language through which that content is manifested. Learning science means learning the language that expresses science. The linguistic challenges this chapter addresses highlight the kind of discipline-specific academic support in language and literacy development that would enable ELLs to be more successful in their reading of science.

ACKNOWLEDGMENT

The author wishes to thank Shu-Wen Lan for her valuable help with data collection and analysis.

REFERENCES

August, D., & Shanahan, T. (Eds.). (2006). *Developing literacy in second-language learners: Report of the National Literacy Panel on Language-Minority Children and Youth.* Mahwah, NJ: Erlbaum.

Bruna, K. R., & Gomez, K. (Eds.). (2009). *The work of language in multicultural classrooms: Talking science, writing science.* New York: Routledge.

Chall, J. S., & Jacobs, V. A. (1983). Writing and reading in the elementary grades: Developmental trends among low-SES children. *Language Arts,* 60(5), 617–626.

Chall, J. S., Jacobs, V. A., & Baldwin, L. E. (1990). *The reading crisis: Why poor children fall behind.* Cambridge, MA: Harvard University Press.

de Oliveira, L. C. (2007). Academic language development in the content areas: Challenges for English language learners. *INTESOL Journal, 4*(1), 22–33.

de Oliveira, L. C. (2010). Nouns in history: Packaging information, expanding explanations, and structuring reasoning. *The History Teacher, 3*(2), 191–203.

de Oliveira, L. C., & Dodds, K. N. (in press). Beyond general strategies for English language learners: Language dissection in science. *Electronic Journal of Literacy Through Science.*

Faltis, C., & Hudelson, S. (1994). Learning English as an additional language in K–12 schools. *TESOL Quarterly, 28*(3), 457–468.

Fang, Z. (2006). The language demands of science reading in middle school. *International Journal of Science Education, 28*(5), 491–520.

Fang, Z., & Schleppegrell, M. J. (2008). *Reading in secondary content areas: A language-based pedagogy.* Ann Arbor, MI: The University of Michigan Press.

Gibbons, P. (2006). *Bridging discourses in the ESL classroom: Students, teachers, and researchers.* London: Continuum.

Good, R. (1993). Editorial: Science textbook analysis. *Journal of Research in Science Teaching, 30*(7), 619.

Halliday, M. A. K. (1993). Towards a language-based theory of learning. *Linguistics and Education, 5*(2), 93–116.

Halliday, M. A. K., & Martin, J. R. (1993). *Writing science: Literacy and discursive power.* Pittsburgh PA: University of Pittsburgh Press.

Hansen, L. (2006). Strategies for ELL success. *Science and Children, 43*(4), 22–25.

Harcourt. (2005). *Harcourt Science: Indiana Edition.* Orlando, FL: Author.

Keenan, S. (2004). Reaching English language learners: Strategies for teaching science in diverse classrooms. *Science and Children, 42*(2), 49–51.

Lee, O. (2005). Science education with English language learners: Synthesis and research agenda. *Review of Educational Research, 75*(4), 491–530.

Martin, J. R. (1993). Literacy in science: Learning to handle text as technology. In M. A. K. Halliday, & J. R. Martin, *Writing science: Literacy and discursive power.* (pp. 166–202). Pittsburgh PA: University of Pittsburgh Press.

Pearson/Scott Foresman. (2006). *Scott Foresman Science (Indiana).* Glenview, IL: Pearson Education.

Saul, E. W. (Ed.). (2004). *Crossing borders in literacy and science instruction.* Newark, DE: International Reading Association.

Scarcella, R. (2002). Some key factors affecting English learners' development of advanced literacy. In M. J. Schleppegrell, & M. C. Colombi (Eds.), *Developing advanced literacy in first and second languages: Meaning with power.* (pp. 209–226). Mahwah, NJ: Erlbaum.

Schleppegrell, M. J. (2004). *The language of schooling: A functional linguistics perspective.* Mahwah, NJ: Erlbaum.

Schleppegrell, M. J., & Colombi, M. C. (Eds.). (2002). *Developing advanced literacy in first and second languages: Meaning with power.* Mahwah, NJ: Erlbaum.

Schleppegrell, M. J., & de Oliveira, L. C. (2006). An integrated language and content approach for history teachers. *Journal of English for Academic Purposes, 5*(4), 254–268.

Stevens, L. P., Jefferies, J., Brisk, M. E., & Kaczmarek, A. S. (2009). Linguistics and science learning for diverse populations: An agenda for teacher education. In K. Richardson Bruna & K. Gomez (Eds.), *The work of language in multicultural classrooms: Talking science, writing science.* (pp.291–315). New York: Routledge.

Tyson, H., & Woodward, A. (1989). Why students aren't learning very much from textbooks. *Educational Leadership, 47,* 14–17.

Unsworth, L. (1997a). Explaining explanations: Enhancing science learning and literacy development. *Australian Science Teachers Journal, 43*(1), 34–46.

Unsworth, L. (1997b). Scaffolding reading of science explanations: Accessing the grammatical and visual forms of specialized knowledge. *Reading, 313,* 30–42.

Unsworth, L. (1998). "Sound" explanations in school science: A functional linguistic perspective on effective apprenticing texts. *Linguistics and Education, 9*(2), 199–226.

Unsworth, L. (1999). Developing critical understanding of the specialised language of school science and history texts: A functional grammatical perspective. *Journal of Adolescent and Adult Literacy, 42*(7), 508–521.

Veel, R. (1997). Learning how to mean-scientifically speaking: Apprenticeship into scientific discourse in the secondary school. In F. Christie & J. R. Martin (Eds.), *Genres and institutions: Social processes in the workplace and school* (pp. 161–195). London: Cassell.

Walker, C. L., Ranney, S., & Fortune, T. W. (2005). Preparing preservice teachers for English language learners: A content-based approach. In Tedick, D. J. (Ed.), *Second language teacher education: International perspectives.* (pp. 313–333). Mahwah, NJ: Erlbaum.

Weiss, L., Banilower, E., McMahon, K., & Smith, P. (2001). *Report of the 2000 national survey of science and mathematics education.* Chapel Hill, NC: Horizon Research.

CHAPTER 7

A FRAMEWORK FOR THE EFFECTIVE SCIENCE TEACHING OF ENGLISH LANGUAGE LEARNERS IN ELEMENTARY SCHOOLS

Trish Stoddart, Jorge Solis, Sara Tolbert, and Marco Bravo

ABSTRACT

This chapter presents a framework for effective science teaching for English language learners (ESTELL) based on two bodies of sociocultural research—the CREDE Five Standards for Effective Pedagogy and the integrated science, language, and literacy instruction literature—which provide converging lines of empirical evidence for a set of socially, culturally, and linguistically responsive instructional practices that have been demonstrated to improve the achievement of English language learners (ELLs). ESTELL is an instructional approach integrating the teaching of scientific inquiry, science discourse, and language and literacy development in a contextualized curriculum that is culturally, socially, and linguistically responsive. This chapter presents a review of the theoretical framework for ESTELL, empirical evidence of impact on ELLs' learning, and a set of instructional exemplars of ESTELL pedagogy.

Teaching Science with Hispanic ELLs in K–16 Classrooms, pages 151–181

INTRODUCTION

The primary goal of science education reform is to improve student learning of science and make rigorous science content and high expectations accessible to all students, especially students from groups whose achievement has traditionally lagged behind that of majority culture students (American Association for the Advancement of Science [AAAS], 1989, 1993; National Research Council [NRC], 1996). Despite two decades of "science for all" reforms, however, significant achievement gaps persist between Anglo European students and cultural and linguistic minority students (Lynch, 2001; Grigg, Lauko, & Brockway, 2006). Of particular concern is the rapidly growing population of students who do not speak English as a first language. By 2010, it is expected that 40% of the U.S. school-age population will be ELLs (NGA Center for Best Practices, 2000). In 2000, 68% of ELLs were concentrated in six states—California, Texas, New Mexico, New York, Florida, and Illinois—with the largest share being in California (Capps, Fix, Murray, Ost, Passel, & Hernandez, 2005). The number of ELL students, however, is growing rapidly in other parts of the country: Nevada (+354%), Nebraska (+350%), and South Carolina, South Dakota, Georgia, Alabama, Arkansas, and Oregon (+200%) (Batalova, Fix, & Murray, 2005). The under-achievement of ELLs is of increasing concern in school districts across the United States.

For at least thirty years, the achievement of ELLs has lagged behind that of native English speakers, and the gap continues to grow (Rodriguez, 2004). The 2005 National Assessment of Educational Progress (NAEP) showed an average 48-point difference in science scores between ELLs and native speakers of English: only 28% of fourth grade ELLs scored at or above the basic level in science, while more than double that number (71%) of native English speakers reached this achievement level (NAEP, 2005). The most recent statewide assessment of science knowledge on the California Standardized Testing and Reporting (STAR) exams shows that ELL students' performance was the lowest of any subgroup. In 2007, 89% of fifth grade ELLs scored as below proficient on the STAR test, while only 10% scored at the proficient level. Additionally, ELL students are significantly less likely than their Anglo European counterparts to pursue advanced degrees in science (Commission on Professionals in Science and Technology, 2007) or to perceive science as relevant to their lives outside of school (Aikenhead, 2001, 2006; Buxton, 2006; Hammond, 2001; Lee & Luykx, 2006; Lemke, 1990; Lynch, 2001; Rodriguez, 1997, 2004; Stanley & Brickhouse, 2001).

Part of the problem is that many ELLs do not have access to rigorous science instruction and often are relegated to remedial instructional programs focusing on the acquisition of basic literacy skills and facts aimed at improving student English proficiency levels instead of teaching high quality science content (Garcia, 1988, 1993; McGroaty, 1992; Moll, Amanti, Neff, & Gonzalez, 1992; Pease-Alvarez & Hakuta, 1992; Valdes, 2001). Understanding the powerful relationship between language, literacy, and science learning for ELLs is fundamental to the development of instructional programs that improve their science achievement. There is, however, currently a limited knowledge base on how to teach science to ELLs and how to prepare the teachers that serve them. (Lee & Luykx, 2004; Lynch, 2000; Stoddart, Pinal, Latzke & Canaday, 2002). Many science educators view language, literacy, and equity issues as beyond the scope of their work and assume that they will be addressed by others in the broader educational reform arena (Lee & Luykx, 2004). Language diversity and equity researchers, on the other hand, primarily attend to issues related to language development and social and cultural context and overlook the teaching of school subjects. As long as these two research agendas continue to operate independently, we cannot achieve the ultimate goal of improving science achievement for all students. There is a critical need to integrate research on the teaching of subject matter with research on student diversity (Darling-Hammond, 1996). In the context of science education, therefore, this will require developing a theoretical and practical knowledge base that integrates knowledge about the effective teaching of science content with knowledge about student language and diversity (Lee & Luykx, 2004). We find that there is much potential in such integration, as it can be mutually beneficial to both domains. As a first step, this chapter draws upon multiple sources of empirical evidence that outlines instruction proven to be effective science teaching for ELLs.

EFFECTIVE SCIENCE TEACHING FOR ELLS

The framework for the pedagogy presented in this chapter draws from sociocultural and Vygotskian theory (Bakhtin, 1981; Rogoff, 1990, 1995; Rogoff & Wertsch, 1984; Tharp, 1997; Tharp & Gallimore, 1988; Vygotsky, 1978; Wertsch, 1985, 1991). Sociocultural theory rests on the principle that learning is social activity, and that it is through the social interaction between the teacher and students and between students—more knowledgeable others—that learning occurs. Learning is enhanced when it occurs in contexts that are culturally, linguistically, and cognitively meaningful and relevant to students (Au, 1980; Deyhle & Swisher, 1997; Ladson-Billings, 1994; Lee and Fradd, 1998; Lemke, 2001; Rosebery, Warren, & Conant, 1992; Tharp & Gallimore, 1988; Warren & Rosebery, 1995). Two bodies of research, based on sociocultural theory, provide converging lines of empirical evidence for a set of socially, culturally, and linguistically responsive instructional practices that are effective in teaching science to ELLs. The first set of studies were produced by researchers from the U.S. Department of Education-funded Center for Research on Education Diversity and Excellence (CREDE) (Doherty & Pinal, 2004; Estrada & Imhoff, 2001; Hilberg, Tharp, & DeGeest, 2000; Saunders & Goldenberg, 1999; Saunders, O'Brien, Lennon & McLean, 1998; Tharp & Dalton, 2007). The second line of evidence is based on the work produced by researchers from five National Science Foundation (NSF)-funded research and development projects: Language Acquisition through Science Education in Rural Schools (LASERS); Seeds of Science, Roots of Reading; the Imperial Valley Project in Science; Science Instruction for All (SIFA); and Promoting Science among English Language Learners (P-SELL). These projects focused on integrated science, language, and literacy instruction for ELLs (Amaral, Garrison, & Klentschy, 2002; Cervetti, Pearson, Barber, Hiebert, & Bravo, 2007; Holliday, Yore, & Alvermann, 1994; Lee, Maerten-Rivera, Penfield, LeRoy, & Secada, 2008; Stoddart, 1999, 2005; Stoddart, Abrams, Gasper, & Canaday, 2000; Stoddart, Pinal, Latzke, & Canaday, 2002; Ku, Bravo, & Garcia, 2004). The authors of this chapter have integrated the findings of these two bodies of research to describe effective science teaching for ELLs.

The CREDE Five Standards for Effective Pedagogy

Researchers from CREDE identified five instructional practices, the CREDE Five Standards for Effective Teaching (CFSEP), which sociocultural theory indicated would improve the teaching and learning of cultural and language minority students and conducted a set of research studies to investigate the relationship between the CFSEP and student achievement. The CFSEP include (1) Joint Productive Activity, (2) Language and Literacy, (3) Connecting School to Students' Lives, (4) Complex Thinking, and (5) Instructional Conversation. These studies demonstrated that cultural and linguistic minority students in classrooms using the practices show significant gains in reading and mathematics achievement. In all the studies, teachers' use of the practices was recorded with the Standards Performance Continuum (Doherty, Hilberg, Pinal, & Tharp, 2002), and student achievement gains were estimated from standardized test scores (SAT-9) from two consecutive years. Teachers' overall use of the practices reliably predicted achievement gains in comprehension, language, reading, spelling, and vocabulary (Doherty et al., 2002). Students whose teachers used the practices extensively in their classroom organization showed significantly greater achievement gains on all SAT-9 tests than students whose teachers had not similarly transformed their teaching. These findings were replicated by Doherty, Hilberg, and Lee (2004). Doherty et al. (2002), in a quasi-experimental design that used a school in an adjacent catchment area as an untreated control group, showed the same patterns of vocabulary gains, exceeding a half a standard deviation in normal curve equivalent scores.

A set of studies by Estrada over a four-year period showed a positive relation between use of the CREDE practices and positive outcomes in first- and fourth-grade reading. Stronger implementation of the pedagogy produced higher student scores on tests of reading and the language of instruction. The vast majority of students in strong implementers' classrooms reached grade level in reading, whereas less than half did so in weaker implementers' classrooms (Estrada & Imhoff, 2001). Hilberg, Tharp, and DeGeest (2000) examined the efficacy of the CREDE practices in mathematics instruction. Two groups of Native American eighth grade students were randomly assigned to either CFESP or traditional conditions for a one-week unit on fractions, decimals, and percents. Students in the experimental condition outperformed controls on tests of conceptual learning at the end of the study and demonstrated better retention of unit content two weeks later.

Several highly successful and well-researched instructional models incorporate three or more CFSEP practices, including Opportunities through Language Arts (OLA), a language arts program for grades 3–5 developed by CREDE researchers in southern California (Saunders & Goldenberg, 2001); the Sheltered Instruction Observation Protocol (SIOP) program, also developed by CREDE researchers (Echevarria, Vogt, & Short, 2000); and Creating Sacred Places for Children (CSPC), a program for Native American Schools. Positive effects on student achievement have been reported and replicated for OLA (Saunders, 1999; Saunders & Goldenberg, 1999), and reported for SIOP (Echevarria et al., 2000; Echevarria, Short, & Powers, 2004).

Teachers' use of CREDE pedagogy also has been linked to factors critical to school performance, such as motivation, perceptions, attitudes, and inclusion. Predominantly Latino ELL students in classrooms where the CFSEP instructional practices were used only slightly or moderately spent more time on-task, perceived greater cohesion in their classrooms, and perceived themselves as better readers having less difficulty with their work, as compared to classrooms where the practices were not present at all (Padron & Waxman, 1999). Native American students in mathematics classes integrating the practices reported more positive attitudes toward mathematics (Hilberg et al., 2000). Findings, replicated over two years with two cohorts of students (Estrada & Imhoff, 2001), indicated that, across language programs, peer inclusion in social choices was greater in classrooms in which students participated in more peer joint productive activities (or peer collaboration).

Although the CFSEP have been shown to be effective in the teaching of reading and mathematics, they have not been articulated in science instruction. Other lines of research, however, that focus more specifically on teaching science to ELLs have identified instructional approaches that parallel those of the CREDE standards (Amaral et al., 2002; Baker & Saul, 1994; Casteel & Isom, 1994; Lee and Fradd, 1998; Lee et al., 2008; Rosebery et al., 1992; Stoddart, 1999; Ku et al., 2004). The research on science for ELLs points to the need for an integration of the teaching of scientific inquiry, science discourse, and language and literacy development in a contextualized curriculum that is culturally, socially, and linguistically responsive.

Integrated Science, Language, and Literacy Pedagogy for ELLs

The promotion of an integrated pedagogy for ELLs is particularly important because the teaching of school subjects, such as science, to ELLs is typically separated from the teaching of language and literacy (Collier, 1989; Cummins, 1981; Met, 1994). It is assumed that ELLs need to be proficient in English before being introduced to more rigorous instruction in the content areas. This is problematic because it may take as long as seven years to acquire a level of language proficiency comparable to native speakers (Collier, 1989; Cummins, 1981). ELLs fall behind academically if they do not learn the content of the curriculum as they acquire English. The amount of time it takes to acquire grade-level English proficiency, however, can be accelerated with the integration of content and language teaching for language minority students (Thomas & Collier, 2003). Research on second language immersion programs finds that contextualized, content-based instruction in students' second language can enhance the language proficiency of ELLs with no detriment to their academic learning (Cummins, 1981; Genesee, 1987; Lambert & Tucker, 1972; McKeon, 1994; Met, 1994; Swain & Lapkin, 1985). The subject matter content provides a meaningful context for the learning of language structure and functions, and the language processes provide the medium for analysis and communication of subject matter knowledge. Inquiry science, therefore, is an excellent context for learning language and literacy.

The context of language use refers to the degree to which language provides learners with meaningful cues that help them interpret the content being communicated; visual cues, concrete objects, and hands-on activities (Krashen,1985). Inquiry science instruction engages students in the exploration of scientific phenomena, and language activities are explicitly linked to objects, processes, hands-on experimentation, and naturally occurring events in the environment, that is, they are contextualized (Baker & Saul, 1994; Casteel & Isom, 1994; Lee and Fradd, 1998; Rodriguez & Bethel, 1983; Rosebery et al., 1992; Stoddart et al., 2002). Thus, learners engage in authentic communicative interactions—describing, hypothesizing, explaining, justifying, arguing, and summarizing—which promote purposeful language (Lee & Fradd, 1998; Warren, Ballenger, Ogonowski, Rosebery, & Hudicourt-Barnes, 2001). They can communicate their understanding in a variety of formats, for example, in writing, orally, by drawing, and by creating tables and graphs (Lee & Fradd, 1998). The contextualized use of language in inquiry science instruction also promotes the understanding of science concepts (Rosebery et al., 1992). An analysis of the language of science serves to understand how language structures the way science concepts are developed, organized, adapted, and communicated (Baquedano-López, Solís, & Kattan, 2005; Kaplan, 1986; Lemke, 1990; Newman & Gayton, 1964). Inquiry involves more than hands-on activities; it also involves active thinking and discourse around activities. Rosebery et al. (1992), in their work with language minority students, emphasize the role of language and discourse in content learning by using the processes of argumentation and collaborative inquiry to guide students into examining scientific claims and the nature of proof.

The relationship between science learning and language and literacy learning, therefore, is reciprocal and synergistic. Through the contextualized use of language in science inquiry, students develop and practice complex language forms and functions. Through the use of language functions such as description, explanation, and discussion in inquiry science, students enhance their conceptual understanding (Stoddart et al., 2002). This is a synergistic approach to teaching and learning in which language and literacy development is contextualized in scientific inquiry projects that promote understanding through collaborative work and discourse between teachers and students. As discussed above, this integrated pedagogy brings together instructional practices—language and literacy contextualized in inquiry science that through discourse and cooperative learning supports the development of scientific understanding—that are aligned with the CFSEP.

Over the past decade, five NSF-funded research and development projects—LASERS; Seeds of Science, Roots of Reading; the Imperial Valley Project in Science; SIFA; and P-SELL—have produced research on the relationship between the integration of science, language, and literacy instruction and ELL student achievement in science, language development, reading, and writing. These studies all have reported significant improvements in ELL science and literacy achievement as a result of the interventions.

Language Acquisition through Science Education in Rural Schools (LASERS), an NSF-funded local systemic change project with seven school districts in central California with large numbers of ELL students, used inquiry science as a context for implementing pedagogy that integrated language and literacy development into cognitively demanding science learning using an instructional approach that emphasized cooperative learning and cultural and linguistic contextualization (Stoddart 1999, 2005; Stoddart et al., 2002). In LASERS, the development of scientific understanding is promoted through the integration of contextualized science inquiry and discourse supported by the teacher through hands-on science activities and science talk. In a series of studies using both performance and standardized assessment, ELL students in LASERS classrooms showed significant achievement gains. In three consecutive summer schools, 1,200 limited English proficient students made significant gains in academic language and science concepts measured on the Woodcock Munoz standardized assessment of academic language and concept maps (Stoddart, 1999). Students also were tracked over three years in two participating school districts. Students (n = 1,300) who were in a LASERS-trained teacher's classroom for one or two years scored significantly higher on the SAT-9 in reading, language, mathematics, and science than students who were not in a LASERS-trained teacher's classroom (Stoddart, 2005).

The NSF-funded Seeds of Science, Roots of Reading project involved science educators and literacy educators in creating and testing an integrated literacy–science curriculum. Reading instruction, including texts, routines for reading, word level skills, vocabulary, and comprehension instruction, was integrated into inquiry-based science (Cervetti et al., 2007). The integrated curriculum was tested in 20 second- and third-grade classrooms over the course of either four or eight weeks against 24 comparison classrooms (12 where science was taught alone and 12 where literacy was taught alone). Students were assessed pre- and post-instruction on science understanding, science vocabulary, and reading comprehension in science. The researchers found positive outcomes for ELLs not only in the area of science knowledge, but also in literacy and vocabulary development, when measured against the comparison groups. Equivalent gains were made by ELLs on all science measures and most literacy measures in comparison to their English-speaking counterparts.

Analysis of the Science Instruction For All (SIFA) project data describes the impact of an instructional intervention designed and implemented to promote achievement of science and literacy among culturally and linguistically diverse students located in the greater San Francisco Bay Area (Baquedano-López et al., 2005; Bravo & Garcia, 2004; E. E. García & Baquedano-López, 2007; Ku, Bravo, & Garcia, 2004; Ku, Garcia, & Corkins 2005; Solís, 2005). Over the course of three years, the SIFA study implemented a curricular intervention in six schools with twenty-one teachers in third- and fourth-grade classrooms, in which each classroom received a year of literacy and science integrated instruction. The study focused on two science units at grades 3 and 4 in which science and literacy assessments were administered at the onset and at the end of the intervention. The results indicate participating students, regardless of linguistic and cultural background, experienced significant growth in their science achievement and understandings of scientific writing. The curriculum had a positive effect on the studentsí achievement and learning of science among students whose home language was either Chinese or Spanish.

The Imperial Valley Project in Science was also an NSF-funded Local Systemic Change initiative in a large school district in southern California. Working primarily with fourth- and sixth-grade students, Amaral et al. (2002) studied the effects of instruction that allowed students to conduct first-hand science investigations and keep a science journal to reflect on science activities and develop writing proficiency. The instructional focus was based on the idea that hands-on science activities establish an authentic purpose and offer increased opportunities for the development of writing skills (Holliday et al., 1994). The study led to significant gains among ELLs in both science knowledge and literacy abilities. Each subsequent year, over the four-year period in which students remained in this intervention, resulted in significant gains in science achievement. Using results from the state's science assessment, these researchers noted mean result increases for both fourth- and sixth-grade ELLs.

The P-SELL project implemented an integrated science and literacy curriculum in Florida for third-grade ELLs in urban elementary schools within an environment increasingly driven by high-stakes testing and accountability; the project examined students' science achievement at the end of the first-year implementation. The study involved 1,134 third-grade students at seven treatment schools and 966 third-grade students at eight comparison schools. Students who received the integrated science and language curriculum showed a statistically significant increase in science achievement over students in the comparison group (Lee et al., 2008). The treatment students also showed a higher score on a statewide mathematics test, particularly on the measurement strand emphasized in the intervention, than did comparison students. The NSF-funded P-SELL project also focused on integrating the teaching of science with English language development for ELLs in seven urban elementary schools in the southeast United States. Pre- and post-instruction assessment of science learning of 818 ELLs in P-SELL classrooms showed significant improvements in science understanding compared to control group students (Lee et al., 2008).

Findings from these five research and development projects all indicate that these integrated curriculum has a positive impact on the science learning and language and literacy development of ELL students.

INSTRUCTIONAL EXAMPLES OF ESTELL PEDAGOGY

In developing the framework for ESTELL, the authors drew on the practices described in the CFSEP and science, language, and literacy integration literature. Both of these approaches emphasize the importance of cultural contextualization in the effective teaching of diverse students and the importance of discourse and cooperative learning in promoting cognitive development. The ELL science literature also emphasizes the synergistic relationship between science inquiry, language and literacy learning, and the development of children's scientific understanding. The framework described in this chapter draws on both of these bodies of literature to describe Effective Science Teaching for English Language Learners (ESTELL). In an ESTELL classroom, teachers and students work together on real scientific inquiry. Activities are rich in language, with teachers developing students' capacity to speak, read, and write English and develop the special language of science. The curriculum is taught through meaningful activities that relate to the students' lives and experiences in their families and communities. Teachers challenge students to think in complex ways and to apply their learning to solving meaningful problems. Teachers and students converse: The basic teaching interaction is conversation, not lecture. A variety of activities are in progress simultaneously, including individual work; teamwork; practice and rehearsal; and mentoring in side-by-side, shoulder-to-shoulder, teacher–student work. Students have systematic opportunities to work with all other classmates.

Below, we present instructional exemplars of the integrated ESTELL practices. Each section foregrounds a specific approach—for example, language and literacy or scientific discourse—but each of these examples reveals a synergistic, integrated (not discrete "add-on") approach or strategy for instruction. The exemplars include: (1) Integrating Science, Language, and Literacy Development; (2) Engaging Students in Scientific Discourse; (3) Developing Scientific Understanding; (4) Collaborative Inquiry in Science Learning; and (5) Contextualized Science Learning.

1. Integrating Science, Language, and Literacy Development

The development of English language and literacy for ELLs involves learning to speak, comprehend, read, and write in a second language. This includes the learning of vocabulary, syntax, and lexical grammar, and the use of language and literacy for both social and academic functions. Research on second language development has emphasized the importance of the contextualized use of language (Cummins, 1981; Genesee, 1987;

Lambert & Tucker, 1972; McKeon, 1994; Met, 1994; Swain & Lapkin, 1985). Contextualization of language use refers to the degree to which language provides learners with meaningful cues that help them interpret the content being communicated: visual cues, concrete objects, and hands-on activities. In primary language development, children begin to understand utterances by relating them to sensory motor activities and the physical context (Krashen, 1985). In the development of a second language, this relationship needs to be explicitly communicated during instruction. By integrating language and literacy with the exploration of scientific phenomena, language activities are explicitly linked to objects, processes, hands-on experimentation, and naturally occurring events in the environment, that is, they are contextualized (Baker & Saul, 1994; Casteel & Isom, 1994; Lee & Fradd, 1998; Rodriguez & Bethel, 1983; Rosebery et al., 1992; Stoddart,1999). The development of science literacy is a social process and part of recognizable cultural expectations for communicating about the natural world (Roth & Lee, 2003).

ESTELL instruction around language and literacy development works to provide students with opportunities for written or verbal language expression *and* development in a contextualized science activity. Students have opportunities to collaborate with peers and the teacher, and the teacher assists students' language development by questioning, listening, rephrasing, or modeling. There is a particular focus on promoting authentic science literacy (graphing data, recording observations, reading and writing expository texts, illustrations, etc.) using science reading materials/references/illustrations for learning science; science language, including science discussion; and the systematic use of scientific vocabulary. Opportunities for literacy practices germane to science provide a context for authentic uses of literacy and increase the likelihood that students will build fluency in these literacy practices. Teachers of ELLs also use the integrated science, language, and literacy lessons as an opportunity for native language development and primary language support.

Example 1a: Integrating Science and Language Development: Life Science, Second Grade

The following example describes an elaborated implementation of the ESTELL approach. All of the ESTELL elements for the integration of science and language development are covered, including attention to authentic science literacy, oral science discourse, science vocabulary use, and the primary language support.

> Ms. D. engages the students in a discussion about bees and pollination. Ms. D. sits next to a large color drawing of a flower and a bee, and the students are seated on a rug in front of her. Ms. D. asks students to look at the picture

> and tell her what they notice about the bee. She repeats students' responses and asks them for more information. For example, one student says, "There is orange stuff on the back of his legs." Ms. D. replies, "Yes, it's a yellow-orange powdery stuff called pollen." She proceeds to tell students about a bee's role in pollination as she points to parts of the flower and bee. Students respond with questions and comments about this process. Ms. D. then tells students that they are going to write a "morning message" together about bees, pollination, and flower parts. Throughout the writing exercise, Ms. D. points out the language structure of the message as they write it; for example, pointing to the beginning of the first sentence, she asks students, "Why did I begin writing here instead of here?" A student replies, "Because it's a story," and Ms. D. says, "Yes, it's the beginning of a paragraph." The message they write describes the pollination process and the function of the various parts of flowers in this process (e.g., the colors of the petals attract bees, the stamen contains pollen, etc.). Ms. D. asks students to read the paragraph in English and Spanish, stopping to ask and answer questions as needed. Ms. D. also leads students through a game in which she asks them to identify words within the words in the "morning message" (e.g., "men" in "stamen").
>
> Ms. D. then transitions into a hands-on activity: dissecting a flower. She tells students what the goals are for the activity and models the process for them. Ms. D. places a large diagram of a flower and points to the flower parts on the diagram as she describes what she would like students to do. Students are told to create sections on their paper to place the flower parts on. Each student is given one flower, a hand lens, and paper. Ms. D. and her teaching assistant help students as they work. Ms. D's interactions with students incorporate substantial inquiry discourse; she asks students to talk about the parts they identify, asking what the function of the part is, where they would find pollen, and so forth. Ms. D. also helps students to make discoveries that extend beyond the assignment. One student, for example, finds a "baby flower" and seed, and takes them apart, and another uses her hand lens to compare the parts of her flower with a flowering plant in a corner of the room. After completing the dissection of their own flowers, placing the parts into categories, and labeling them, students take their hand lenses and examine other students' flowers. Students appear to be very engaged in the activity. After completing the primary assignment, students are seen using their hand lens to examine other students' flower parts and are heard discussing what they have found.

This teacher has developed a thorough understanding of how to design and implement an integrated science and language lesson. The design and implementation of the lesson uses a substantial amount of science inquiry and a range of language activities designed to engage students and advance their learning in science and language. The language and literacy activities are contextualized by being related to observations of pictures and examination of flowers. The lesson covers in-depth science and language content, and the implementation provides students with an opportunity to reflect on their learning. Students are provided with tools to participate in both science inquiry and writing about that inquiry. While there was an initial focus on writing a "message" that does not correspond to scientific forms of data observation and recording, the teacher related authentic science literacy tasks by having students write about their pollination observations.

Example 1b: Integrating Science and Primary Language Development: Life Science, Third Grade

The following example describes how the role of the primary language can be instrumental to maintaining student engagement during science activities that often rely on technical or new science language. The use of the primary language is important when appropriate; it facilitates the development of conceptual understanding and provides a link to the development of English as an additional language. In the following example, a third-grade classroom is involved in the examination of plant life through an experiment using seed pods. This observation takes place mid-way through implementation of a lesson. The teacher begins this observation by eliciting several types of science knowledge and observations, but principally by (1) having students review what they have learned so far in this lesson, and (2) having students make conceptual connections to observations, including making predictions of what they might observe next. The following exchange occurs after several previous student contributions.

> **Teacher:** After they were pollinated, what changes did we see with the plants?
> **Student:** It start... um, the () the () petals they start to getting long they start to () I can't–
> **Teacher:** [points to board] Fall off?
> **Student:** No seca... I don't know how to say it in Eng–
> **Teacher:** Well, tell me in Spanish.
> **Student:** Um, se secaron, secaron [translation: *they dried, dried*].
> **Teacher:** Se secaron [translation: *they dried up*].
> **Student:** Secaron [translation: *dried*].
> **Teacher:** Muy bien, se secaron [translation: *good, they dried up*]. It dried and it fell off, right? [writes on board]
> **Student:** Yeah. [a few students] It dried and fell off.
> **Teacher:** OK, then we started to see a part of the plant we'd never seen before.

The student in this example is encouraged by the teacher to switch from speaking English to Spanish, which allows the student to participate in the sharing of observations about seed pods. The teacher repeatedly uses key science vocabulary while eliciting information and observations from students, including such key words as *petals, pollen/pollinate/pollination/pollinated, buds, leaves, flowers, stems, roots,* and *seeds.*

2. Engaging Students in Scientific Discourse

In addition to being a discipline, science activities are achieved through a social process where the language used for competent participation requires specialized ways of talking, writing, and thinking about the world in scientific ways (Cervetti et al., 2007). Learning and doing science is not just a process of acquiring a set of facts, principles, and procedures; it also involves using the language of science in ways of talking and representing the natural world through discourse, interaction, and collaboration. Science is a discourse about the natural world: "Biology is not plants and animals. It is language about plants and animals.... Astronomy is not planets and stars. It is a way of talking about planets and stars" (Postman, 1979, p. 165). Learning science and talking about science are, therefore, interrelated. The discourse of science has its own vocabulary and organization that are embodied in the ways scientists think and communicate about their work. Language mediates and structures the ways in which scientists think about and investigate problems. These processes include formulating hypotheses, proposing alternative solutions, describing, classifying, using time and spatial relations, inferring, interpreting data, predicting, generalizing, and communicating findings (Chamot & O'Malley, 1986; National Science Teachers Association, 1991). The use of these language functions is fundamental to the process of inquiry science (NRC, 1996). By engaging in scientific discourse, students learn how to think about science, how to "do" science, and, consequently, develop their own scientific understanding.

Instructional conversations (ICs) are an example of an effective instructional arrangement for teaching students through dialogue (Dalton, 1998; Tharp & Dalton, 2007). These conversations can be achieved when the teacher organizes the classroom to accommodate conversation, articulating a clear academic goal for guiding conversation, ensuring student talk is more prevalent than teacher talk, guiding all talk to incorporate students' contributions, monitoring student comprehension of their talk, and by carefully scaffolding dialogue. The goals of ICs are to lead students to develop more complex and elaborated levels of understanding of academic concepts, activities, tasks, and practices. Effective ICs are those that are responsive to a range of both student comprehension levels and the types of contributions they make. Practices in ESTELL, therefore, highlight the role of the teacher in scaffolding students within their zone of proximal development to encourage scientific reasoning and dialogue. The teacher elicits and models conversation that requires scientific reasoning to involve students in sustained discussion on science topics. The teacher elaborates, recasts, and connects student ideas and invites students to follow up on others' talk. Students have opportunities to interact with peers and the teacher, while the teacher assists students' language development by questioning, listening, rephrasing, or modeling. (Chapin, O'Connor, & Anderson, 2003; O'Connor & Michaels, 1996). Through these group discussions, students begin to examine and reformulate a range of ideas and develop more complex understandings (Baquedano-López et al., 2005).

Example 2a: Instructional Conversation: Nature of Science, Fifth Grade

Students have just completed an investigation into gravity and acceleration. Their question was, "Do balls of different weights, masses, and sizes fall at different rates?" Their results were inconsistent. The teacher leads the students in a discussion on experimental error related to the investigation.

Teacher: (to Student 1) Tell us what happened during your group's investigation.

Student 1: Well, we got different results for each trial. During two trials, all the balls fell at the same rate. During one trial, the tennis ball fell first.

Teacher: (to all students) Did anyone else experience this?

Student 2: Yes, the same thing happened in our group.

Students 3: We didn't.

Student 4: Our group found that all the balls fell at the same rate.

Teacher: Can anyone explain what might have been going on here? How is it possible that you could have gotten different results for each trial? Students 1 and 2 got different results, but Students 3 and 4 found that the balls fell at the same rate each time. What does that mean?

Student 5: Maybe they conducted the experiment wrong?

Teacher: What do you mean? How could the experiments have been conducted differently? You all have the same materials.

Student 6: Yes, but we went outside to conduct our investigation, and it was kind of hard to tell which ball was falling first. The ping pong ball was taken by the wind when we dropped it, which made it fall slightly after the other two.

Student 2: Well, it seemed like in our group we kind of expected the tennis ball to land first, so maybe we could have let that one go like half a second before the ping pong ball, you know? It's like we wanted it to win.

Teacher: So what could you do differently to eliminate these sources of experimental error next time?

The teacher in this example involves students in a scientific discussion on experimental error. His approach to questioning is such that he encourages students to respond to each other and engage each other in a discussion on the importance of precision and accuracy of measurement in a scientific investigation. He uses instructional conversation to encourage the students to reach this conclusion without giving them the answer directly.

Example 2b: Instructional Conversation: Life Science, First Grade

Students have been studying the animal kingdom and learning about the various things that animals need for survival (i.e., food, shelter, water, etc.). Today, Ms. H. is leading a discussion on the types of food animals eat to help students understand that differences among animals are related to differences in their means of survival: In this case, that the shape of animals' teeth is related to the type of food that they eat. Ms. H. gives each group two models of an animal jaw bone (one herbivore and one carnivore) and asks them to make three observations about the differences between the two jaw bones. As students work in their groups, Ms. H. walks around to each group and facilitates dialogue among the student groups, engaging each group in a thoughtful discussion of the differences in animal eating habits. As she travels from group to group, Ms. H. connects student responses, facilitates clarification of claims, and revoices student ideas. Ms. H. intervenes as a facilitator, not as a knowledge producer. Students generate science knowledge through dialogue tied to a hands-on investigation.

Lin: This guy—he likes to eat plants!
Ms. H.: Lin thinks that Model A is from an animal that eats plants. Do you agree with her?
Jon: Yea, because his teeth are more smoother.
Ms. H.: Okay, Jon says the teeth on Model A are smoother than the teeth on Model B. Is that what you are saying, Jon?
Jon: Yes, this one is smoother than that one.

In this excerpt, the teacher's role is to revoice, encourage participation, and connect student–student responses. Notice that she does not intervene to correct scientific reasoning, that is, students' use of anthropomorphic reasoning, but allows students to engage in their own sense-making dialogue to understand scientific concepts.

3. Developing Scientific Understanding

ELLs can and need to be challenged to think critically about science concepts and topics to develop higher-order understandings. Too often, ELLs are relegated to remedial instructional programs focusing on the acquisition of basic skills that supposedly match their English-proficiency level and are not engaged in intellectually challenging activities (Garcia, 1988, 1993, 1997; Moll, 1992; Valdes, 2001). ESTELL integrated language, literacy, and inquiry science practices promote the development of English language and literacy while simultaneously promoting the development of students' scientific understanding. According to Padilla, Muth, and Padilla (1991), the same problem-solving processes are used whether students are conducting science experiments or reading assigned science texts. The cognitive strategies they use in both include making inferences, drawing conclusions, making predictions, and verifying predictions. The teaching of language arts with science, therefore, engages students in the development of thinking processes as they predict, classify, and interpret (Carin & Sund, 1985). Baker (1991) has talked about this as developing metacognitive skills (e.g., formulating conclusions, analyzing critically, evaluating information, recognizing main ideas and concepts, establishing relationships, applying information to other situations). The integration of scientific inquiry with contextualized scientific discourse promotes the development of students' understanding and promotes habits of mind inherent to science work.

In ESTELL, the teacher designs activities that promote complex reasoning of science concepts by having students make judgments about the value of data and consistency of individual and collective thinking. Students have opportunities to reflect and evaluate their own and others' scientific reasoning. The teacher designs and promotes student-led inquiry by having students share and evaluate their research design, findings, and implications of their investigations, and the teacher provides feedback.

Example 3a: Developing Scientific Understanding: Physical Science, Third Grade

The following example from third grade describes a lesson on the scientific method using an activity with paper airplanes. The example scores high in promoting several ESTELL practices, including collaborative inquiry and language development. It is an exemplary case of promoting complex scientific processes and thinking through guided inquiry.

> The teacher opens the lesson by showing a six-minute video that he created that (1) describes the concepts of air pressure and lift; (2) introduces the lesson in which students will develop and test out paper airplanes; and (3) introduces the three steps of the scientific method (i.e., hypothesis, experiment, and conclusion) and describes why it was important for students to use it in this lesson. In the video, keywords are presented both orally and visually. After the video, the teacher creates a list on the board with students about the processes of scientific inquiry and what each means for their inquiry activity about flight. The teacher also explains how students should record their findings on their method worksheet, which includes sections for them to record their hypotheses, findings, and

> conclusions. Students work in small groups to discuss, design, and create three paper airplanes. When they complete their airplanes, they go into the hallway to test out each airplane twice and measure how far they flew. Students record their results and observe the flight tests of other groups. Students write down their observations on their method sheet and are asked by the teacher to, within their small groups, compare their hypotheses with their findings and generate some conclusions about why their airplanes flew those distances. Finally, students are asked to discuss within their groups if their hypotheses were correct, and tie their findings back to the scientific concepts of air pressure and lift.

This lesson supports ESTELL practices that challenge student thinking because it orients students to more complex engagement of science concepts and supports the examination of student investigations through repeated feedback. The teacher provides clear expectations for testing out the merits of their observations, connects the scientific method to the activity, and structures time for students to discuss and evaluate their findings based on the initial standards for evaluation.

Example 3b: Developing Scientific Understanding: Life Science, Sixth Grade

Students in sixth grade are studying a unit on single-celled organisms. The teacher asks students to investigate bacteria levels in the school. In small groups, students formulate hypotheses about where they think the most bacteria collects, and identify three places from which they will collect bacteria to be grown in three separate petri dishes. During the next week, students chart the growth of their bacteria samples. Students then measure their samples and analyze their results to determine which sample grew the most bacteria and decide whether or not their hypotheses were correct. Finally, students present their findings to the rest of the class. Then, individual group findings are compiled to generate class findings, and groups are asked to evaluate their group findings in light of class findings. Each group then reports their new conclusions to the whole class with an explanation of why they changed their views, if they changed their views.

In this example, the teacher engages students in an open-to-guided inquiry investigation in which students are required to formulate their own hypotheses and determine their sources of data collection. Students engage in all aspects of the inquiry cycle. Their investigation is tied to what they have been learning and is designed to facilitate increased understanding of the unit on monerans.

4. Collaborative Inquiry in Science Learning

Research in effective instruction for ELLs has demonstrated that students and teachers working together in groups on a joint product increases content understanding, language acquisition, and literacy development (Brown, Metz, & Campione, 1996; Doherty, Hillberg, Pinal, & Tharp, 2003; Genessee, 1999; Johnson & Johnson, 1989; Kagan, 1989; Slavin, 1987; Dalton, 1998; Strong, 1983, 1984; Tharp & Gallimore, 1998). Working in groups also provides more opportunities for ELLs to use language in authentic social contexts. When effectively monitored by the teacher, group members can help scaffold each other, allowing for each individual to participate at a level appropriate to their language development (Herrell & Jordan, 2004; Kagan, 1989). Additionally, scientists do not conduct their work in isolation. Scientific inquiry is conducted in communities of practice. Science is a social endeavor, whereby scientists collaborate to collect data and write up findings, as well as submit findings to the larger scientific community for further evaluation and review.

Collaborative inquiry combines principles of cooperative learning, distributed expertise, and legitimate peripheral participation with principles of inquiry science instruction so that students work together in small groups, where each group member is expected to perform a specific task while all students collaborate in a science inquiry investigation or related activity (Brown, Ash, Rutherford, Nakagawa, Gordon, & Campione, 1993; Lave & Wenger, 1991; Meyers, 1993). Students are expected to create a tangible or intangible learning product through collaborative inquiry. Some examples of tangible products include an investigation proposal, a lab report, a data chart, etcetera. Examples of intangible products include sustained participation in a discussion involving scientific reasoning, an informal oral report to the class about a group's progress or preliminary findings in an inquiry task, etcetera.

ESTELL focuses on promoting effective learning communities through inclusive and collaborative student engagement. Collaborations occur between teacher and students, but more emphasis is placed on student–student interactions in small groups or pairs. However, the role of teacher in collaborative inquiry is not passive. The teacher works closely with each group to ensure that all students are participating within their appropriate zones of proximal development; s/he monitors students' engagement with the task and scaffolds them through questioning and prompting to keep them actively interacting with both content and language. The teacher promotes the creation of learning products, including artifacts, processes, procedures, or findings about science. The teacher supports student sharing, evaluation, and feedback of class products. Additionally, the teacher regulates the quality of the product, tangible or intangible, to ensure that students have accomplished both the science learning and the literacy and language learning goals at hand. Collaborations can take many forms: shared ownership, authorship, use, or responsibility for a collaborative product or tasks. In ESTELL, however, there is a particular disciplinary focus with regard to collaborations in science that promote collaboration, sharing of science authority, and specific science productions.

Arranging students in groups or promoting collaboration alone is insufficient to increase learning outcomes. Particular attention must be given to the learning task at hand, and the promotion of English-proficient students and teachers as scaffolds for less English-proficient students is crucial (Genessee, 1999; Genessee, Lindholm-Leary, Saunders, & Christian, 2005; Jacob, Rottenberg, Patrick, & Wheeler, 1996). Organizing students in groups to engage in the completion of a carefully prepared joint science learning task can also fail to achieve the desired outcome if equitable participation among all group members is not monitored by the teacher (Lee & Luykx, 2006; Kurth, Anderson, & Palinscar, 2002). Social stereotypes and other inequities can negatively affect the nature of student–student interactions. Therefore, the teacher must take measures to ensure each group member is encouraged to engage with the group and provided with opportunities for substantial participation in the creation of the joint product.

Example 4a: Collaboration and Production: Physical Science, Second Grade

Second-grade students have been studying simple machines and recently completed an investigation with levers. Using lab materials, they explored the fulcrum, effort, and load of a class-1 lever. Students investigate the levers in small groups. Each group was asked to detail their findings from tests they conducted using spring scales to determine the relationship between the fulcrum, effort, and load of a class-1 lever. The groups enter their findings on a chart. As the groups work, the teacher goes around to each group to probe for more detailed responses, asking them to think through their ideas, to make predictions and justify them. Each group presents their chart to the whole class and discusses their findings. All group members participate in the presentation. The teacher leads the class in a discussion and helps them to identify shared conclusions across groups. Following this, the teacher asked the students go back into their small groups to further investigate how class-1 levers are used to form a see-saw, hammer claws, scissors, and pliers. Each group is given a different item: a model see-saw, a hammer, scissors, or pliers. Then groups present their findings, explaining how levers are used to form the item their group investigated.

This example shows how the teacher promotes student collaboration in which all students are involved in the levers activity, with the clear expectation that students themselves can generate scientific observations and conclusions. Students produce collaborative products and shared understandings of the relationship between fulcrum, effort, and load of a class-1 lever. Concepts are then tied to common household items so that students participate in a contextualized application of knowledge produced in the first activity.

Example 4b: Collaboration and Production: Earth and Physical Science, Fifth Grade

Fifth grade students have been studying different types of water contamination and pollutants. They go on a field trip to a local watershed. In small groups, students collect water samples from different areas in the watershed, note the turbidity of the water in each sample, and record the temperature of the water samples at the time of collection. Students bring the data back to the classroom and conduct tests on the samples, including chlorine, copper, hardness, iron, nitrate, pH, and phosphate tests to determine levels of the various pollutants in each sample. The students then decide how they will represent the data—as a chart, a graph, a table, etcetera—and then create the representation and enter the data. Then students in their groups discuss the data and preliminary findings and prepare to present their work to the rest of the class.

In this activity, students collaborate to contribute to the collection of the data set and work together on the analysis of the findings. Students are working together on all aspects of the investigation, from sample collection, to testing, to presentation of findings. Collaborative products include the results and data table generated and the group presentation to the rest of the class.

5. Contextualized Science Instruction

Effective science instruction for ELLs requires that complex concepts are connected to hands-on investigations or familiar cultural models through inquiry-based learning activities. ESTELL pedagogy advances teaching beyond physical hands-on activities or isolated inquiry investigations and extends it to include the purposeful integration of students' *funds of knowledge* from home, school, or community (Gonzalez & Moll, 2002; Hammond, 2001; Ladson-Billings, 1995; Moll et al., 1992). As such, contextualized science instruction provides a framework for integrating the other four components of ESTELL pedagogy. Implementing the first four components in decontextualized and discrete ways prevents optimal implementation of the ESTELL pedagogy.

Contextualization, in this context, means the systematic incorporation of sociocultural resources, existing prior knowledge from everyday life experiences, or funds of knowledge into science practice. The teacher relates science learning to the world that surrounds students and makes connections to local, regional, and global science issues and investigations.

The teacher may also initiate and develop science projects that promote student expertise and leadership in issues related to local physical, geographic, and/or ecological science phenomena (e.g., leadership/ activism in local environmental contexts). Whenever possible, teachers involve family and community members as science experts and/or in the investigation of community-related science issues. ESTELL practices are also responsive to both continuities and discontinuities between students' social, linguistic, and cultural backgrounds and the Western scientific worldview. In ESTELL, science learning is also contextualized through the use of the tools, inquiry processes, and discourse used by scientists in their work (Barad, 2007; Lemke, 1990; NRC, 1996).

Contextualization is crucial to establishing the relevance of particular science topics as students participate in meaningful activities that are tied to their own questions, experiences, and concerns, and build on their own cultural, linguistic, and intellectual resources.

Example 5a: Contextualization: Environmental Science, Sixth Grade

Students in a predominantly agricultural community have completed an investigation on household pesticides and are now learning about how some pesticides used in farming pose a serious health threat to migrant farm workers. Using data collected and published by the California Department of Health Services (CDSH), students create pie graphs to represent the percentage of pesticide-related illnesses that were attributed to various industries (e.g., agricultural industry, 54.3%) and the ethnicity of agricultural workers affected (e.g., 85% Latino). Students then participate in a jigsaw reading of the peer-reviewed journal article, "Pesticide-related Illness among Migrant Farm Workers in the United States" (Das, Steege, Baron, Beckman, & Harrison, 2001) in the *International Journal of Occupational Environmental Health.* Using the assigned section of the article, each group member is required to research one of the following subtopics in an expert group: demographic characteristics of migrant workers, causes of pesticide-related illness, reporting incidences of pesticide-related illness, and pesticide-related illness prevention. After expert groups meet, group members return to their original group to teach their home group members about their topic of expertise. The following day, their teacher, Mr. Z., has invited representatives from the Environmental Protection Agency, the Environmental Working Group, and the United Farm Workers to come and talk to the class about how the chemical makeup of the more harmful pesticides affect the human body and to share their current organizing efforts to protect workers from such pesticides. Then groups create a public service announcement and poster to educate people in their community about pesticide-related illness prevention and reporting.

The students in this agricultural community research a science topic that is both socially and culturally relevant. In this case, although they are not involved in an empirical investigation, they must distill information from a scholarly journal and present their findings to both group members and, later, to their community in the form of a poster and public service announcement. Also, the teacher brings in local experts so that students both further develop their expertise on the topic of pesticides and human health and at the same time increase their awareness of community resources in the area.

Example 5b: Contextualization: Physical Science, Fourth Grade

Students in Ms. T.'s classroom will begin to learn about the Doppler Effect the following week. Ms. T. and her students live in a large urban community. Throughout the day, it is common to hear sirens or honking horns while the students are in class. The students are accustomed to waiting for the noise to fade and then continuing on with the lesson. Ms. T. asks her students to listen carefully three times to passing noises, such as sirens, over the next few days and draw a picture or keep a journal log to record what they hear. This activity is assigned in preparation for the upcoming unit.

The following week, Ms. T. asks her students to share their entries with each other in small groups and discuss their experiences listening to passing sounds in their neighborhoods. She asks them to make hypotheses about why it is that sounds seem to get louder and then fade. Then she introduces an activity in which students work in groups to make Doppler balls using a buzzer, a battery holder, a foam-filled baseball, and a 9-volt battery. Once students make their Doppler balls, they record observations about the sound of the buzzer when it is stationary, when it is moving toward them, when it is moving away from them, and when the they run to and from the stationary ball. After students have completed the activity, they share their results with the class. Ms. T. then introduces the official scientific term for what it is that the students have now discovered: "the Doppler Effect."

In this example, the teacher uses experiences familiar to her students to introduce a scientific concept. The example is culturally appropriate for the community, since the students live in a large urban area in which sirens are common street noises. The teacher might have chosen other types of examples, such as airplanes taking off, but many of her students may not have had a personal experience with this sound, since there is no major airport in the community. To introduce the new science topic, Ms. T. deliberately chooses an example that she knows most or all of her students can relate to, and then she provides opportunities for students to offer their own examples after conducting a community investigation. Additionally, the teacher engages students in a direct investigation of the Doppler Effect using familiar materials. Students participate in an investigation of the science topic using Doppler balls before the topic is officially introduced. In this way, Ms. T contextualizes her lesson on the Doppler Effect through the use of a concrete, hands-on experience and involves students as community researchers.

CONCLUSION

The research and instructional practices discussed above demonstrate that the integration of ESTELL pedagogy into science teaching is a powerful model for improving ELL achievement. The challenge is to prepare novice teachers to effectively use this instructional approach in their classrooms. Most teachers, however, are not prepared to teach academic content to diverse learners (Bryan & Atwater, 2002; Lee & Luykx, 2004; Rodriguez & Kitchen, 2005). The majority of teacher education programs do not model an integrated approach to instruction, and in the coursework there is little connection between learning to teach science and the use of culturally responsive pedagogy. Subject matter teaching methods are taught with little emphasis on integrating the language and culture of the student population being served (Dalton, 1998; Fradd & Lee, 1995; Stoddart, 1993a). Issues relating to cultural and linguistic diversity, when taught, are presented in separate courses and often focus on social conditions, not pedagogy (Met, 1994; Zeichner, 2003). Finally field experiences are often disconnected from, and not well coordinated with, the university-based components of teacher education (Wilson, Floden & Ferrini-Mundy, 2001), and contradictory views of teaching and learning are frequently manifested by the schools and teacher education program (Stoddart, 1993b). If novice teachers are to learn to effectively teach science to diverse learners, there must be coherence between their own learning experiences of science content, the pedagogy taught and modeled in science teacher education methods courses, and the models they observe in their field placements.

The next steps in the elaboration of ESTELL pedagogy is to develop a coherent model of teaching and coaching for novice teachers that integrates ESTELL pedagogy at every stage of teacher preparation and induction, from prerequisite science content courses, to the science teaching methods courses in the credential programs, to the clinical setting of student teaching and the first year of teaching. It is only through such coherent, integrated programs of teacher preparation that novice teachers will develop the knowledge and skills to effectively teach science to ELLs.

REFERENCES

Aikenhead, G. (2001). Students' ease in crossing cultural borders into school science. *Science Education, 85*(2), 180–188.

Aikenhead, G. (2006). *Science Education for Everyday Life: Evidence-Based Practice.* New-York: Teachers College Press.

Amaral, O. M., Garrison, L., & Klentschy, M. (2002). Helping English Learners increase achievement through inquiry-based science instruction. *Bilingual Research Journal,* 29(2), 213–239.

American Association for the Advancement of Science. (1989). *Science for all Americans.* New York: Oxford University Press.

American Association for the Advancement of Science. (1993). *Benchmarks for science literacy.* New York: Oxford University Press.

Au, K. H. (1980). Participation structures in a reading lesson with Hawaiian children: Analysis of a culturally appropriate instructional event. *Anthropology and Education Quarterly, 11*(2), 91–115.

Baker, L. (1991). Metacognition, reading, and science education. In C.M. Santa and D.E. Alvermann (Eds.), *Science learning: Processes and applications* (pp. 2–13). Newark, DE: International Reading Association.

Baker, L. & Saul, W. (1994). Considering science and language arts connections: A study of teacher cognition. *Journal of Research in Science Teaching , 31,* 1023–1037.

Bakhtin, M. M. (1981). *The dialogic imagination,* trans. and ed. M. Holquist and C. Emerson. Austin, TX: University of Texas Press.

Baquedano-López, P., Solís, J. L. , & Kattan, S. (2005). Adaptation: The language of classroom learning. *Linguistics and Education, 6,* 1–26.

Barad, K. (2007). *Meeting the universe halfway: Quantum physics and the entanglement of matter.* Durham, North Carolina: Duke University Press.

Batalova, J., Fix, M., & Murray, J. (2005). *English language learner adolescents: Demographics and literacy achievements.* Report to the Center for Applied Linguistics. Washington, DC: Migration Policy Institute.

Brown, A. L., Ash, D., Rutherford, M., Nakagawa, K., Gordon, A., & Campione, J.C. (1993). Distributed expertise in the classroom. In G. Salomon (Ed.), *Distributed Cognitions: Psychological and educational considerations* (pp. 188–228). New York: Cambridge University Press.

Brown, A. L., Metz, K. M. & Campione, J. C. (1996). Social interaction and individual understanding in a community of learners: The influence of Piaget and Vygotsky. In A. Tryphon & J. Vonèche (Eds.), *Piaget–Vygotsky: The social genesis of thought.* (pp. 145–170). Mahwah, New Jersey: Lawrence Erlbaum.

Bryan, L. A., & Atwater, M. M. (2002). Teacher beliefs and cultural models: A challenge for science teacher preparation programs. *Science Education, 86*(6), 821–839.

Buxton, C. A. (2006). Creating contextually authentic science in a "low-performing" urban elementary school. *Journal of Research in Science Teaching, 43*(7), 695–721.

Capps, R., Fix, M., Murray, J., Ost, J., Passel, S., Hernandez, S. H. (2005). *The new demography of America's schools: Immigration and the No Child Left Behind Act.* Washington, D.C.: Urban Institute. Retrieved January 4, 2008, from http://www.urban.org/publications/311230.html.

Carin, A. A., & Sund, R. B. (1985). *Teaching modern science (4th ed.).* Columbus, OH: Merrill.

Casteel, C. P., & Isom, B.A. (1994). Reciprocal processes in science and literacy learning. *The Reading Teacher, 47*(7), 538–545.

Cervetti, G.N., Pearson, P. D., Barber, J., Hiebert, E., & Bravo, M. A. (2007). Integrating literacy and science: The research we have, the research we need. In M. Pressley, A. K. Billman, K. Perry, K. Refitt & J. Reynolds (Eds.), *Shaping literacy achievement.* New York: Guilford.

Chamot, A., & O'Malley, M. J. (1986). *A cognitive academic language learning approach: An ESL content based curriculum.* Wheaton, MD: National Clearinghouse for Bilingual Education.

Chapin, S., O'Connor, C., & N. C. Anderson. (2003). *Classroom discussions: Using math talk to help students learn, grades 1–6.* Sausalito, CA: Math Solutions Publications.

Collier, V. P. (1989). How long? A synthesis of research on academic achievement in a second language. *TESOL Quarterly, 23,* 509–531.

Commission on Professionals in Science and Technology (2007). CPST Professional Women and Minorities: A Total Human Resources Data Compendium.

Cummins, J. (1981). *Bilingualism and minority-language children.* Toronto: OISE Press.

Dalton, S. S. (1998.) *Pedagogy matters: Standards for effective teaching practice.* Santa Cruz, CA: University of California, Center for Research on Education, Diversity, & Excellence.

Darling-Hammond, L. (1996). The right to learn and the advancement of teaching: Research, policy, and practice for democratic education. *Educational Researcher, 25*(6), 5–17.

Das, R., Steege, A., Baron, S., Beckman, J., & Harrison, R. (2001). Pesticide-related illness among migrant farm workers in the United States. *International Journal of Occupational Environmental Health, 7,* 303–312.

Deyhle, D., & Swisher, K. (1997). Research in American Indian and Alaska Native education: From assimilation to self-determination. In M. W. Apple (Ed.), *Review of Research in Education, Vol. 22* (pp. 113–194). Washington, DC: American Educational Research Association.

Doherty, R. W., & Pinal, A. (2004). Joint productive activity and cognitive reading strategy use. *TESOL.*

Doherty, R. W., Hilberg, R. S., & Lee, V. (2004). *Observing the five standards in practice: Development and application of the standards performance continuum* (CREDE Research Brief #11). Retrieved on March 28, 2010 from Center of Applied Linguistics website: www.cal.org/crede/pdfs/rb11.pdf

Doherty, R., Hillberg, R., Pinal, A. & Tharp, R. (2003). Five standards and student achievement. *NABE Journal of Research and Practice, 1,* 1–24.

Doherty, R. W., Hilberg, R. S., Pinal, A., & Tharp, R. G. (2002). *Transformed pedagogy, organization, and student achievement.* Paper presented at the annual conference of the American Education Research Association, New Orleans, LA.

Echevarria, J. E., Vogt, M. E., & Short, D. J. (2000). *Making content comprehensible for English language learners.* Boston: Allyn and Bacon.

Echevarria, J., Short, D., & Powers, K. (2004). Using sheltered instruction to improve the achievement of English language learners. *Manuscript in preparation.*

Estrada, P., & Imhoff, B. D. (2001). *Patterns of language arts instructional activity: Excellence, inclusion, fairness, and harmony in six first grade classrooms.* Paper presented at the annual meeting of the American Education Research Association, Seattle, WA.

Fradd, S. H., & Lee, O. (1995). Science for all: A promise or a pipe dream for bilingual students? *Bilingual Research Journal, 19,* 261–278.

García, E. (1988). *Effective schooling for language minority students.* In National Clearinghouse for Bilingual Education (Ed.), *New Focus.* Arlington, VA: Editor.

García, E. (1993). Language, culture and education. In L. Darling-Hammond (Ed.), *Review of Research in Education* (pp. 51–98). Washington, DC: American Educational Research Association.

García, E. (1997). Multilingualism in U.S. schools: Treating language as a resource for instruction and parent involvement. *Early Child Development and Care,* 127–128, 141–155.

García, E., & Baquedano-López, P. (2007). Science instruction for all: An approach to equity and access in science education. *Language Magazine 6*(6), 24–31.

Genesee, F. (1999). *Program alternatives for linguistically diverse students* (Educational Practice Rep. No. 1). Washington, D.C.: Center for Research on Education, Diversity, & Excellence.

Genesee, F. (1987). Considering two-way bilingual programs. *Equity and Choice, 3*(3), 3–7.

Genesee, F., Lindholm-Leary, K., Saunders, W., & Christian, D. (2005). *Educating English Language Learners: A synthesis of research evidence.* Cambridge, UK: Cambridge University Press.

Gonzalez, N., & Moll, L. (2002). Cruzando el puente: Building bridges to funds of knowledge. *Educational Policy, 16,* 623–641.

Grigg, W. S., Lauko, M. A., and Brockway, D. M. (2006). *The Nation's Report Card: Science 2005* (NCES 2006–466). U.S. Department of Education, National Center for Education Statistics. Washington, DC: U.S. Government Printing Office.

Hammond, L. (2001). Notes from California: An anthropological approach to urban science education for language minority families. *Journal of Research in Science Teaching, 38,* 983–999.

Herell, A. & Jordan, M. (2004). *Fifty strategies for teaching English Language Learners,* 2nd ed. New Jersey: Pearson Education.

Hilberg, R. S., Tharp, R. G., & DeGeest, L. (2000). Efficacy of CREDE's standards-based instruction in American Indian mathematics classes. *Equity and Excellence in Education, 33*(2), 32–40.

Holliday, W.G., Yore, L. D., & Alvermann, D.E. (1994). The reading–science learning–writing connection: breakthroughs, barriers, and promises. *Journal of Research in Science Teaching, 31,* 877–893.

Jacob, E., Rottenberg, L., Patrick, S., & Wheeler, E. (1996). Cooperative learning: Context and opportunities for acquiring academic English. *TESOL Quarterly, 30*(2), 253–280.

Johnson, D. W., & Johson, R. T. (1984). *Circles of learning: Cooperation in the classroom.* Alexandria, VA: Association for Supervision and Curriculum Development.

Kagan, S. (1989). *Cooperative learning: Resources for teachers.* San Jan Capistrano, CA: Resources for Teachers.

Kaplan, R. B. (1986). Culture and the written language. In J. M. Valdes (Ed.), *Culture bound: Bridging the culture gap in language teaching* (pp. 8–19). New York: Cambridge University Press.

Krashen, S. D. (1985). *Inquiries and insights: Second language teaching immersion & bilingual education literacy.* Englewood Cliffs, NJ:Alemany Press.

Ku, Y. M., Bravo, M., & Garcia, E. E. (2004). Science instruction for all. *ABE Journal of Research and Practice, 2*(1), 20–44.

Ku , Y. M. , Garcia, E. E. , & Corkins , J. (2005). *Impact of the instructional intervention on science achievement of culturally and linguistically diverse students.* Paper presented at the American Educational Research Association, Montreal, Canada.

Kurth, L., Anderson, C., & Palinscar, A. (2002). The case of Carla: Dilemmas of helping all students to understand science. *Science Education, 86,* 287–313.

Ladson-Billings, G. (1995). But that's just good teaching! The case for culturally relevant pedagogy. *Theory into Practice, 34,* 159–165.

Ladson-Billings, G. (1994). *The dreamkeepers: Successful teachers of African American_children.* San Francisco, CA: Jossey-Bass Publishers.

Lambert, W. E., & Tucker, G. R. (1972). *Bilingual education of children: St. Lambert experiment.* Rowley, MA: Newbury House.

Lave, J. & Wenger, E. (1991). *Situated learning: Legitimate peripheral participation.* Cambridge, UK: Cambridge University Press.

Lee, O., & Fradd, S. H. (1998). Science for all, including students from non-English-language backgrounds. *Educational Researcher, 27*(4), 12–21.

Lee, O. & Luykx, A. (2004). *Science Education and Student Diversity: Synthesis and Research Agenda.* Research report produced by A joint project of the Center for Research on Education, Diversity, and Excellence (CREDE) and the National Center for Improving Student Learning and Achievement (NCISLA) in Mathematics and Science.

Lee, O., & Luykx, A. (2006). *Science education and student diversity: Synthesis and research agenda*: Cambridge University Press.

Lee, O., Maerten-Rivera, J., Penfield, R., LeRoy, K., & Secada, W. G. (2008). Science achievement of English language learners in urban elementary schools: Results of a first-year professional development intervention. *Journal of Research in Science Teaching, 45*(1), 31–52.

Lemke, J. (1990). *Talking science: Language, learning, and values.* New Jersey: Ablex Publishing Corporation.

Lemke, J. L. (2001). Articulating communities: Sociocultural perspectives on science education. *Journal of Research in Science Teaching, 38*(3), 296–316.

Lynch, S. (2000). *Equity and science education reform.* Mahwah, NJ: Erlbaum Publishers.

Lynch, S. (2001). "Science for All" is not equal to "One Size Fits All": Linguistic and cultural diversity in science education reform. *Journal of Research in Science Teaching, 38*(5), 662–627.

McGroaty, M. (1992). The societal context of bilingual education. *Educational Researcher, 21,* 7–10.

McKeon, D. (1994). Language, culture and schooling. In F. Genesee (Ed.), *Educating second language children: The whole child, the whole curriculum, the whole community* (pp. 15–32). New York: Cambridge University Press.

Met, M. (1994). Teaching content through a second language. In F. Genesee (Ed.), *Educating second language children: The whole child, the whole curriculum, the whole community* (pp. 159–182). Oakleigh: Cambridge University Press.

Meyers, M. (1993). *Teaching to diversity.* Toronto, Canada: Irwin Publishing.

Moll, L. C. (1992). Bilingual classroom studies and community analysis. *Educational Researcher, 21,* 20–24.

Moll, L., Amanti, C., Neff, D., & Gonzalez, N. (1992). Funds of knowledge: A qualitative approach to developing strategic connections between homes and classrooms. *Theory Into Practice, 31,* 2, 132–141.

National Assessment of Educational Progress. (2005). *National and State Reports in Science.* Retrieved October 2, 2007 from http://nationsreportcard.gov/.

National Research Council. (1996). *National science education standards.* Washington, DC: National Academy Press.

National Science Teachers Association. (1991). *Scope, sequence, and coordination of secondary school science.* Washington, DC: Author.

Newman, S. S., & Gayton, A. H. (1964). Yokuts narrative style. In D. Hymes (Ed.), *Language and culture in society* (pp. 372–381). New York: Harper-Row.

NGA Center for Best Practices. (2000). *Teacher supply and demand: Is there a shortage?* Issue Brief. Retrieved April 20, 2005, from http://www.nga.org/eda/files/ooo125TEACHERS.PDF.

O'Connor, M. C., & Michaels, S. (1996). Shifting participant frameworks: Orchestrating thinking practices in groups discussion. In D. Hicks (Ed.), *Discourse, learning and schooling* (pp. 63–103). New York: Cambridge University Press.

Padilla, M. J., Muth, K. D., Padilla, R. K. L. (1991). Science and reading: Many process skills in common? In C. M. Santa & D. E. Alvermann (Eds.), *Science learning—Processes and applications* (pp. 14–19). Newark, Delaware: International Reading Association.

Padron, Y. N., & Waxman, H. C. (1999). Classroom observations of the Five Standards of Effective Teaching in urban classrooms with English language learners. *Teaching and Change, 7*(1), 79–100.

Pease-Alvarez, L. & Hakuta, K. (1992). Enriching our views of bilingualism and bilingual education. *Educational Researcher,* 21, 4–6.

Postman, N. (1979). *Teaching as a conserving activity.* New York: Delacorte.

Rodriguez, A. (1997). The dangerous discourse of invisibility: A critique of the NRC's National Science Education Standards. *Journal of Research in Science Teaching, 34,* 19–37.

Rodriguez, A. J. (2004). *Turning despondency into hope: Charting new paths to improve students' achievement and participation in science education.* Southeast Eisenhower Regional Consortium for Mathematics and Science Education @ SERVE. Tallahassee, Fl. www.serve.org/Eisenhower www.serve.org/Eisenhower.

Rodriguez, A. J. and Kitchen, R. (2005). *Preparing prospective mathematics and science teachers to teach for diversity: Promising strategies for transformative pedagogy.* Mahwah, New Jersey: Lawrence Erlbaum Associates.

Rodriguez, I. & Bethel, L. J. (1983). An inquiry approach to science and language teaching. *Journal of Research in Science Teaching, 20,* 291–296.

Rogoff, B. (1995). Observing sociocultural activity on three planes: Participatory appropriation, guided participation, and apprenticeship. In J. V. Wertsch, P.

del Rio, & A. Alvarez (Eds.), *Sociocultural studies of mind* (pp. 139–164). Cambridge, UK: Cambridge University Press.

Rogoff, B. (1990). *Apprenticeship in thinking: Cognitive development in social context.* New York: Oxford University Press.

Rogoff, B., & Wertsch, J. V. (1984). *Children's learning in the zone of proximal development.* San Francisco: Jossey-Bass.

Rosebery, A. S., Warren, B., & Conant, F. R. (1992). Appropriating scientific discourse: Findings from language minority classrooms. *The Journal of the Learning Sciences, 21,* 61–94.

Roth, W.-M., & Lee, S. (2004). Science education as/for participation in the community. *Science Education, 88*(2), 263–291.

Saunders, W., & Goldenberg, C. (2001). The effects of an instructional conversation on transition students' concepts of friendship and story comprehension. In R. Horowitz (Ed.), *Talking texts: Knowing the world through instructional discourse.* Newark, DE: International Reading Association.

Saunders, W., & Goldenberg, C. (1999). *The effects of comprehensive Language Arts/Transition Program on the literacy development of English learners* (Technical Report). Santa Cruz, CA: Center for Research, Diversity & Excellence, University of California.

Saunders, W., O'Brien, G., Lennon, D., & McLean, J. (1998). Making the transition to English literacy successful: Effective strategies for studying literature with transition students. In R. G. R. Jimenez (Ed.), *Promoting learning for culturally and linguistically diverse students* (pp. 99–132). Monterey, CA: Brooks Cole Publishers.

Slavin, R. (1987). Ability grouping and student achievement in elementary schools: A best-evidence synthesis. *Review of Educational Research, 57*(3), 293–336.

Solís, J. L. (2005). Locating student classroom participation in science inquiry and literacy activities. In J. Cohen, K. McAlister, K. Rolstad & J. MacSwan (Eds.), *ISB4: Proceedings of the 4th International Symposium on Bilingualism.* Somerville, MA: Cascadilla Press.

Stanley, W. B., & Brickhouse, N. W. (2001). Teaching sciences: The multicultural question revisited. *Science Education, 85*(1), 35–49.

Stoddart, T. (1993). The professional development school: Building bridges between cultures. *Educational Policy, 7*(1), 5–23.

Stoddart, T. (1999). *Language Acquisition Through Science Inquiry.* Paper presented at the annual meeting of the American Educational Research Association, Montreal.

Stoddart, T. (2005). *Improving student achievement with the CREDE Five Standards Pedagogy.* Technical Report No. J1 (pp. 2143–2151). Santa Cruz, CA: University of California, Center for Research on Education, Diversity and Excellence.

Stoddart, T, Abrams, R., Gasper, E., & Canaday, D. (2000) Concept maps as assessment in science inquiry learning – a report of methodology. *International Journal of Science Education,* 22(12), 1221–1246.

Stoddart, T., Pinal, A., Latzke, M., & Canaday, D. (2002). Integrating inquiry science and language development for English Language Learners. *Journal of Research in Science Teaching, 39*(8), 664–687.

Strong, M. (1983). Social styles and the second language acquisition of Spanish-speaking kindergartners. *TESOL Quarterly, 17,* 241–258.

Strong, M. (1984). Integrative motivation: Cause or result of successful second language acquisition? *Language Learning, 34,* 1–13.

Swain, M., & Lapkin, S. (1985). *Evaluating bilingual education: A Canadian case study.* Clevedon, England: Multilingual Matters.

Tharp, R. G. 1997. *From at-risk to excellence: Research, theory, and principles for practice.* Santa Cruz, CA: Center for Research on Education, Diversity & Excellence. Retrieved from http://www.crede.ucsc.edu/products/print/reports/rr1.html.

Tharp, R. G. & Dalton, S. S. (2007). Othodoxy, cultural compatibility, and universals in education. *Comparative Education, 43(*1), 53–70.

Tharp, R. G. & Gallimore, R. (1988). *Rousing minds to life: Teaching, learning and schooling in social context.* New York: Cambridge University Press.

Thomas, W. P., & Collier, V. P. (2003). *A national study of school effectiveness for language minority students' long-term academic achievement.* Santa Cruz, CA: Center for Research on Education, Diversity & Excellence, University of California, Santa Cruz.

Valdés, G. (2001). *Learning and not learning English: Latino students in American schools.* New York: Teachers College Press.

Vygotsky, L. S. (1978). *Mind in society: The development of higher psychological processes,* trans. and ed. M. Cole, V. John-Steiner, S. Scribner, and E. Souberman. Cambridge, MA: Harvard University Press.

Warren, B., Ballenger, C. Ogonowski, M. Rosebery, A., & Hudicourt-Barnes, J. (2001). Rethinking diversity in learning science: The logic of everyday sense-making. *Journal of Research in Science Teaching 38,* 1–24.

Warren, B., & Rosebery, A. S. (1995). Equity in the future tense: Redefining relationships among teachers, students, and science in linguistic minority classroom. In W. G. Secada, E. Fennema, & L. B. Adajian (Eds.), *New directions for equity in mathematics education* (pp. 298–328). New York: Cambridge University Press.

Wertsch, J. V. (1985). *Culture, communication, and cognition: Vygotskian perspectives.* New York: Cambridge University Press.

Wertsch, J. V. (1991). A sociocultural approach to socially shared cognition. In L. B. Resnick, J. M. Levine, & S. D. Teasley, eds., *Perspectives on socially shared cognition* (pp. 85–100). Washington, DC: American Psychological Association.

Wilson, S. M., Floden, R. E., & Ferrini-Mundy, J. (2001). *Teacher preparation research: Current knowledge, gaps, and recommendations.* Document R-1-3. 1–83. Center for the Study of Teaching and Policy

Zeichner, K. M. (2003). The adequacies and inadequacies of three current strategies to recruit, prepare, and retain the best teachers for all students. *Teachers College Record, 105*(3), 490–519.

CHAPTER 8

PRE-SERVICE ENGLISH LANGUAGE LEARNER SCIENCE TEACHER PREPARATION IN THE SOUTHEAST UNITED STATES

Teresa J. Kennedy, Jason T. Abbitt, and Michael R. L. Odell

ABSTRACT

Over the past two decades, the Hispanic population in the southeastern United States has grown by as much as 300% in some states, placing pressure on teacher preparation programs and pre-school through grade 12 schools to meet the needs of Spanish-speaking English language learners (ELLs). This descriptive study examines teacher education programmatic materials and science methods course syllabi to determine the extent to which ELLs are being addressed in higher education institutions in the Southeastern states. In the materials examined, it was found that pre-service teachers are provided with limited preparation to address the needs of ELLs. Of the 29 teacher preparation programs examined, few required pre-service educators to offer specific courses targeting ELLs, and even fewer included ELLs as a major

Teaching Science with Hispanic ELLs in K–16 Classrooms, pages 183–199

topic covered during science methods courses. Recommendations for pre-service science education course design targeted at preparing linguistically responsive teachers are included.

INTRODUCTION

The National Research Council's (NRC) national science education standards state that the commitment to science for all implies inclusion of those who traditionally have not received encouragement and opportunity to pursue science, specifically naming students with limited English proficiency within this context (NRC, 1996). In the last decade, the enrollment of students who speak languages other than English has grown at a phenomenal rate across the nation. According to the U.S. Department of Education's *Title IX–General Provisions* (2004), ELLs, formerly referred to as limited English proficient (LEP) students, are categorized as individuals who:

- are between 3 and 21 years of age;
- are enrolled or preparing to enroll in K–12 academic programs;
- were born in a country other than the United States or speak a native language other than English; and
- have difficulty in speaking, reading, writing, or understanding English, resulting in not being able to meet the state's proficiency level of achievement, unsuccessfully participating in English-only classrooms, and having limited opportunities to fully participate in society.

ELLs are one of the fastest-growing student groups in the United States, currently numbering over 5 million and representing 10 percent of all K–12 students (Suro & Singer, 2002; U.S. Department of Education, 2008). Between 1990 and 2000, ELL enrollment increased by 105 percent, a significant increase, given that overall student enrollment had only increased 12 percent during this time period (Kindler, 2002). The numbers of ELLs in our nation's schools have jumped by 57 percent over the past ten years (National Clearinghouse for English Language Acquisition, 2007). Clearly, ELLs constitute the largest growing segment of the educational system in the U.S.A. This is a segment consistently demonstrating low levels of achievement and accounting for the highest dropout rates in our nation's schools.

While dropout rates in the U.S.A. have declined from 15 percent in 1972 to 9 percent in 2006, and from 11 to 9 percent between the years of 2000 and 2006, the rates of ELLs ranging from age of 16 to 24 years old who are not enrolled in high school and lack a high school credential continue to persist (U.S. Department of Education, National Center for Education

Statistics, [NCES] 2008). While the NCES 2008 report stated that the status dropout rates for Whites, Blacks, and Hispanics each declined between 1972 and 2006, it also documented that for each year between 1972 and 2006, the status dropout rate was lowest for Whites and highest for Hispanics. High school completion rates for ELLs fall 20 percentage points behind proficient English speakers (Ballantyne, Sanderman, & Levy, 2008).

In 2000, one in five students in U.S. classrooms came from a home in which a language other than English was spoken (Crawford, 2000). At that time, over 400 languages were represented in U.S. schools, with Spanish-speaking students representing 77 percent of this total. These numbers are growing, and Spanish continues to be the most common language, other than English, spoken across the nation. Given this data, it is highly likely that every teacher, at some point in his or her career, will have ELLs in the classroom and will need to be ready to assist these students to fully access academic content to enable their educational success.

The Hispanic population is documented as the most rapidly increasing segment of our nation's population (NRC, 2006). Between 1990 and 2000, the Hispanic population grew by at least 250 percent in many southeastern states. Since 1990, Georgia, North Carolina, South Carolina, and Tennessee reported increases of over 300 percent in their Latino population (Fry, 2005). Table 8.1 illustrates the percent change in the Hispanic population in the southeast from 2000–2006. Six Southeastern states ranked in the top ten of Hispanic growth rate. Texas and Florida ranked lower, in part due to their already large Hispanic populations; however, East Texas and Central

TABLE 8.1 Percent Change in the Hispanic or Latino Population by States: 2000–2006

State	Population Growth	Rank
Arkansas	60.9	1
Georgia	59.4	2
South Carolina	57.4	3
Tennessee	55.5	4
North Carolina	54.9	5
Alabama	48.8	8
Florida	34.3	28
Mississippi	33.7	29
Texas	24.4	38
Louisiana	14.6	46

Source: U.S. Census Bureau, Population Estimates, July 1, 2000–July 1, 2006. Internet Release Date: February 8, 2008.

Florida have counties reporting significant increases in Hispanic population growth (U.S. Census Bureau, 2008).

Before the 1990s, the majority of Hispanics in Southeastern states were temporary migrant laborers. Recent census data show that Hispanics are becoming permanent residents (Fry, 2008). Until two decades ago, the Hispanic population in the U.S.A. was concentrated in established areas of Hispanic settlement. Beginning in the 1990s, the Hispanic population started to disperse, establishing population centers in the South, a region that historically had very few Hispanic residents (Fry & Gonzales, 2008). As a result, the South accounts for a greater share of overall Hispanic population growth than any other region in the United States. Since 2000, the bulk of the Hispanic population growth has occurred in Southern states, accounting for almost half of the increase of 5.1 million in the Hispanic population (Fry, 2008).

Unlike the rest of the country, data indicate that the Hispanic population in the Southeast is much more likely to have been born abroad, particularly in Mexico. Many among this population do not speak English well and have limited formal education. The recent and fast growth of Hispanics in the region is beginning to have dramatic impacts on the infrastructure of communities, particularly on the schools (Kochhar, Suro, & Tafoya, 2005).

Although school intervention programs are beginning to spring up across the U.S.A., and funding sources targeted at supporting Spanish-speaking ELLs also are growing, significant academic performance gaps persist between ELLs and their non-ELL peers (Abedi, 2006; Abedi, Leon, & Mirocha, 2003; Government Accountability Office, 2006; Solano-Flores & Trumbull, 2003).

Today's science teachers must be prepared to not only teach their specific subject area(s), but also address the socio-cultural struggles of ELLs attempting to learn another culture, acquire a new language, perform at a competitive level with their non-ELL peers, and integrate their worldviews and their personal experiences within the American classroom.

Two common educational intervention options for ELLs are variations of transitional bilingual education and English as a second language (ESL) pull-out programs. Unfortunately, ELLs are too often treated as remedial learners, grouped separately from their English-speaking peers, and in many schools without funding for formal intervention programs, special education staff assume responsibility for ELLs (Kennedy, 2001). The National Clearinghouse for English Language Acquisition (Ballantyne, Sanderman, & Levy, 2008) reported that research on teacher training and preparedness suggests that "teachers who do not hold bilingual or ESL certification are not well prepared to meet the needs of these children" (p. 9). Highlights of their report include the following statistics:

- Only 29.5 percent of mainstream teachers with at least one ELL in their classroom have the training to work effectively with ELLs;
- only 20 states require that all teachers receive specific training in working with ELLs;
- less than 1/6th of higher education institutions offering pre-service teacher preparation include training for working with ELLs;
- only 26 percent of teachers have received training related to meeting the needs of ELLs in their staff development programs; and
- over half, 57 percent, of teachers believe that they need more training to be able to provide effective education for ELLs.

Considering the number of students in our schools representing language communities other than English, the current reform movement must surpass previous change efforts in regard to linguistically diverse students if we are to truly serve all students well (Kennedy, 2001). This requires reform in university pre-service programs and in-service district staff development programs. The Committee on Multicultural Education for the American Association of Colleges for Teacher Education challenged all teacher education institutions in 2002 to prepare teachers capable of ensuring an "equitable education for students whose primary language is not English," further stating that "school educators, especially classroom teachers, must be better prepared for the changing fabric of the American classroom" (Brisk, Barnhardt, Herrera, & Rochon, 2002, p. 6). The Committee stressed the need for teacher preparation programs to focus on accommodative classroom practices for ELLs in order to ensure their academic success. Suggesting that the preparation of non-English speakers to survive in English-dominant schools and society is more complex than current policy debates would suggest, the Committee further called for drastic action in institutions of higher education in order to prepare future teachers as well as provide professional development to practicing teachers.

As schools and communities work to meet the educational needs of their ELLs, examinations into how pre-service teachers are prepared to meet the needs of ELLs has become increasingly important. Are pre-service elementary and secondary teachers being adequately prepared in their teacher preparation programs to address the needs of ELLs in the context of learning science and being prepared to ultimately enter the science, technology, engineering, and mathematics (STEM) workforce? There is a need to investigate the existing status of STEM teacher preparation to understand current practices and identify promising practices in regard to how pre-service science teacher education programs can better prepare pre-service teachers to enter the teaching force equipped to address the needs of Hispanic ELLs.

A recent study conducted by the National Clearinghouse for English Language Acquisition (Ballantyne, Sanderman, & Levy, 2008, p. 120) divided states into five categories in respect to the "amount of training or expertise required for all new teachers" across the U.S.A., finding that:

- only four states have specific coursework or certification requirements for all teachers (Arizona, California, Florida, and New York; noting that Pennsylvania has begun to require teachers graduating by 2011 from approved teacher education programs in the state to have completed specific coursework focusing on ELLs);
- only 17 states have teacher certification standards for all teachers containing reference to the special needs of ELLs (Alabama, Colorado, Idaho, Illinois, Iowa, Louisiana, Maryland, Massachusetts, Michigan, Minnesota, Nevada, New Jersey, North Dakota, Rhode Island, Tennessee, Vermont, and Virginia; noting that seven additional NCATE states will soon likely fall into this category: Alaska, Connecticut, Delaware, Georgia, Kansas, Mississippi, and South Carolina);
- only 8 states have made reference to language as an example of diversity in their teacher certification standards for all teachers (Arkansas, Montana, New Mexico, North Carolina, Ohio, Oregon, West Virginia, and Wyoming); and
- to date, 15 states have no requirement whatsoever for teachers to have expertise or training in working with ELLs (District of Colombia, Hawai'i, Indiana, Kentucky, Maine, Missouri, Nebraska, New Hampshire, Oklahoma, Pennsylvania, South Dakota, Texas, Utah, Washington, and Wisconsin).

This descriptive study expands on prior research and is an attempt to develop a snapshot to determine how pre-service teachers are prepared in their science methods courses at large teacher-producing universities across the southeastern United States, since the Southeast accounts for a greater share of recent overall Hispanic population growth than any other region in the U.S.A.

RESEARCH QUESTIONS

This study examines the status of pre-service teacher preparation in the Southeast in regard to preparation to teach science to Hispanic ELLs, and specifically:

1. Which teacher preparation courses address teaching ELLs?
2. How do established science methods courses address ELLs?

METHODOLOGY

A survey of programmatic information and syllabi from science methods courses was conducted. The researchers contacted teacher preparation programs in the southeastern United States at the four largest teacher-producing institutions in each state, as identified by U.S. Department of Education statistics (2006) regarding the number of recent graduates from teacher preparation programs. An e-mail message requesting syllabi for science methods courses was sent to faculty members from each institution that was currently teaching science methods courses at the elementary, middle, and secondary levels. The e-mail request indicated that a study was being conducted on the topics included in science methods courses.

While many faculty members responded to the e-mail request with their syllabi, several of those contacted expressed a reluctance to provide their syllabi without further information about the specific topics being reviewed. In order to reduce a possible bias or the addition of information prior to submission, the study's focus on ESL/ELL topics was not specifically identified in the request, nor in any follow-up communication. The e-mail request was sent out on two separate occasions.

For the purposes of this study, the Southeast was operationally defined as the states of Alabama, Arkansas, Florida, Georgia, Louisiana, North Carolina, Mississippi, South Carolina, Tennessee, and Texas (eastern region only). Institutions in Texas were limited to those in East Texas because the demographics and past history of this region are typical of other southeastern states. Forty total institutions were contacted and thirty-four syllabi were received. Of the syllabi obtained, most were from courses focusing on science methods in the elementary grades. The syllabi represent twenty-nine institutions and include all targeted states. Many institutions did not have separate middle/high school science methods courses. In a number of cases, the middle/high school science methods courses were taught by adjunct faculty, which may have impacted the response rate. A number of institutions only offered elementary science methods courses in the fall semester, which also may have affected the response rate.

For institutions returning syllabi, catalogue course descriptions and program requirements also were examined to place the methods course syllabi in the context of each teacher preparation program. Programs, course descriptions, and science methods syllabi were analyzed by the researchers to determine if the topic of addressing the needs of ELLs specifically was addressed.

A rubric (see Appendix) was developed with which the researchers analyzed each syllabus for topics that specifically addressed ELLs in science content areas. Programs were analyzed by examining requirements and course descriptions to determine if information related to working with ELLs was addressed in the program and, if so, in which courses. Although all of the institutions returning syllabi had teacher preparation programs with options specifically targeted to prepare ESL/bilingual teachers, this study primarily focused on programs resulting in an elementary generalist.

RESULTS

Twenty-five (74%) out of thirty-four syllabi contained a statement concerning teaching diverse students. Eleven (32%) syllabi included teaching diverse students as a topic, though ELLs were not specifically addressed. Five syllabi (15%) included either statements regarding teaching of ESL/ELL students or course topics regarding the relationship of ESL/ELL to P–12 science teaching. Of these five syllabi, four contained statements that reference ESL/ELL or language diversity, and three included ESL/ELL in a course topic or assignment. Tables 8.2 and 8.3 provide an overview of the data gathered.

The five syllabi including either statements regarding teaching strategies for ESL/ELL students or course topics regarding the relationship of ESL/ELL to P–12 science teaching, and those including course topics or class sessions regarding the relationship of ESL/ELL to P–12 science teaching were the focus of further examination to clarify the context in which these topics were present in the courses. One course syllabus contained statements regarding "language diversity" but no course topics focusing specifically on the issue of ESL/ELL learners in P–12 science. In this syllabus, the statements regarding language diversity were limited to a list of benchmarks for pre-service educators. Among these benchmarks was one statement: "The teacher must demonstrate knowledge and awareness of varied cultures and linguistic backgrounds." The list of benchmarks was an attachment to the syllabus and was not otherwise referenced in the course description or topics.

Another course included two class sessions on equity and diversity topics, but did not reference ESL/ELL elsewhere in the syllabus. In the description

TABLE 8.2 Frequency of Syllabi by Grade Level

Grade Level	Frequency
Elementary	13
Elementary/Middle	2
Middle	6
Middle/Secondary	1
Secondary	7
Unidentified	5

TABLE 8.3 Science Methods Syllabi Analysis ($N = 25$)

Number of Syllabi	Percentage	Indicator*
25/34	74%	Course syllabus contains *statements relating to course content or learning objectives* regarding diversity (cultural, ethnic, gender, etc.), multiculturalism, or addressing special needs of students in P–12 classroom teaching.
11/34	32%	Course syllabus identifies *specific course topics or class sessions* regarding diversity (cultural, ethnic, gender, etc.), multiculturalism, or addressing special needs of students in P–12 classroom teaching.
4/34	12%	Course syllabus contains *statements regarding relationship of ESL/ELL or language diversity* to teaching of P–12 science.
3/34	9%	Course syllabus contains syllabus identifies *specific course topics or class sessions regarding the relationship of ESL/ELL or language diversity to teaching of P–12 science.*

of these two class sessions on diversity, "ELL" was a bulleted item for class discussion that also included addressing special needs in the classroom, student equality, and "how to reach all students." There was no course assignment or other activity indicated that specifically noted ESL/ELL.

The two courses in which ESL/ELL were most prominently mentioned included statements regarding specific state English to speakers of other languages (ESOL) standards. These two syllabi also integrated the competencies associated with the ESOL standards into course activities and assignments. ESOL competencies addressed included teacher knowledge of: curriculum, curriculum materials, resources, instructional materials, instructional technology, and assessment.

The courses referred to above were offered at universities in the same state, though not at the same institution. One university offered a course for elementary pre-service teachers and the other university offered a course for middle school pre-service teachers.

In the course for elementary educators that prominently mentioned ESL/ELL students and their needs, instruction addressing ESL/ELL was described in context of "modifying science lessons and activities to meet the needs of various student populations." Course objectives additionally emphasized integrating ESOL strategies as one of the "skills related to the process of science, as they apply to the learning of science by elementary school children." Further emphasis on ESOL instructional strategies was indicated in those course objectives relating to assessment strategies and the development of lesson plans aligned with local, state, and national science education standards. Overall, ESOL strategies were emphasized in three of the seven course objectives. In examining the schedule of class sessions, teaching ESL/ELL students was a topic included at two points in the course, during the first week of the course, which focused on standards for science education, and during a week that focused on diversity in the classroom.

In the course for pre-service middle school educators, ESOL was listed among the "experiences that promote effective science teaching in grades 5–9." Specific state ESOL standards also were referenced in the description of the course goal and objectives. Although a detailed description of class sessions was not included in the syllabus, state ESOL standards were indicated as applying to two of the items listed in the course outline, including activities related to graphing and representing data and another laboratory-based activity. This particular syllabus also included one exam and three assignments that referenced ESOL strategies and standards. An overview of an exam to be given during the course specified ESOL strategies and activities that met ESOL standards among the topics to be included in the exam. Two assignments, one in which students would review a middle school science website and another requiring students to evaluate science textbooks, also noted the state ESOL standards. A more detailed description of a lesson planning assignment also was provided in this syllabus. The assignment included explicit directions that "your written lesson plan, teaching strategies and methods must incorporate ESOL strategies." Further mention was made of ESL/ELL in this assignment in a list of questions for the students to consider, including "What ESOL strategies are you using?" and "How will you assess student learning, including those with special needs, ESOL, etc.?" The inclusion of ESOL strategies was a requirement for the written objectives for this lesson planning assignment. This course represented the most notable integration of teaching methods relating to ESL/ELL in a science methods course of any of the syllabi that were received.

An examination of programmatic material describing the contents of a teacher education program and catalogue course descriptions showed that 64 percent of the institutions addressed ELLs in courses with a Special Education prefix. Almost half (48%) of programs addressed ELLs as part of a foundations course such as "School and Society" or a course specific to Multicultural Education. These course descriptions did not indicate that ELLs were addressed in the context of academic disciplines such as history, science, or mathematics. Sixteen percent of the programs included a stand-alone generic ELL course in the educational plan of their preservice teachers.

DISCUSSION AND RECOMMENDATIONS

Population projections cite a dramatic demographic transformation in the United States over the next 45 years, driven by continuing immigration and fast population growth among Latino communities. The Pew Hispanic Center projects that the Hispanic population in the United States will triple to represent 30 percent of the total population by 2050 (Passel & Cohn, 2008). The increasing number of ELLs in the mainstream classroom demands that pre-service science education programs adequately prepare new teachers with skills that will enable them to meet the challenges of teaching academic content in the science classroom.

Schools and their teachers are ultimately accountable for ensuring that all students master the same academic content in reading, mathematics, and science as non-ELLs. Teacher preparation programs need to meet this challenge by supporting faculty to increasing their capacity to infuse diversity issues related to educating ELLs throughout their curriculum. Science education teacher preparation programs need to provide opportunities for pre-service teachers to participate in courses or experiences related to teaching ELLs in the mainstream classroom. Science educators that prepare pre-service teachers also need to be prepared so they can integrate instructional strategies that target ELLs into the science education methods courses.

The National Clearinghouse for English Language Acquisition and Language Instruction Educational Programs 2008 Roundtable report on professional development for all mainstream teachers of ELLs (Ballantyne, Sanderman, & Levy, 2008) presents a vision of teacher education and professional development that is effective and relevant for all education personnel, including teacher educators, para-educators, in-service teachers, pre-service teachers, principals, and district staff. The report describes the need for collaborative, active learning within professional learning communities, further stressing the need to focus on multiple dimensions of diversity, fostering cross-cultural learning, and stressing disciplinary standards and pedagogical content knowledge. Courses, furthermore, should be driven by research and continually evaluate programmatic needs in order to improve student outcomes that ultimately narrow the achievement gap for ELLs.

All schools discussed in this paper were accredited by either the National Council for Accreditation of Teacher Education (NCATE), the *Teacher Education Accreditation Council* (TEAC), or their state accreditation body. It is recommended that science education programs also consider following National Staff Development Council (NSDC) context, process, and content standards (NSDC, 2001) related to ELLs. These standards support establishing professional learning communities for content area teachers that include ELL experts in the community who model responsibility for ELL learning, as well as encourage schools and districts to assign adequate resources enabling teachers to learn how to interpret and access research of relevance to ELLs.

More specifically, we recommend that pre-service courses aimed at preparing linguistically responsive pre-service teachers in the science classroom should consider providing strategies and opportunities to accomplish the following:

- Gain essential understandings of second language learning, including distinctions between conversational language proficiency development and academic language proficiency development in a non-threatening learning environment. This applies to native language speakers as well as ELL students. Pedagogical approaches in science classrooms should place a significant amount of attention on ensuring that students have access to language and vocabulary just beyond current levels of competence, commonly referred to as monitoring comprehensible input levels through explicit attention to linguistic form and function (Lucas, Villegas, & Freedson-González, 2008). Pre-service teachers should be capable of conforming to instructional congruence or "the process of merging academic disciplines with students' linguistic and cultural experiences to make the academic content accessible, meaningful, and relevant for all students" (Lee, 2004, p. 72).
- Use varying degrees of scaffolding techniques according to ELLs' linguistic and academic backgrounds to actively involve all students in the classroom while guiding them through the learning process (Canney, Kennedy, Shroeder & Miles, 1999; Kennedy, 2001).
- Create a classroom environment that facilitates both language development and content learning among ELLs by gaining experience in the identification of linguistic tasks leading to success in the classroom (Brisk & Zisselsberger, 2007).
- Incorporate the use of the students' native languages in their teaching, allowing students to build proficiencies in both their native language and in English (Crawford, 2000; Kennedy & Canney, 2000) through allowing speakers of the same language to work together and to discuss scientific concepts in their native language before they are required to communicate them, orally and in writing, in English.
- Collaborate with parents and community members to contribute to and be included in instructional processes that promote negotiation of meaning within natural learning contexts, including modified speech, providing visual support, and planned meaningful redundancy of content (Kennedy, 2001).
- Engage in experiences designed to conceptualize and implement systemic ways to address challenges related to promoting classroom practices that are equitable and rigorous (Buxton, Lee, & Santau, 2008);
- Work with practicing teachers to develop science curricula integrating literacy development with science learning, as well as engage in action research agendas that consider teachers' perspectives in promoting science inquiry with students learning English (Fradd, Lee, Sutman, & Saxton, 2001).

- Take into account national and state science standards and know how to teach in ways that facilitate student learning of the standards (NRC , 1996), as well as follow Teachers of English to Speakers of Other Languages (TESOL) standards for English language proficiency (TESOL, 2006).
- Take into account national and state established discipline assessments, as well as follow the U.S. Department of Education *Framework for High-Quality English Language Proficiency Standards and Assessments* (2008) during all phases of curriculum design.

Pre-service science methodology courses covering the topics above may enable new teachers to plan classroom instruction aimed at improving the English proficiency of their students. Such courses would also provide teachers with the necessary tools to promote ELLs' academic achievement levels so they would be comparable with the achievement of their non-ELL peers. Research suggests that attending to student language competencies, specifically their academic language competencies, leads to academic success, implying that raising students' English proficiency could help to close the achievement gap prevalent in today's schools. Addressing the needs of ELLs as a common pedagogical practice is critical, and instruction toward this end must be included in every science teacher methods course.

FUTURE RESEARCH

The data for this research were limited, due to the rates of return. Some probable trends, however, were identified. More science methods course syllabi need to be collected to develop a better understanding of how ELLs are addressed in the science methods course. All of the syllabi examined contained common topics, including instructional models, science instruction, lesson planning, student misconceptions, assessment, and diversity. Given that the ELL population is increasing, research into science-specific strategies would be beneficial for developing methods courses, methods textbooks, and other instructional materials. The research community would benefit from a syllabus-sharing archive similar to the one created by the Association of Science Teacher Educators (ASTE) to look for trends in instruction and coverage of English language learning or other science education topics.

Based upon the finding that many teacher preparation programs include English language learning as part of educational foundations and/or special education courses, the authors recommend that pre-service faculty and teachers be surveyed to ascertain attitudes towards ELL students. If pre-service teachers receive their primary preparation in a special education course, for example, do they see ELL students as being disabled? Do science educators see ELL preparation as the responsibility of ELL specialists, special educators, or foundations faculty?

Acknowledging that these questions remain unanswered, we believe that the research presented here offers a starting point for further investigations. Teacher education currently lags behind the realities of today's classrooms. If science educators in the Southeast hope to reduce the science achievement gap among Hispanic students, ELL research specific to the learning of science has to be included as an important part of the research agenda, and the findings need to be incorporated into science teacher preparation courses. The challenge we have before us is to integrate research-based practices that provide experiences to better prepare pre-service science teachers to meet the needs of all students.

ACKNOWLEDGEMENTS

A special thank you to Patricia Stout, journalism student at the University of Colorado, who helped copyedit this chapter.

REFERENCES

Abedi, J. (2006). Language issues in item-development. In S. M. Downing & T. M. Haladyna (Eds.), *Handbook of test development* (pp. 377–398). Mahwah, NJ: Erlbaum.

Abedi, J., Leon, S., & Mirocha, J. (2003). *Impact of student language background on content-based performance: Analyses of extant data* (CSE Technical Report No. 603). Los Angeles: University of California, National Center for Research on Evaluation, Standards, and Student Testing.

Ballantyne, K. G., Sanderman, A. R., Levy, J. (2008). *Educating English language learners: Building teacher capacity.* Washington, DC: National Clearinghouse for English Language Acquisition. Retrieved January 10, 2009, from http://www.ncela.gwu.edu/practice/mainstream_teachers.htm.

Brisk, M.E., Barnhardt, R., Herrera, S., & Rochon, R. (2002). Educators' preparation for cultural and linguistic diversity: A call to action. American Association of Colleges for Teacher Education, Committee on Multicultural Education. Retrieved January 10, 2009, from http://escholarship.bc.edu/lynch_facp/11/.

Brisk, M.E., & Zisselsberger, M. (2007, April). We've let them in on the secret: Using SFL theory to improve teaching of writing to bilingual learners. Paper presented at the annual meeting of the American Educational Research Association (AERA), Chicago.

Buxton, C, Lee, O. & Santau, A. (2008). Promoting science among English language learners: Professional development for today's culturally and linguistically diverse classroom. *Journal of Science Teacher Education, 19*(5), 495–511.

Canney, G., Kennedy, T.J., Shroeder, M., & Miles, S. (February, 1999). Instructional strategies for K–12 limited English proficiency (LEP) students in the traditional classroom. *The Reading Teacher*, 539–544.

Crawford, J. (2000). *At war with diversity: USA language policy in an age of anxiety*. Clevedon, England: Multilingual Matters.

Fradd, L., Lee, O., Sutman, F. X., & Saxton, M. K. (2001). Promoting science literacy with English language learners through instructional materials development: A case study. National Association for Bilingual Education. Retrieved January 10, 2009, from http://findarticles.com/p/articles/mi_qa3722/is_200110/ai_n8997309.

Fry, R. (2008). *The role of schools in the English language learner achievement gap*. Washington, DC: Pew Hispanic Center.

Fry, R.(2005). *The high schools Hispanics attend: Size and other key characteristics*. Washington, D.C.: Pew Hispanic Center.

Fry, R. & Gonzales, F. (2008). *One-in-five and growing fast: A profile of Hispanic public school students*. Washington, DC: Pew Hispanic Center.

Government Accountability Office (GAO). (2006). *No child left behind act: Assistance from education could help states better measure progress of students with limited English proficiency*. Washington, DC: United States Government Accountability Office.

Kennedy, T. J. (2001). Preparing teachers for educating linguistically diverse students. *Northwest Passage: NWATE Journal of Education Practices, 1*(1), 51–55.

Kennedy, T. J. & Canney, G. (2000). Collaborating across language, age, and geographic borders. In K. Risko and K. Bromley (Eds.), *Collaboration for diverse learners: Viewpoints and practices* (pp. 310–329).

Kindler, A.L. (2002). *Survey of States' limited English proficient students and available educational programs and services: 2000–2001 summary report*. Washington DC: National Clearinghouse for English Language Acquisition.

Kochhar, R., Suro, R., & Tafoya, S. (2005). *The new Latino south: The context and consequence of rapid population growth*. Washington, DC: Pew Hispanic Center.

Lee, O. (2004). Teacher change in beliefs and practices in science and literacy instruction with English language learners. *Journal of Research in Science Teaching, 41*(1), 65–93.

Lucas, T., Villegas, A.M., & Freedson-González, M. (2008, October). Linguistically responsive teacher education. *Journal of Teacher Education, 59*(4), 361–373.

National Clearinghouse for English Language Acquisition (NCELA). (2007). *The growing numbers of limited English proficient students*. Washington, DC: Author. Retrieved January 10, 2009, from http://www.ncela.gwu.edu/policy/states/reports/statedata/2005LEP/GrowingLEP_0506.pdf.

National Council for Accreditation of Teacher Education (NCATE). (2006). *Professional standards for the accreditation of schools, colleges, and departments of education*. Retrieved January 10, 2009, from http://www.ncate.org/documents/standards/unit_stnds_2006.pdf.

National Council for Accreditation of Teacher Education (NCATE). (2007). *The NCATE Unit Standards*. Retrieved January 10, 2009, from http://www.ncate.org/documents/standards/unitstandardsMay07.pdf.

National Research Council (NRC). (2006). *Hispanics and the future of America*. In Tienda, M. and Mitchell, F. (Eds.). Panel on Hispanics in the United States

[and] Committee on Population Division of Behavioral and Social Sciences and Education. Washington, DC: National Academies Press.

National Research Council (NRC). (1996). *The national science education standards.* Washington DC: National Academy Press.

National Staff Development Council (NSDC). (2001). *Standards for staff development: Revised.* Oxford, OH: Author.

Passel, J. S. & Cohn, D.V. (2008). *USA population projections: 2005–2050.* Washington, DC: The Pew Hispanic Center. Retrieved January 10, 2009, from http://pewhispanic.org/reports/report.php?ReportID = 85.

Solano-Flores, G., & Trumbull, E. (2003). Examining language in context: The need for new research and practice paradigms in the testing of English-language learners. *Educational Researcher, 32*(2), 3–13.

Spillane, J. P., Diamond, J. B., Walker, L. J., Halverson, R., & Jita, L. (2001). Urban school leadership for elementary science instruction: Identifying and activating resources in an undervalued school subject. *Journal of Research in Science Teaching, 38,* 918–940.

Suro, R. & Singer, A. (2002). *Latino growth in metropolitan America: Changing patterns, new locations.* Washington, DC: Brookings Institution and The Pew Hispanic Center.

Teachers of English to Speakers of Other Languages (TESOL). (2006). *PreK–12 English language proficiency standards.* Alexandria, VA: Author.

U.S. Census Bureau. (2008). *Population estimates, July 1, 2000 to July 1, 2006.* Washington, DC: Author.

U.S. Department of Education. (2008, October). *Framework for high-quality English language proficiency standards and assessments: Brief.* Washington, DC: Author. Retrieved January 10, 2009, from http://www.aacompcenter.org/cs/aacc/print/htdocs/aacc/resources_sp.htm.

U.S. Department of Education. (2006). *Digest of statistics, 2006.* Washington, DC: Author. Retrieved January 10, 1009, from http://nces.ed.gov/pubSearch/pubsinfo.asp?pubid = 2007017.

U.S. Department of Education, National Center for Education Statistics. (2008). The condition of education 2008 (NCES 2008-031). Washington, DC: Author. Retrieved January 10, 2009, from http://pewhispanic.org/reports/report.php?ReportID = 19.

U.S. Department of Education. (2004). *Title IX–General Provisions, Limited English Proficient.* Section 9109, Part A, Section 25. Washington, DC: Author. Retrieved January 10, 2009, from http://www.ed.gov/policy/elsec/leg/esea02/pg107.html.

APPENDIX
Rubric: ESL/ELL in Science Education

State: ______________________

Inst. ID: ____________________

Teaching Methods Level:

☐ Elementary
☐ Elem/Middle
☐ Middle
☐ Middle/Secondary
☐ Secondary
☐ All K–12

Diversity/ESL/ELL Course topics:
Check all that apply

____ **Diversity Statements included in syllabus**
Course syllabus contains *statements relating to course content or learning objectives* regarding diversity (cultural, ethnic, gender, etc.), multiculturalism, or addressing special needs of students in P–12 classroom teaching.

____ **Course Topics relating to Diversity included in syllabus**
Course syllabus identifies *specific course topics or class sessions* regarding diversity (cultural, ethnic, gender, etc.), multiculturalism, or addressing special needs of students in P–12 classroom teaching.

____ **Statements regarding ESL/ELL or language Diversity**
Course syllabus contains *statements regarding relationship of ESL/ELL or language diversity* to teaching of P–12 science.

____ **Course Topics regarding ESL/ELL included in syllabus**
Course syllabus contains syllabus identifies *specific course topics or class sessions regarding the relationship of ESL/ELL or language diversity to teaching of P–12 science*.

CHAPTER 9

USING INQUIRY TO TEACH INQUIRY

A Pre-Service Science Education Model with Possibilities for Developing Hispanic English Language Learners' Academic Discourse

Elsa Q. Villa and Kerrie Kephart

ABSTRACT

There is emerging evidence that inquiry-oriented approaches to science teaching and learning show promise for meeting the needs of an increasingly culturally and linguistically diverse U.S. student population. Despite calls from national associations of science educators and educational researchers for adopting inquiry methods in science teaching, however, there is evidence of a lingering "teaching gap" (Stigler & Hiebert, 1999). Science teaching in the U.S. is still largely focused on developing students' procedural knowledge, while it neglects their conceptual development. Teacher preparation programs face the challenge of preparing the next generation of teachers to adopt methods that they may have never experienced as learners. This chapter presents findings of a study of K–4 teacher candidates' development of

Teaching Science with Hispanic ELLs in K–16 Classrooms, pages 201–231

understanding of inquiry pedagogy in science. In their growing understanding of inquiry, they also realized its potential for supporting English language learners' conceptual and linguistic development. We highlight the reflections of five of the study participants who, through immersion in inquiry teaching and reflection on their teaching experiences, demonstrated transformation in their perspectives of pedagogical methods and their relationship with linguistic development.

INTRODUCTION

> What is vital in becoming proficient learners is being able to understand how to get and make sense of the collection of knowledge acquired through critical thinking. For example, in the fall semester, my partner and I worked with the Full Option Science System (FOSS) Balance and Motion science kit. At first, I thought the lessons had no closure. *I felt uncomfortable teaching them.* I felt I was doing something wrong and questioned my abilities as a teacher. We would complain about their lack of connection from one lesson to the other. To our surprise, at the end of the kit, where the students build the foam ramps and make a marble roll from one end to the other, *the students pulled all the concepts of balance and motion together to make it work! They used the vocabulary presented in the lessons to express their ideas and that is when I realized that inquiry was at work all along.* I allowed my inexperience to get in the way of seeing the real value of these lessons while I was presenting them. (Gracie, personal reflection, 2008; emphasis added)

Gracie is a pseudonym for one of the teacher candidates in the study that is the focus of this chapter. The findings of this study demonstrate how inquiry pedagogy transformed teacher candidates' perspectives on science teaching and learning and, in so doing, enabled them to embrace the use of inquiry pedagogy, a form of teaching that the literature suggests promotes and develops science literacy and academic language skills (Amaral, Garrison, & Klentschy, 2002; Lee, 2002; Lee & Fradd, 1998; Rosebery, Warren, & Conant, 1992). Gracie is representative of the students who attend the university in which the study was conducted: a university located on the U.S./Mexico border where the majority of students are Hispanic. Gracie is married, with school-aged children, and she attends a university where students studying to become teachers are as diverse as their ages. They include students who work off-campus, are single, married, parents, grandparents, immigrants, non-native English speakers, and first-generation U.S. citizens. In the quotation above, taken from Gracie's personal reflection on the two semesters during which she studied mathematics and science methods, she touched on some of the themes that emerged from this study, which include non-traditional pedagogical methods that challenge traditional teacher-centered instruction, the time-intensive nature of inquiry pedagogy, students' acquisition of science discourse, and development of conceptual understanding through inquiry.

This chapter presents the findings of a study that immersed K–4 teacher candidates who were pre-service teachers in their last semester of undergraduates studies in teaching inquiry in their field-based classrooms, where the vast majority of classrooms were traditional ones in which the teacher perceived him/herself as transmitting knowledge and students as passive receivers of that knowledge. To counter this perception, Villa, first author and the instructor in the university science methods course, required teacher candidates to use research-based, commercial science curricula developed through National Science Foundation grants. Findings from this study, conducted in a low socio-economic elementary school on the U.S./Mexico border, show how teacher candidates—represented here in their own voices, as drawn from interviews, focus groups, and written reflections—acquired an understanding of inquiry methods and how these methods promoted and developed academic discourse in science and English.

BACKGROUND AND NEED

In a nation with a growing immigrant population (U.S. Census, 2004), the need to support students' academic language development, especially that of English Language Learners (ELLs), is a growing concern. According to the U.S. Census, between 1990 and 2000 the percentage of people over age five who spoke a language other than English in their home increased by 47%, from 32 million (or 12% of the total U.S. population) in 1990 to 47 million (or 18% of the total U.S. population) in 2000 (Meyer, Madden, & McGrath, 2004). Concomitantly, the percentage of students designated as ELLs in U.S. schools also rose over the same time period, from 5.1% of all students in 1990 to 6.7% in 2000 (Meyer et al., 2004). By 2004–2005, according to the Educational Testing Service (ETS) (2008a), the ELL population had risen to 10.5% of all students in U.S. schools. Meanwhile, there is evidence of a widening gap between ELLs and their native English-speaking counterparts on certain measures of academic success. For example, the Educational Testing Service (2008b) reports ELLs did not show improved performance from 2005 to 2007 on the National Assessment of Educational Progress (NAEP) report. Moreover, the achievement gap between ELLs and native-English speaking students widened in all subjects as measured by the NAEP from 2005 to 2007. Bilingual education experts, nevertheless, caution that assessing ELLs through standardized tests may not be the most accurate measure of their achievement and abilities (ETS, 2008b).

Genesee, Lindholm-Leary, Saunders, and Christian (2005) also note the dearth of research on effective curriculum and methods for working with

ELLs, especially in science. For this reason, we seek to corroborate emerging evidence that learning science through inquiry methods of teaching promotes and develops both science literacy and academic language skills (Amaral, Garrison, & Klentschy, 2002; Hampton & Rodriguez, 2001; Lee, 2002; Lee & Fradd, 1998; Rosebery, Warren, & Conant, 1992).

Historical Perspective of Inquiry Science

In the early 1960s, school districts across the nation introduced innovative science curriculum to elementary classrooms. These programs were Science Curriculum Improvement Study (SCIS), Science-A Process Approach (SAPA), and Elementary Science Study (ESS), which featured hands-on teaching materials and curricula. This marked the genesis of national science reform efforts, which have been underway for almost fifty years. In the mid-1980s, the American Association for the Advancement of Science (AAAS) founded Project 2061, with a mission to inform and educate the general population about science, mathematics, and technology. Their landmark publication, *Science for All Children*, led to the creation of national science benchmarks and influenced the development of national and state standards for K–12 science and mathematics content and teaching (AAAS, 2008). Moreover, federal funding agencies such as the National Science Foundation contributed millions of dollars to develop effective materials and professional development for educators. Despite these massive efforts over a long period of time, however, science teaching in elementary schools is still a rare occurrence. Where elementary science is happening, inquiry methods are generally not being used, which Newman et al. (2004) suggest is due primarily to "teachers lacking the necessary skills and knowledge to teach inquiry" (p. 259). While gains in student science achievement have been made since 1990, the gap between minorities and White students remains relatively constant, and the lowest achievement levels are among those living in poverty (Science and Engineering Indicators [SEI], 2008). Compounding the challenges facing elementary science programs has been the massive influx of ELLs across the nation. Teachers are typically uninformed of effective methodologies and strategies for teaching this diverse population of learners, a deficiency in their preparation that can and should be addressed (Menken & Antuñez, 2001).

The Need to Prepare Teachers for Inquiry Science

In a meta-analysis of science teacher education research, Clift and Brady (2005) examined 21 studies between 1995 and 2001 and found that many of them focused on teaching for conceptual change. In examining the impacts of science methods courses and field experiences on teacher candidates' beliefs and practices, many of the studies found that pre-service teachers accepted new practices, yet struggled with applying them in actual classrooms. Newman et al. (2004) found that among the challenges in teaching science methods to pre-service elementary education teachers was their lack of experience as learners in inquiry classrooms. Stigler and Hiebert (1999) describe this as a "teaching gap," whereby socialization into teaching results in resistance to change. Their video study of teachers from several industrialized nations found that, while U.S. teachers desired to change their teaching methodology to align with current reform efforts, there was little evidence that change actually occurred.

Teacher candidates tend to have pre-conceived ideas of what a teacher does and how a teacher acts, notions that are acquired through their prior experience as students and that are strongly resistant to change. Lortie (1975) coined the term "apprenticeship of observation" to describe the phenomenon in which "those who teach have normally had sixteen continuous years of contact with teachers and professors" (p. 61). This experience informs their mental schema, which shapes their ideas of what it means to be a teacher as they enter the teaching profession (Darling-Hammond, Hammerness, Grossman, Rust, & Shulman, 2005; Lortie, 1975; Oakes & Lipton, 2003; Zeichner & Gore, 1989). Thus, due to their past experiences as learners in traditional teacher-centered classrooms, there is a tendency for teacher candidates to hold a mental model that teaching is about transmission of knowledge from teacher to student through direct instruction. Such a model is in conflict with the precepts of inquiry instruction. A challenge for teacher preparation programs, therefore, is to interrupt teacher candidates' teaching schemata.

About Inquiry

We define inquiry as the process of making sense of, and deriving meaning from, phenomena through observation, research, questioning, and discourse with peers and experts. In a pedagogical context, inquiry creates a classroom environment in which children are guided to explore various phenomena in a specific domain to gain conceptual understanding (Carter, 2004; Moscovici & Nelson, 1998) in an environment that is not structured. Teachers do not impose rules in a top-down fashion; children obtain results that may vary from teacher-expected results, generate and answer their own questions, and are self-directed and self-regulated learners (Carter, 2004). Wells (1995) described inquiry as having three fundamental features: research, interpretation, and communication. These features are motivated by the learner's curiosity; are characterized by messy, ill-defined problems or questions; involve evaluation; and require communication with others about the state of knowledge and understanding as the inquirer researches and investigates. Wells further defines inquiry as the "interdependence of action and reflection in the attempt to achieve common knowledge and shared understanding" (p. 251). Using inquiry achieves what Wells identifies as the goal of education: "to equip [a culture's] maturing members to become self-directed, creative inquirers" (p. 251).

Inquiry practices support an idea John Dewey (1916) expounded over a century ago: the notion that real learning occurs when learners are intrinsically motivated to make sense and meaning of their world while learning by doing. Inquiry science is doing science, or as Lopez and Tuomi (1995) stated, "As any scientist knows, the best way to learn science is to do science" (p. 78). That is, children ask questions, collect data, make conjectures, conduct experiments, make connections, and seek answers. Jean Piaget theorized that learners learn by making connections between what is experienced, actively or passively, and then by creating a schema or concept of what is understood (Brooks & Brooks, 1993; Fosnot & Perry, 2005). It is by being immersed in this type of classroom environment that teacher candidates can learn the pedagogy of inquiry. Yet the challenge remains to find such classrooms in a nation where they are few and far between.

Science Language and Scientific Practice

While all students, native English-speaking and non-native English-speaking alike, face challenges in learning complex and abstract ways of using academic language, the difficulties for ELLs are compounded because they have fewer linguistic resources in English to draw on in order to make sense of academic texts. Moreover, discursive practices valued in mainstream science classrooms may conflict with the values and practices ELLs bring with them from home. Thus, for ELLs, learning English in order to do science, mathematics, or history means more than simply learning grammar and vocabulary. As Hart and Lee (2003) point out, learning science while learning English increases the cognitive complexity for ELLs, compared to that experienced by their native-English speaking peers. They note that "ELLs frequently confront the demands of science learning through the vehicle of a yet-unmastered language. In academic genres and registers, such as those related to science, language is interpreted and expressed in fundamentally different ways from the commonsense understandings embedded in informal everyday language" (p. 477). This means that in academic contexts, ELLs are exposed to, and are likely to struggle with, new and more complex ways of using English. In science, these ways of using language are associated with such scientific practices as planning and conducting investigations, gathering data, using tools and technologies, thinking critically and logically about relationships between evidence and explanations, constructing and analyzing alternative explanations, and communicating scientific arguments.

Linguists and scholars of academic literacy have described scientific ways of using language and the challenges such usage poses for second language learners and native English speakers alike. Halliday (1993), for example, describes aspects of the grammar of scientific English that encode scientific values, processes, and dispositions. He argues that these features of the language have historically co-evolved with scientific activity and serve to enable scientists to express the complex, abstract phenomena that are at the heart of scientific activity. In order to address the challenges that scientific language poses for learners, Halliday suggests that teachers need to draw learners' attention to aspects of the language of science that go beyond technical vocabulary.

Drawing on the work of Halliday and Martin (1993), Schleppegrell (2002) highlights the types of grammatical and lexical resources a student writer needs to be able to control in English in order to understand and write scientific texts. She demonstrates how incomplete control of such linguistic resources as modality, verb tense, logical connection, complex noun phrases, and embedded clauses may limit an English learner's ability to elaborate significant aspects of the results of an experiment or relate those results to other researchers' findings.

Gee (2005) provides analysis of science talk in an elementary school classroom, showing how subtle but significant aspects of scientific meaning and expression may not be available to students if their experience with scientific discourse never transcends the language of everyday conversation. Gee argues that students in elementary science classes need to have "embodied experiences" with scientific ideas, combined with opportunities for "active assembly of meaning" (p. 28) in order to be able to appropriate scientific discourse. Moreover, he argues that this appropriation may cause identity conflicts for some learners, especially those for whom mainstream English is not the primary language. Such conflicts arise whenever participation in scientific discourse forces them to abandon ways of using language that, although they may serve them well outside of academics, do not function well in expressing complex scientific ideas.

Lee (2002) and Lee and Fradd (1998) discuss how incongruities between ELLs' home cultures, the mainstream academic culture of school, and specifically the values and practices embodied in modern Western science may pose challenges for ELLs' participation in science activities. Students who have had limited schooling prior to entering an English language classroom or whose native culture is based in oral language traditions may, for example, struggle with forms of thinking and expression valued in modern Western science, such as hypothesizing, inferencing, and making arguments based on empirical evidence.

Inquiry and Language Development

Educational researchers have argued that inquiry-oriented science pedagogy may help to make science content and practices more accessible to ELLs, while at the same time helping them to develop academic language (Amaral, Garrison, & Klentschy, 2002; Ballenger, 1997; Lee, 2002; Lee & Fradd, 1998; Rosebery, Warren, & Conant, 1992; Stoddart, Pinal, Latzke, & Canaday, 2002). Since inquiry focuses on meaning-making from direct experience and builds on children's backgrounds and interests, inquiry-based science instruction has the potential to help address many of the challenges that learning science in English poses for ELLs. First, since inquiry is initiated from learners' direct, often hands-on—or in Gee's (2005) terms, "embodied"—experiences with phenomena in their immediate environment, it creates common ground—a shared context—from which all learners in the classroom may draw as they collaborate and attempt to describe and develop explanations for what they perceive. In such a context, learners may initially use everyday conversational language to express their ideas (Ballenger, 1997). A learner who is struggling with language may point to objects in the environment or imitate certain phenomena with his or her body in ways that circumvent the need for words. Then, as an inquiry lesson progresses, when a peer or teacher uses new vocabulary or more complex forms of expression to discuss experiences that all members of the class have participated in, the ELL is better able to make sense of the language being used, which is the first step toward appropriating it into his or her own linguistic repertoire (Rosebery, Warren, & Conant, 1992).

Adopting an inquiry approach to science instruction in and of itself, however, may not be sufficient to support the science learning and language development of all learners of diverse cultural and linguistic backgrounds. Teachers may also need to adopt certain stances and employ certain other types of scaffolding in order to support ELLs. For example, teachers need to develop cultural sensitivity and knowledge of the cultural bases of varying discourse patterns their students may exhibit. Lee (2002) suggests that teachers need to draw on their understanding of students' cultural backgrounds, and perhaps use students' native languages, in order to create a bridge between the language, practices, and values they bring from home and those that are valued in science at school. She argues that this needs to be done in such a way that it affirms the beliefs, values, and ways of using language students bring from their homes and communities, building on them whenever possible.

To date, a handful of studies have sought to document the effects of using inquiry approaches with second language learners. Amaral, Garrison, and Klentschy (2002) conducted a quantitative study of achievement of K–6 ELLs learning science through inquiry-based instruction in a southern California school district. Their findings indicated that the longer students had participated in such instruction (from zero to four years), the higher their scores on standardized achievement tests in science, writing, reading, and mathematics.

Several other studies have analyzed classroom discourse and have described specific types of practices and discursive evidence that such practices support ELLs to begin to develop conceptual understanding, scientific habits of mind, and scientific forms of expression. Rosebery, Warren, and Conant (1992), for example, studied middle- and high-school native speakers of Haitian Creole learning science in a collaborative, interdisciplinary, inquiry environment as part of the Cheche Konnen project. Instruction in science, mathematics, and language development were integrated in this project. Through analysis of classroom discourse and think aloud protocols, Rosebery et al. determined that after a year of participation in the project, these language minority students were able to appropriate scientific tools, reasoning, and discourse. Ballenger (1997), a colleague of Rosebery et al. who also investigated the Cheche Konnen project, conducted a microanalysis of a science discussion in which ELLs contributed personal narratives in order to anchor their knowledge claims. Ballenger argues that allowing students to personalize the discussion and use everyday sense-making practices provided them a "way in" to scientific reasoning and discourse. Further investigations by the project team (Warren, Ballenger, Ogonowski, Rosebery, & Hudicourt-Barnes, 2001) corroborate and expand on Ballenger's initial findings.

While Ballenger does not discuss whether the teachers in the classroom she studied engaged in explicit instruction to scaffold the Cheche Konnen students into scientific discourse, others have argued that such instruction is essential to making scientific meanings, values, ways of thinking, and "ways with words" (Heath, 1983) accessible to language minority students (Gee, 2005; Lee, 2002; Lee & Fradd, 1998; Schleppegrell, 2002). Lee (2002) further argues that direct instruction in science processes is a necessary element of effective instructional scaffolding for students whose cultural and linguistic backgrounds may be to some degree incompatible with the values and practices of Western science. Stoddart et al. (2002) take a less strident approach to the issue of explicit teaching of science discourse and direct instruction in science processes, calling instead for "integration" of language and science instruction.

In sum, theories of second language literacy development suggest that the type of embodied, hands-on experiences with natural phenomena and authentic science problems and practices that inquiry science pedagogy creates provide a rich context for both science content and academic literacy development for ELLs. As the empirical studies of linguistic minority students in inquiry science classrooms demonstrate, the nature of inquiry is such that it encourages learners to negotiate meaning and engage in discursive practices that have the potential to mimic expert scientific practice. It is still not clear, however, to what extent, how, and when teachers might need to engage in explicit instruction about language and discourse processes in order for ELLs to make maximum benefit of inquiry science pedagogy. Nevertheless, evidence is emerging that not only are ELLs capable of participating in inquiry science activities, but doing so supports their achievement in science and their development of scientific discourse and dispositions.

OUR PRE-SERVICE SCIENCE EDUCATION MODEL: USING INQUIRY TO TEACH INQUIRY

As previously mentioned, teacher candidates are seldom, if ever, exposed to inquiry lessons either in their university coursework or field experiences. At the university where the study described in this chapter was conducted, K–4 teacher candidates in their last semester of studies are required to take a science methods course while participating in an internship, or field experience, at an elementary school. The aim of the science methods course is for teacher candidates to learn inquiry pedagogical methods. In addition to learning about inquiry methods, the teacher candidates then teach a series of connected lessons in actual classrooms with one or two other teacher candidates. They prepare, teach, and reflect on these lessons using a lesson study process that has them engage in dialogic reflection with team members. They also write critical reflections to connect their teaching experiences to course readings in inquiry, learning theories, and assessment. In the lesson study process, team members rotate teaching and observing each other teach. As they observe children interacting and engaging in the lesson, they refer to guiding questions and take field notes in a journal. One such question may be, "How are children talking about the lesson?" Following each lesson, the team debriefs what they saw, felt, and experienced. Each team member follows up by writing a critical reflection on his or her experiences and observations in his or her journal.

An essential component of this model is a cooperating teacher, who can mentor and guide these teaching teams in appropriately implementing inquiry instruction. However, teachers in the cooperating schools rarely, if ever, teach science or any other subject matter using inquiry methods. Thus, a challenge for implementing this model was to find a surrogate in the form of apposite inquiry materials for teacher candidates to use. The Full Option Science System (FOSS) and Science and Technology for Children (STC) kits, National Science Foundation (NSF) supported science curricula, were selected as the framework through which the teacher candidates would experience inquiry. Each FOSS and STC kit contains materials and connected activities that allow children to discover a big idea or phenomenon in science, such as the notions that plant reproduction is systematic and recursive and that caterpillars have a predictable life cycle. Big ideas are amalgamated concepts derived from domains in various science disciplines (AAAS, 2008). The kits are designed to step a teacher through the various connected activities with embedded assessments and provide the teacher with appropriate prompts to allow children to explore and investigate.

Both FOSS and STC represent a form of scripted curriculum, as these science kits prompt the teacher with questions for students and provide worksheets for students to record their observations and conjectures. Opponents of scripted curriculum might question the use of these inquiry kits as prescriptive, ignoring children's prior knowledge and culture, dehumanizing them, and rendering them automatons (Delpit & White-Bradley, 2003). At the same time, scripted curriculum has been criticized as stifling teachers' creativity and relegating them to the role of technicians rather than intellectuals (Giroux, 1988; Smyth, 2000). While we also reprove the use of such materials, we argue that the FOSS science kits are not like archetypal prescriptive, scripted curricular units that are often found in reading programs, which tend to deskill teachers (Crawford, 2004). These kits, on the contrary, support the teacher or teacher candidate in learning new skills by providing them with embedded guidelines for implementing an inquiry environment. Within such an environment, children are allowed to freely explore and navigate the science phenomena under investigation as they make sense of and derive meaning from their experiences through participating in a dialogic process in which teachers guide them to question, doubt, ponder, research, and argue. The kits, furthermore, bring "student lives, perspectives, cultures, and experiences into the center of the curriculum in a way that involves students as coconstructors and cocreators (rather than passive consumers) of that curriculum, along with teachers" (Smyth, 2000, p. 496). Teachers and students are, thus, co-learners and active agents as they navigate the messiness of inquiry classrooms.

In this model, the kits function as the missing "expert" to help "focus, build upon, or, when necessary, redirect the meanings that are being constructed" (Wells, 1995, p. 258). And, just as the kits provide materials that intrigue the children and guide them to ask probing questions in a messy and ill-defined way, the teacher candidates also venture into this messy, ill-defined undertaking to explore and investigate inquiry pedagogy. In a sense, then, in this model we use inquiry to teach inquiry.

CONTEXT, PARTICIPANTS, AND METHODOLOGY

The Texas county in which the study was conducted is an urban county on the U.S.–Mexico border where the number of families living in poverty is among the highest in the state (Texas State Data Center & Office of the State Demographer [TSDC], 2008), and the population is predominantly Hispanic. In significant ways, the community in which this study was set typifies emerging immigrant communities across the nation. The study was conducted at a public elementary school located in a low socioeconomic neighborhood where the vast majority of the children are ELLs. Specifically, the school serves 800 children from pre-kindergarten through sixth grades. Ninety-seven percent of the student population is Hispanic, and 90% are on free and reduced lunch. Over 30% are classified as ELLs.

In partnership with the school principal, 16 teacher candidates in a university science methods course, held on the school site, were all assigned to cooperating teachers at the school site, an atypical condition for education majors at this university, where teacher candidates in any given cohort are usually assigned to different schools for their field experience. Each team of teacher candidates taught one FOSS unit for a particular K–4 grade level during the semester, the length of time for the study. These lessons were taught during a segment of their required field experience hours. The pre-service teacher teams taught from one to two times a week, depending on the classroom demands and teacher flexibility. They attended the science methods course each week and wrote weekly critical reflections on their readings relating what they were reading with their classroom teaching experiences. These written reflections provided one source of data for the study, collected over the course of the semester. Another source of data was journals the candidates maintained in which they recorded their observations and reflections each time they taught on a weekly basis. Yet another source of data was notes Villa made immediately following each class of what the teacher candidates were saying during the class. These notes were based on recall and were written the same day as the class meeting. While note-taking based on recall may overlook some details, this procedure has the advantage of putting research participants at ease (Bogdan & Biklen, 1998), as note-taking on the spot could risk loss of trust and confidence. Twice during the semester Villa asked the class to have online discussions that were asynchronous. Data also were drawn from these electronic conversations.

Of the 16 teacher candidates in the science methods course, five underwent what we characterize as transformations of their perceptions of teaching inquiry. These women, whom we call Gracie, Maribel, Amparo, Marisol, and Dee Dee, demonstrated through comments they made and questions they asked in class, or in their reflective writing, that they understood the purpose of inquiry, were noticing how children were reacting to it, were reflecting on how it differed from traditional teaching, and were developing epistemological curiosity about the nature of inquiry learning. In a reading reflection, for example, Maribel distinguished between the uses of assessments in traditional teaching—for categorizing students and meeting system-mandates—and their use in inquiry, where they play a crucial role in the formative assessment of students' developing understanding.

The other 11 candidates in the science methods course did not appear to undergo such deep transformations in their perceptions of teaching inquiry. No systematic data were collected that would allow us to determine all of the reasons why they did not undergo such a transformation; however, there were no obvious identifying characteristics that applied to all of them. A few commented that they were unable to complete an inquiry unit because their cooperating teacher had them doing other things in the field-based classroom or they "ran out of time." We suspect also that there may have been varying degrees of passive or active resistance to the approach among these teacher candidates. Nevertheless, we believe that the fact that some candidates did not appear to undergo major transformation in their perceptions of inquiry does not diminish in any way the importance of the findings from those who did.

Thus, it is from the five "transformed" teacher candidates' comments in interviews and in their reflective writing that the examples discussed in the findings below are drawn. Villa conducted one-on-one interviews with two of these women, Maribel and Amparo. All five participated in two focus groups on the candidates' experiences with their teaching and how children were reacting to the lessons. Descriptive information for these four women and Gracie, whose quote began this chapter, is summarized in Table 9.1.

TABLE 9.1 Description of Study Participants

Pseudonym	Description	Status
Gracie	Bilingual, married Hispanic female with three children	Graduated; employment status unknown
Maribel	Bilingual, single Hispanic female	Graduated and in first year of teaching
Amparo	Bilingual, married Hispanic female with four children	Graduated and attending graduate school
Marisol	Bilingual, married Hispanic female with two children	Graduated; employment status unknown
Dee Dee	Bilingual, single Hispanic female with four children	Graduated and in first year of teaching

These teacher candidates, like the majority of their cohort, have ethnic and linguistic backgrounds similar to those of the majority of elementary students in the school where they had their field experience: All were Hispanic and raised in homes that were either bilingual (Spanish–English) or monolingual Spanish. Several had, themselves, been enrolled in bilingual education programs at the pre-college level.

The research questions for this study dealt with connections between theory and practice, transformation of perspectives on science teaching and learning, and learners' science language development. The specific questions were:

1. *Transformation of perspectives on science teaching and learning*: How did the candidates initially perceive of science teaching and learning, and what kinds of transformations in their perceptions occurred through participation in the methods class?
2. *Connecting theory and practice*: To what extent did the teacher preparation model enable candidates to connect learning theories with classroom practice and what types of connections did they make?
3. *Learners' language development*: What kinds of developments in children's academic language do the candidates perceive to have occurred, and what kinds of connections do the candidates make between learners' language development and their engagement in inquiry science?

To study the effectiveness of the model, we used discourse analysis (Gee, 1999) to analyze samples of the teacher candidates' written journal entries about their science teaching experiences and transcripts of the interviews. Analysis of the data followed a grounded theory approach (Glaser & Strauss, 1967; Strauss & Corbin, 1998) in which salient themes emerged as we read and discussed the texts multiple times. These themes were coded and grouped around such topics as theory and practice, autonomy, teaching and learning, assessment, inquiry, and language development. Key themes emerged that addressed our initial research questions and illustrated the teacher candidates' understandings about appropriate pedagogies for teaching science to ELLs.

FINDINGS

To effect change, individuals must find "dissatisfaction, inconsistency, or intolerability in [a] current situation" (Fullan, 2007, p. 22). Such transformation can cause ambivalence. The teacher candidates in the study initially resisted the project; however, once resistance dissipated, the teacher candidates emerged with new understandings, and it is how they expressed these understandings that we will examine closely. This ambivalence must be addressed, however, because it represents the beginning of the journey toward transformation.

Ambivalence Toward Messy, Ill-Defined Problems

When the study project was introduced to the study cohort, many willingly agreed to move forward, albeit in indecisive ways. As one student wrote: "I have to admit that in the beginning of the semester I was not sure how I felt about inquiry. I thought, well, if in the past the traditional method seemed to work, why change it now?" Another student expressed similar sentiments: "You have been taught so long to do things in a particular way that for you to be able to change them is hard.... One thing that I have to admit is that, in the end, having difficulties at the beginning made it worth it because it allowed me to focus more at getting better at it." Because we used inquiry to teach inquiry, these initial thoughts relayed the mixed emotions that they had proceeding into this project. Another student described it as feeling like being "thrown into the wolves." Once they understood inquiry and accepted its inherent messiness, however, the teacher candidates were able to critically examine more traditional approaches to teaching and their limitations.

Thus, at the end of the semester, one student described the entrenched nature of the traditional ways of schooling as follows:

> Maybe at the beginning it will be a little bit difficult for us letting go of what we have been taught to be correct. Now we know that... sticking to old traditional ways will not facilitate students' learning. Instead it will only create more followers. We as individuals were not encouraged to become autonomous thinkers. My experience as a student was one where the teacher had total control of the classroom, like in most of the classrooms in today's schools. The teacher would not allow interaction, neither with other students nor with herself. She was the one writing the rules for everyone to follow, and ideas and opinions were not valued in the classroom.

This student had made connections to the course readings of a Piagetian scholar, Constance Kamii, who describes a controlling teacher as one who uses heteronomy as opposed to autonomy (Kamii & Joseph, 1989). *Heteronomy* is the governing of behavior and thought through external forces like teacher-created rules, while *autonomy* is self-governance. Piaget advocated autonomy "as the aim of education" (Kamii & Joseph, 1989, p. 46) and the "necessary consequence of constructivism" (p. 50). In constructivist classrooms, children are allowed the freedom to explore and make sense of their world. In contrast, the teacher-controlled, heteronomous classroom leads a child to "uncritically believe what he is told, including illogical conclusions, propaganda, and slogans" (Kamii & Joseph, 1989, p. 49). For the teacher candidates in this study, observing this duality between the classroom environments they were attempting to create and those they had experienced in their own education, and, most noticeably, observing the children's reactions as they became engaged, contributed to transforming their identities as teachers.

Transformation of Teacher Candidates' Perspectives

Reflection is necessary and essential for thinking critically and solving problems. Dewey (1997/1910) noted the duality of accepting transmitted facts versus critically reflecting:

> If the suggestion that occurs is at once accepted, we have uncritical thinking, the minimum of reflection. To turn the thing over in mind, to reflect, means to hunt for additional evidence, for new data, that will develop the suggestion, and will either, as we say, bear it out or else make obvious its absurdity and irrelevance. (p. 13)

As the teacher candidates read theory, critically reflected, and made those connections between what they experienced implementing their inquiry lessons and observing the traditional classrooms where they were conducting their field experience, they transformed their perspectives on an aspect of teaching, that is, how students acquire knowledge. They experienced a shift in their thinking from understanding knowledge as teacher-transmitted to viewing it as student-generated. One member of the study cohort, Amparo, aptly described how these two perspectives on knowledge development contrasted:

> Most teachers are not used to the inquiry approach to teaching lessons. Teachers do not like the noise level. I observed that one teacher did not like this one student asking so many questions. This teacher's remarks were always "shut up and be quiet, listen to what the teacher has to say." I have come to understand that in a classroom of questioning and a classroom of discourse, noise and excitement is what characterizes and defines inquiry, hands-on lessons.

Another student described a revolution in her teaching philosophy because of her experience with inquiry teaching/learning:

> My teaching philosophy has greatly changed ever since being introduced to inquiry-based instruction. What I have liked about this learning process is

> that inquiry engages students to understanding concepts and straying away from solely drill-based "correct answers." It truly brings back the authenticity of learning. Students become their own teachers and learning becomes that much more significant to them. My idea of teaching made a one-hundred and eighty-degree turn.

Yet another student stated simply: "I realize I must break the pattern of teaching in which I was taught." Statements such as these provide evidence that these teacher candidates were developing a meta-awareness of different pedagogical approaches and the effects they have on student learning, which we argue is an important step toward embracing inquiry learning and, ultimately, affecting changes in school climates.

Creating a Climate of Inquiry

Receiving positive feedback from children who were developing epistemological curiosity further contributed to the changes in teacher candidates' perspectives on teaching and learning, resulting in their desire to use inquiry pedagogy. Epistemological curiosity is a phrase coined by Freire (1998) to describe learners' curiosity to acquire knowledge. It is an "essential of thinking," that occurs when a learner can "maintain the state of doubt and... carry on systemic and protracted inquiry" (Dewey, 1997/1910, p. 13). Teacher candidate Amparo realized this toward the end of the semester as she was teaching the science inquiry materials and wrote the following in one of her reflections:

> Ever since I have been doing my Science lessons with [inquiry] kits, I find myself wanting to incorporate more of the inquiry method and more hands-on with all the lessons that I teach the students. I find myself wanting for children to experience what they are to learn and I feel empty if I do not incorporate some hands-on activity.... I can begin to see how the third grade class has changed in how they view science, because they get to ask whatever is on their mind.

Dewey (1997/1910) asserted that the "business of education" is "to develop a lively, sincere, and open-minded preference for conclusions that are properly grounded, and to ingrain into the individual's working habits, methods of inquiry and reasoning appropriate to the various problems that present themselves" (pp. 27–28). Marisol revealed how such working habits and methods of inquiry developed in children: "Everything they learned from the science, they took to social studies. They took it... everywhere." Amparo also related how her students had developed such habits:

> To my surprise these students seemed to have millions of questions, and what I found interesting is that as the lessons progressed, their own questions were being answered. Therefore, I believe that this is how students make connections with the "Oh, now I understand why that happened."

In this climate of inquiry, the teacher candidates observed how children initiated their own research. Maribel, for example, shared how a second-grade child, inspired by the butterfly cycle, conducted her research:

> And this student at the end of the semester... made a habitat for a ladybug. We didn't even ask her to do that. She brought in this cup with leaves and a rock, and then she put holes in the cup so the ladybug could have air. And she said: "Look what I brought." I put her in front of the class, and she shared with the students what she had created. The students were asking her questions about where she got the butterfly [sic] and where the ladybug is going to get water from. She started explaining all this information. So she was really into it. She did research on her own. We didn't even ask her to do it. It's just incredible how they start writing information. I just saw a great improvement in their writing.

Maribel described another student who also synthesized the concepts as they studied the butterfly cycle:

> This is James, a student from a 3rd grade class [see Figure 9.1]. We're not supposed to introduce the [word] metamorphosis. We started working with butterflies, and he was the first student that did this [drew a diagram of the life cycle]. I never introduced the cycle, because this is something that they should discover on their own. Metamorphosis is something that you do at the very, very end of all the lessons, but he did it himself. He drew the egg, and then he wrote caterpillar, cocoon, and then butterfly. And then he put arrows indicating that it was a cycle. This was the time that I knew that he finally understood.

This newly developed environment of inquiry not only facilitated the children's curiosity about science concepts as they made connections with the inquiry investigations, it also developed their academic language skills.

Academic Language Development

Language development goes hand-in-hand with certain types of direct experience because, as Gee (2005) notes, the meanings we ascribe to words in a particular situation give them a tone or a flavor. Gee calls these "situated meanings," that is, meanings "rooted in embodied experience [such that] one has to 'see' the meaning as a pattern extracted from the concrete data

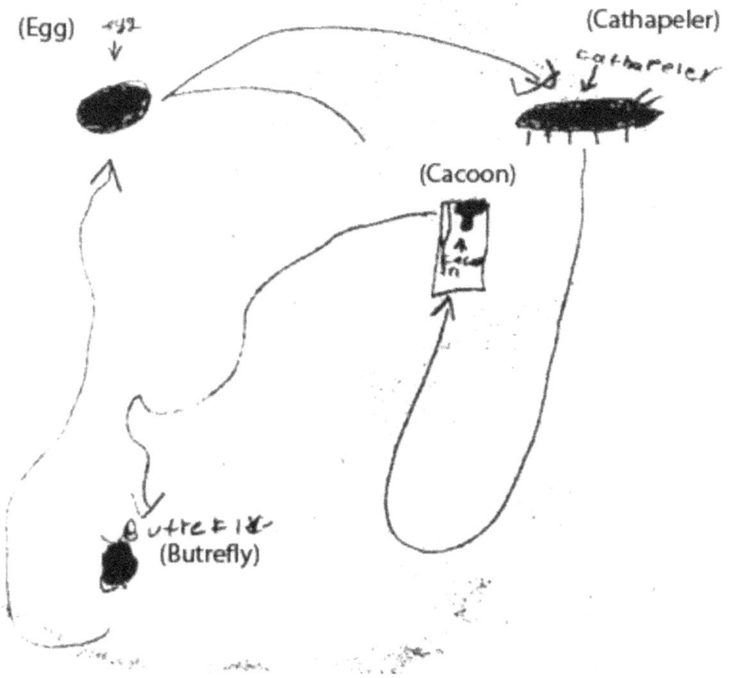

Figure 9.1 James's butterfly cycle.

of experience" (p. 25). He goes on to say that being able to extract such patterns of meaning requires "lots of practice with experiences that trigger the pattern" (p. 25). Gee cites research in cognitive psychology that has elucidated the role that direct, embodied experience plays in learning and the development of language. There is a growing consensus among cognitive psychologists that the meaning of a word is associated in a learner's brain with a rich tapestry of perceptual data—images, smells, sounds, feelings, actions and interactions—that come from the learner's experience with the use of the word. Meanings, thus, are woven together in such a way that use of the words recalls the associated experiences the learner has had with them in the past (Gee, 2005).

It is not surprising, then, that the kinds of experiential learning contexts that the teacher candidates in this study created using the FOSS kits—contexts in which the children were free to explore with physical elements and living creatures of the natural world—yielded fertile ground for the development of academic language among ELLs and native English-speaking children alike. For example, Amparo reflected in writing that in her third-grade bilingual class studying geology using an inquiry approach prepared the children to learn new vocabulary:

> The inquiry-based approach seemed to have these third grade bilingual students ready to learn through hands-on [activities] and ready for new terms to help them describe their findings.... Working with [FOSS] science lessons, I observed that... these students were very eager to express their newly acquired vocabulary.

Amparo went on to describe how this connection between hands-on experience and children's readiness to absorb the meaning of new terms played out in an inquiry unit she taught on rocks:

> In a prior lesson the word "properties" was introduced.... [N]ot only was this term introduced verbally or [in writing], but it was introduced through a hands-on approach. The following day students were being asked to tell me how they could identify rocks. Immediately the students came up with the word "properties." The students knew that a rock has different properties. [Their] discussion ranged from "I can see that one property of a rock is its size. Another property is its color" to questions like, "If it breaks or doesn't break, is that a property?" With these kinds of questions, it is obvious that these students can acquire new vocabulary if introduced within a hands-on and inquiry-based approach. These students got to experience the actual term being connected to the actual investigation.

The teacher candidates also paid attention to the timing of the introduction of new vocabulary and to the ways of talking and writing about the phenomena the children were experiencing in the inquiry lessons. In an interview, Maribel reflected on how the FOSS curriculum discouraged preteaching of vocabulary:

> At the beginning, you would not introduce any of the vocabulary. You would do an observation and, as you went along, you would introduce a word or two. But this was during the investigations. Especially the big terms like metamorphosis. You were not supposed to say that word, because they needed to see the whole process in order to fully understand what that was. So, you wouldn't say the vocabulary at the beginning of the lesson.

She also described how, in a lesson she taught with bilingual second- and third-graders on the growth cycle, students began to notice significant parts of the anatomy of the waxworms they were working with on their own:

> Corry[1] [a hyperactive student with a special needs designation] had the waxworm, and he was playing with it in his little piece of paper, and for some reason he flipped the paper. And the waxworm just started hanging from it. [Maribel demonstrates how the waxworm hung from under the inverted paper.] And then he just yelled: "Miss, miss, look at Spiderman!" And then everybody looked

> at it, and the little thing was right there just hanging upside down. So that was the clasper. The little thing that was holding him from the paper, that was the clasper. So that is when you introduce [the vocabulary word].

Thus, Maribel recognized that it was time to introduce the scientific term for something Corry had noticed and that had captured the attention of the whole class. Once Corry made this discovery, Maribel noted, all the children in the class started trying to replicate what he was experiencing with his waxworm. Prior to this moment of discovery in the inquiry process, Corry and his classmates could not have appreciated the functionality of the clasper for the waxworm, and so they had no need for a technical term to name it. Once they experienced it firsthand, however, the word took on a precise, functional meaning, and students like Kathy (see Figures 9.2a and 9.2b) began to use it as they talked and wrote about their experiences with waxworms.

The teacher candidates also described the language development, in English and in Spanish, of ELLs through engagement in the FOSS inquiry lessons. For example, in an interview, Dee Dee described the process of instruction in Spanish and English in a lesson on trees in her bilingual kindergarten class. She described how she and her co-teacher, also a teacher

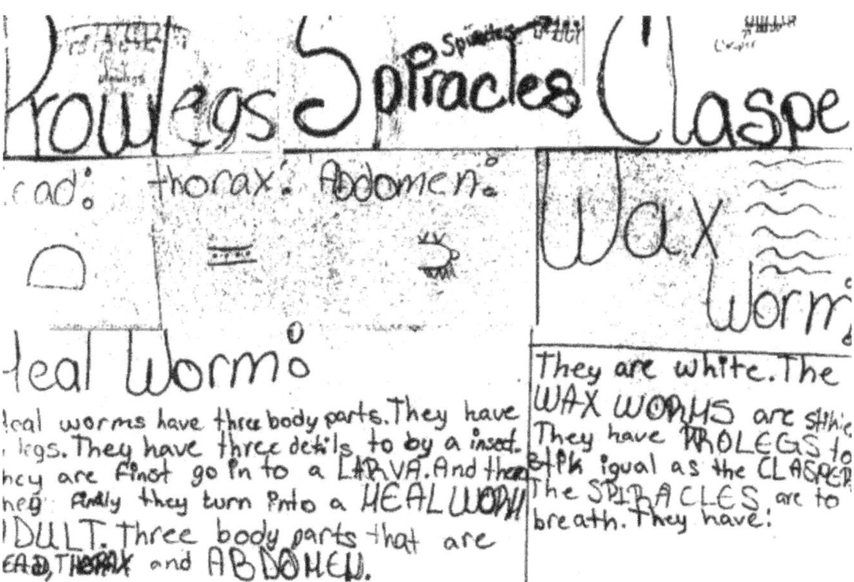

Figure 9.2a Kathy's poster of mealworms and waxworms.

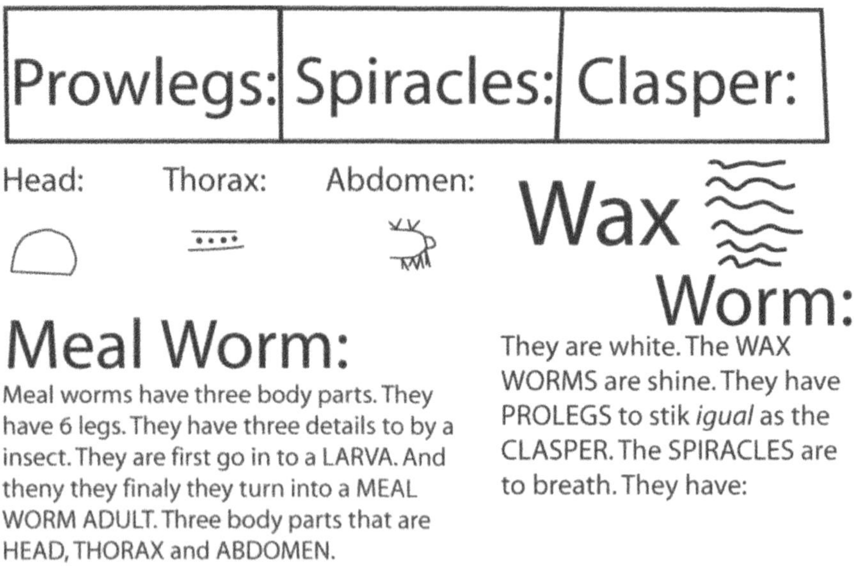

Figure 9.2b Authors' reproduction of Kathy's poster of mealworms and waxworms.

candidate, began the lesson in Spanish, but switched to English as the lesson progressed:

> When we were going to talk about trees, [first] it was done in Spanish: "And we're going to talk about the parts of a tree," you know. And someone will say the words in Spanish. That was like a pre-introduction, and then when we actually got to the lesson, it was done in English. And again you allowed the answers to be in Spanish, so you know that they understood what was being asked. But they just didn't have that transitioning yet in their mind to answer in English. But they were giving the right answers. A lot of them were in Spanish, especially in the beginning. You know, "When we're looking at this tree, what do you see that is different?" "Oh, the trunk is bigger." "Okay, say it in Spanish." We would say it in English. "Oh, this one has three leaves." "Oh, this one has the leaves that are in the pine tree," or whatever. And, a lot of it was in Spanish and then we would repeat what they say in English.

What's important to note about the kindergartners in Dee Dee's class is that the majority had had little exposure to English prior to coming to school, yet through the inquiry lesson they developed passive recognition and understanding of English vocabulary and grammar as it was being modeled by Dee Dee and her co-teacher in the lesson. They clearly demonstrated an understanding of the English vocabulary to describe the parts of the tree, as well as aspects of English grammar relevant to doing science, such as the language of comparison, for instance, "[W]hat do you see that is different? Oh, the trunk is *bigger*." The strategic use of the children's first language in the beginning of the lesson helped to form a bridge to enable them to understand what they were experiencing. Gradual introduction of English into the lesson, after the learners were oriented to the lesson's topic and context through their first language and had had direct, hands-on experience with concrete uses of the language, helped them to gain passive understanding of English. This gradual transition into the second language, where the focus is first and foremost on comprehension, and then on modeling usage in the second language, is recommended at the beginning stages of second language development, especially for very young learners who are not yet literate in their first language (Brisk & Harrington, 2000; Lessow-Hurley, 2005; Perez, 2004). In inquiry instruction, the connections between first and second language are made all the more meaningful through the learners' direct engagement with and inquiry about the world around them.

In an interview, Maribel described how she observed subsequent stages of second language development among children in the second grade bilingual class she worked with: "[A] lot of [the students] started using the vocabulary in English while writing in Spanish. I never told them to write in English or in Spanish. I just felt that, if they were ready, they were going to start writing." Maribel described how students in this second-grade classroom gradually began to incorporate more and more English into their written work, as they were ready, without explicit instruction or encouragement to do so. Such codeswitching at the lexical level is not surprising, given that the language of instruction in this classroom was primarily English. A student from Maribel's class, Pedro, exhibited a different type of codeswitching that is common as bilingual students transition from passive understanding of the second language to active use in their oral and written expression around the topics of their inquiry. Maribel described his tactic of codeswitching at the sentence or topic level in his written descriptions in his observation log:

> This is Pedro. These were the pictures that they started to do [see Figure 9.3]. And he started writing in English. "Today we saw caterpillars. And they have abdomens, thorax, and head. They make webs and ate." And "eggs." And he goes back to Spanish: "*Todas las pupas son mariposas. Ya criecieron; son bellisimas mariposas.*" [*All of the pupas are butterflies. They have already grown; they are beautiful butterflies.*]

Maribel was very impressed by one ELL in the same class who wrote only English during the inquiry lessons: "Veronica started by writing English, which was different. She would speak English a little more than the other students. All the way through, she wrote in English, which I thought was a big deal. She started to do it in her drawings as well as her observations."

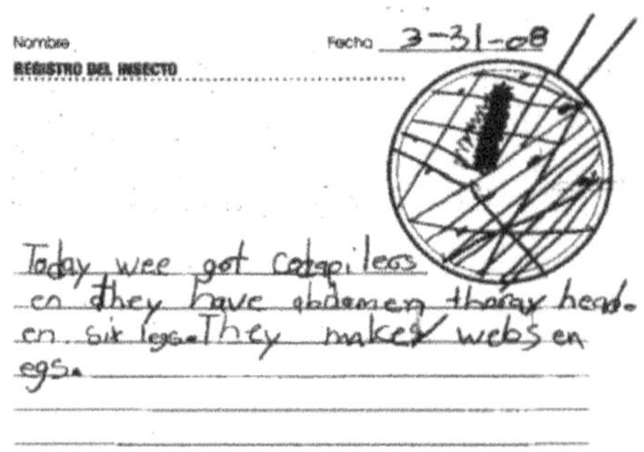

Date: 3-31-2008
Today wee got catapilers en they have abdomen thorax head. en six legs. They makes webs en egs.

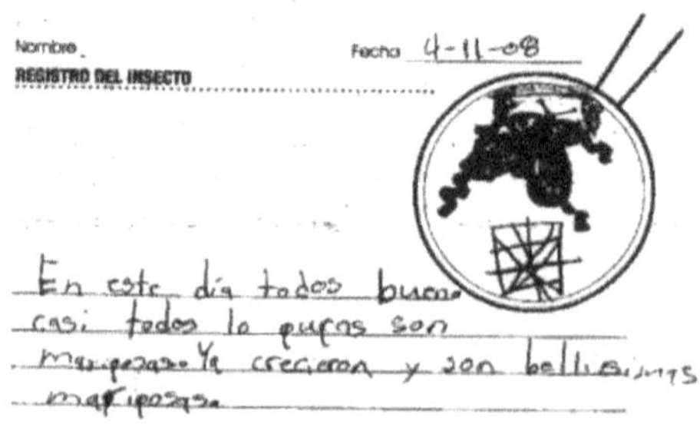

Date: 4-11-2008
En este día todos bueno casi todos lo pupas son mariposas yo crecieron y son bellisimas mariposas.

Figure 9.3 Pedro's observation logs—codeswitching.

Figure 9.4 shows three samples of Veronica's writing from her observation logs, spanning a two-month period.

In the third-grade class in which she taught, Maribel was impressed not only by ELLs' use of English, but also by the quantity of writing they produced around the inquiry lesson:

> This is my 3rd grade class, but it's just impressive on how much they started to write. You don't tell them anything. You just tell them, "On what we've been observing and information that we've seen, go ahead and write any differences that you see, and then if there's something that is alike, put it in the middle [of the Venn diagram]." Then I started circling to see how much vocabulary they were using.

She described another ELL whose language development throughout the inquiry lesson was marked:

> This was Edgar, a student who at the beginning of the year spoke purely Spanish. He would not understand any word in English.... Yeah, he wouldn't use English at all. At the end of the semester or at the middle of the semester he was starting to use a lot of vocabulary in English and then writing also in English [see Figure 9.5].

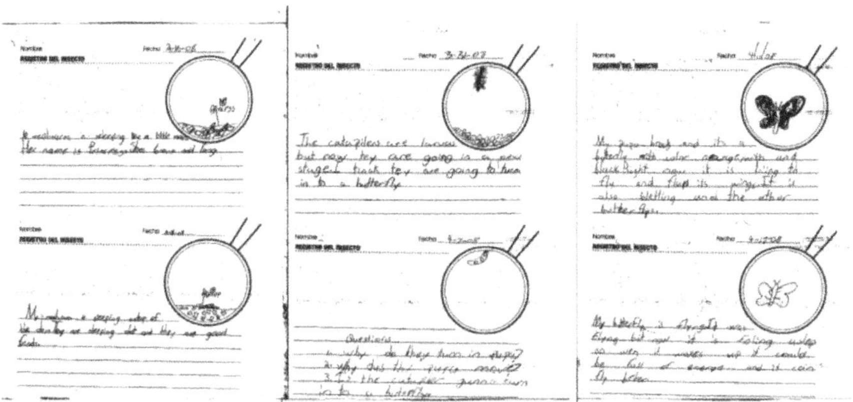

Date: 2-15-2008
The mealworm is sleeping a little cacon.
Her name is Princess. Shes brown and long.

Date: 2-18-2008
My mealworm is sleeping ontop of the oter. They are sleeping alot and they are good friends.

Date: 3-31-2008
The catapilers are larvas but now tey are going in a new stage. I think they are going to turn in to a butterfly.

Date: 4-7-2008
Questions:
1. Why do they turn in pupa?
2. Why dus the pupa move?
3. Is the catapiler gunna turn in to a buterfly.

Date: 4-11-2008
My pupa brock and its a butterfly witin color orange with and black. Right now it is triing to fly and flap its wings. It is also bletting and the other butter flies.

Date: 4-17-2008
My butterfly is flying. It was flyng but now it is folling aslep so wen it wakes up it could be full of energe and it can fly beter.

Figure 9.4 Veronica's observation logs.

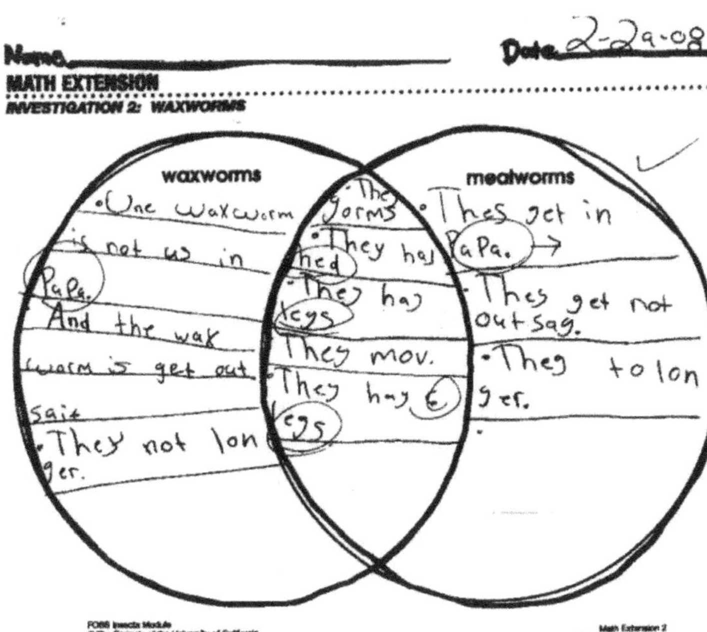

Figure 9.5 Edgar's use of English in his Venn diagram.

As the quotes from teacher candidates and the examples of written work produced by their students suggest, language development and development of science discourse was occurring as the theory predicts it would, corroborating what previous studies with ELLs have shown. Through guided inquiry and copious opportunities to practice using science discourse in meaningful ways, these elementary students were beginning to express understanding of the big ideas in the curricula, first in their native language and then in English. Such development in the students' comprehension and use of English further stimulated the transformations in the teacher candidates themselves. Students' learning of science and their beginning to express new understandings in English acted as a catalyst for this transformation.

DISCUSSION

We have presented compelling evidence that the teacher candidates highlighted in this chapter demonstrated a multi-dimensional perspective of what inquiry classrooms look and feel like. They experienced the potential of inquiry pedagogy for supporting ELLs' development of conceptual understanding and academic discourse. Understanding and using inquiry are effective mechanisms for teachers to guide learners in developing epistemological curiosity. This driving curiosity set the context for the learners' development of domain-specific language.

Having the opportunity to experience inquiry in the process of using inquiry science materials with children enabled these teacher candidates to make theory–practice connections about inquiry pedagogy and how learners develop conceptual understanding. It is not clear, however, if the candidates fully understood the theoretical connections between inquiry-based pedagogy and the development of language and literacy, as such connections were not emphasized in the methods class, due mainly to limitations of time.

Among observations the candidates made in their journals and in interviews about the impact of inquiry on the children's development, they called attention to the linguistic growth children seemed to be achieving in the language-based activities and materials provided in the FOSS kits, such as the observation logs. This is not surprising, given that the FOSS curriculum emphasizes activities that foreground negotiation of meaning and engagement in scientific literacy practices. It is notable, nevertheless, that the curriculum does not make the theoretical connections between inquiry, language use, and academic language development explicit. In fact, there is a dearth of texts available for teacher preparation that help to bridge this gap (Lindfors, 1999, however, is a notable exception), and none that specifically focuses on inquiry science and second language development. Very few teacher preparation programs include courses for mainstream teachers in principles of second language acquisition and techniques for supporting ELLs in the mainstream classroom. In order to make the most of the opportunities that inquiry pedagogy provides for academic language development, the model for teacher preparation described in this chapter needs to be modified to incorporate a more explicit focus on how academic language develops and how best to support it for ELLs. Further investigation of effective inquiry practices with ELLs is also needed in order to describe the process whereby learners develop scientific discourse, as well as what types of practices or guidance from teachers, if any, is necessary to enable them to do so.

A limitation of the model described here is its reliance on the FOSS kits to act as surrogates for a mentor who might guide the teacher candidates early on and support them throughout the process. Having such a mentor might allow transformation to occur sooner and in a greater number of candidates than through using the FOSS kits alone. Observing such a mentor in action while implementing inquiry lessons might enable the teacher candidates to read theoretical papers with a different lens and could help deepen their understanding of what they need to do to create a transformative classroom environment. It is possible that the speed of transformation cannot be affected by outside forces, due to its time-intensive nature (Wells, 1995), a factor that may be related to in-service teachers' general resistance to the open-endedness of inquiry pedagogy (Karplus & Thier, 1967).

Another limitation is the discontinuity of teacher candidates' opportunities to teach inquiry lessons. Their required field experience places them at the elementary school campus for only a few days per week. In spite of these limitations, implications for the model and further study of it abound. Once a critical mass of teacher candidates graduates, gains experience in their own classrooms, and begins to move into leadership roles, classrooms and schools have the potential to be transformed into liberated spaces where children explore and make discoveries that allow them to deeply understand science concepts while acquiring academic language in both English and their native language. This has the potential to springboard them into advanced learning at the secondary and post-secondary levels. Such transformation is particularly important for children and classrooms in low socioeconomic areas where too often schools have failing scores and high dropout rates. Creating a literate science community is the goal for Project 2061 and other initiatives of foundations and government agencies, and it is hoped that this will reduce our dependence on foreign scientists and engineers to fill workforce gaps.

ACKNOWLEDGEMENTS

We would like to thank The Boeing Company for partially funding this research through their community grant program and Elaine Hampton for her leadership and mentoring as this project unfolded. We would also like to thank Christopher C. Villa for providing technical support in re-imaging the children's productions and are especially grateful to Marcela Calderón, Rocio Luna, Gabriela G. Iglesias, and Monica J. Cabral for sharing their insights and experiences with the co-authors in the preparation of this chapter.

REFERENCES

Amaral, O. M., Garrison, L., & Klentschy, M. (2002). Helping English learners increase achievement through inquiry-based science instruction. *Bilingual Research Journal, 26*(2), 213–239.

American Association for the Advancement of Science (AAAS). (2008). *AAAS: Programs: Education: Project 2061.* Retrieved August 25, 2008, from http://www.project2061.org/.

Ballenger, C. (1997). Social identities, moral narratives, scientific argumentation: Science talk in a bilingual classroom. *Language and Education, 11*(1), 1–14.

Bogdan, R. C., & Biklen, S. K. (1998). *Qualitative research in education: An introduction to theory and methods* (3 ed.). Boston: Allyn & Bacon.

Brisk, M. E. & Harrington, M. M. (2000). *Literacy and bilingualism: A handbook for ALL teachers.* Mahwah, NJ: Lawrence Erlbaum.

Brooks, J. G., & Brooks, M. G. (1993). *In search of understanding: The case for constructivist classrooms.* Alexandria, VA: Association for Supervision and Curriculum Development.

Carter, A. (2004). Autonomy, inquiry and mathematics reform. *The Constructivist, 15*(1). Retrieved November 17, 2007, from http://www.odu.edu/educ/act/journal/vol15no1/carter01.pdf.

Clift, R.T., & Brady, P. (2005). Research on methods courses and field experiences. In M. Cochran-Smith & K. M. Zeichner (Eds.), *Studying teacher education: The report of the AERA panel on research and teacher education* (pp. 309–424). Washington, D.C.: American Educational Research Association by Lawrence Erlbaum Associates, Inc.

Crawford, P.A. (2004). "I follow the blue..." A primary teacher and the impact of scripted curriculum. *Early Childhood Education Journal, (32)*3, 205–210.

Darling-Hammond, L., Hammerness, K., Grossman, P., Rust, F., & Shulman, L. (2005). The design of teacher education programs. In L. Darling-Hammond & J. Bransford (Eds.), *Preparing teachers for a changing world: What teachers should learn and be able to do* (pp. 390–441). San Francisco: Fossey-Bass.

Delpit, L., & White-Bradley, P. (2003). Educating or imprisoning the spirit: Lessons from ancient Egypt. *Theory Into Practice, (42)*4, 283–288.

Dewey, J. (1916). *Democracy and education: An introduction to the philosophy of education.* New York: The Macmillan Company.

Dewey, J. (1997/1910). *How we think.* Mineola, NY: Dover Publications.

Educational Testing Service (2008a). Press releases: New ETS report offers highlights from achievement gap symposium. Retrieved July 27, 2008 from: http://www.ets.org/portal/site/ets/menuitem.c988ba0e5dd572bada20bc47c3921509/?vgnextoid = 4209e3dfa430b110VgnVCM10000022f95190RCRD&vgnextchannel = dd2d253b164f4010VgnVCM10000022f95190RCRD.

Educational Testing Service (2008b). Addressing achievement gaps: The language acquisition and educational achievement of English-language learners. *Policy Notes: News from the ETS Policy Information Center, 16(*2), 1–15.

Fosnot, C. T., & Perry, R. S. (2005). Constructivism: A psychological theory of learning. In C. T. Fosnot (Ed.), *Constructivism: Theory, perspectives, and practice* (pp. 8–38). New York: Teachers College Press.

Freire, P. (1998). *Pedagogy of freedom: Ethics, democracy, and civic courage.* Lanham: Rowman & Littlefield Publishers.

Fullan, M. (2007). *The new meaning of educational change* (4th ed.). New York: Teachers College Press.

Gee, J.P. (1999). *An introduction to discourse analysis.* London: Routledge.

Gee, J. P. (2005). Language in the science classroom: Academic social languages as the heart of school-based literacy. In R. Yerrick & W. M. Roth (Eds.), *Establishing scientific discourse communities: Multiple voices of teaching and learning research* (pp. 19–37). Mahwah, New Jersey: Lawrence Erlbaum.

Genesee, F., Lindholm-Leary, K., Saunders, W., & Christian, D. (2005). English language learners in U.S. schools: An overview of research findings. *Journal of education for students placed at risk, 10*(4), 363–385.

Giroux, H. (1988). *Teachers as intellectuals: Toward a critical pedagogy of learning.* South Hadley, MA: Bergin & Garvey.

Glaser, B. and Strauss, A. (1967). *The discovery of grounded theory.* Chicago: Adeline.

Halliday, M.A.K. (1993). Some grammatical problems with scientific English. In Halliday, M. A. K. & Martin, J. M. (eds.) *Writing science: Literacy and discursive power.* (pp. 69–85). Abingdon, UK: Routledge.

Halliday, M.A.K. & Martin, J. M. (1993). *Writing science: Literacy and discursive power.* Abingdon, UK: Routledge.

Hampton, E., & Rodriguez, R. (2001). Inquiry science in bilingual classrooms. *Bilingual Research Journal, (25)*4, 461–478.

Hart, J. E. & Lee, O. (2003). Teacher professional development to improve the science and literacy achievement of English language learners. *Bilingual Research Journal, 23*(3), 475–501.

Heath, S. B. (1983). *Ways with words: Language, life, and work in communities and classrooms.* Cambridge, UK: Cambridge University Press.

Kamii, C., & Joseph, L. L. (1989). *Young children continue to reinvent arithmetic—2nd grade: Implications of Piaget's theory.* New York: Teachers College Press.

Karplus, R., & Thier, H. D. (1967). A new look at elementary school science: Science curriculum improvement study. Chicago: Rand McNally & Company.

Lee, O. (2002). Promoting Scientific Inquiry with Elementary Students from Diverse Cultures and Languages. *Review of Research in Education, 26,* 26–69.

Lee, O., & Fradd, S. H. (1998). Science for all, including students from non-English language backgrounds. *Educational Researcher, 27*(4), 12–21.

Lessow-Hurley, J. (2005). *The foundations of dual language instruction.* Boston: Pearson Education.

Lindfors, Judith. (1999). *Children's inquiry: Using language to make sense of the world.* New York: Teachers College Press.

Lopez, R., & Tuomi, J. (1995). Student-centered inquiry. *Educational Leadership, 52*(8), 78. Retrieved June 25, 2008, from Vocational and Career Collection database.

Lortie, D. C. (1975). *Schoolteacher: A sociological study.* Chicago: University of Chicago Press.

Menken, K. & Antuñez, B. (2001). *An overview of the preparation and certification of teachers working with limited English proficient students.* Washington, DC: National Clearinghouse for Bilingual Education.

Meyer, D., Madden, D., & McGrath, D. J. (2004, August). *English language learner students in U.S. public schools: 1994 and 2000.* (NCES 2004-035). U.S. Department of Education, National Center for Education Statistics. Washington, DC: U.S. Government Printing Office.

Moscovici, H. & Nelson, T.H. (1998, Jan.). Shifting from activitymania to inquiry. *Science and Children,* 14–17.

Newman, W. J., Abell, S.K., Hubbard, P.D., McDonald, J., Otaala, J, and Martini, M. (2004). Dilemmas of teaching inquiry in elementary science methods, *Journal of Science Teacher Education, 15*(4), 257–279.

Oakes, J., & Lipton, M. (2003). *Teaching to change the world.* Boston: McGraw Hill.

Perez, B. (2004). *Becoming biliterate: A study of two-way bilingual immersion education.* Mahwah, NJ: Lawrence Erlbaum.

Rosebery, A. S., Warren, B., & Conant, F. R. (1992). Appropriating scientific discourse: Findings from language minority classrooms. *The Journal of the Learning Sciences, 21,* 61–94.

Schleppegrell, M. (2002). Challenges of the science register for ESL students: Errors and meaning-making. In M. Schleppegrell & C. Colombi (eds.), *Developing Advanced Literacy in First and Second Languages* (pp. 119–142). Mahwah, NJ: Lawrence Erlbaum Associates.

Science and Engineering Indicators 2008 (SEI) (2008). National Science Foundation. Retrieved July 12, 2008 from http://www.nsf.gov/statistics/seind08.

Schleppegrell, M. (2002). Challenges of the science register for ESL students: Errors and meaning-making. In M. Schleppegrell & C. Colombi (eds.), *Developing Advanced Literacy in First and Second Languages,* pp. 119–142. Mahwah, NJ: Lawrence Erlbaum Associates.

Smyth, J. (2000). Reclaiming social capital through critical teaching. *The Elementary School Journal, (100)*5, 491–511.

Stigler, J.W., & Hiebert, J. (1999). *The teaching gap: Best ideas from the world's teachers for improving education in the classroom.* New York: Free Press.

Stoddart, T., Pinal, A., Latzke, M., & Canaday, D. (2002). Integrating inquiry science and language development for English language learners. *Journal of Research in Science Teaching, 39*(8), 664–687.

Strauss, A. and Corbin, J. (1998). *Basics of qualitative research: Techniques and procedures for developing grounded theory* (2nd Ed.). Newbury Park: Sage Publications.

Texas State Data Center & Office of the State Demographer (TSDC) (2008). Retrieved July 12, 2008, from http://txsdc.utsa.edu/maps/thematic.

U.S. Census Bureau. (2004). Projected population of the United States, by race and Hispanic origin: 2000 to 2050. Retrieved March 14, 2007, from http://www.census.gov/ipc/www/usinterimproj/.

Warren, B., Ballenger, C., Ogonowski, M., Rosebery, A., & Hudicourt-Barnes, J. (2001). Re-thinking diversity in learning science: The logic of everyday language. *Journal of Research in Science Teaching, 38*(5), 529–552.

Wells, G. (1995). Language and the inquiry-oriented curriculum. *Curriculum Inquiry, (25)*3, 233–269.

Zeichner, K., & Gore, J. (1989). Teacher socialization. Retrieved October 15, 2006, from http://ncrtl.msu.edu/http/ipapers/html/pdf/ip897.pdf.

NOTE

1. All names of the elementary students referred to by the teacher candidates are pseudonyms.

CHAPTER 10

TRANSFORMATIVE PROFESSIONAL DEVELOPMENT FOR IN-SERVICE TEACHERS

Enabling Change in Science Teaching to Meet the Needs of Hispanic English Language Learner Students

Carla C. Johnson

ABSTRACT

The challenges relating to the science achievement of Hispanic English language learners (ELLs) are a growing cause for concern in the United States. Teachers in the field have had few opportunities to learn about strategies that may make their science instruction more relevant for the increasingly diverse population of first-generation Hispanic students that enter their classrooms. This chapter presents the Transformative Professional Development (TPD) model as a potential approach to structuring professional development for in-service science teachers to improve their practice. Implementation of the TPD model has resulted in increased student learning of science and enhanced science teacher quality. Critical components of science teacher pro-

Teaching Science with Hispanic ELLs in K–16 Classrooms, pages 233–252

fessional development programs will be shared and implications for future research discussed.

INTRODUCTION

There has been continual growth over the last decade in the population of culturally and linguistically diverse students in the United States (National Center for Education Statistics [NCES], 2006). The majority of immigrants over the past two decades are native Spanish speakers (Garcia, 2002) who enter schools with lower literacy skills than their peers in the United States and are more likely than any other ethnic group to live in urban areas and attend low-performing urban schools (Pew Hispanic Center, 2005). Of Hispanic students who live in urban areas, 71% speak Spanish as their primary form of communication (NCES, 2006).

Academically, Hispanic English language learners (ELLs) are significantly behind their native English-speaking peers in the United States, and this is a growing cause for concern in an era when global competitiveness and success in life depend not only on knowledge of science and mathematics, but, more importantly, on literacy (Business Roundtable, 2005; Committee on Prospering in the Global Economy of the 21st Century, 2007). In science, the achievement gap between Hispanic and White students averaged 29 points on the 2005 National Association for Educational Progress Report. Additionally, 70% of Hispanic students in the United States drop out of high school (Laird, DeBell, Kienzl, & Chapman, 2007). Traditional science instruction, which typically places English-speaking, White American students at an advantage, has been revealed as a contributing factor to the achievement gap (Gibbons, 2003; Lee, 2004) because Hispanic students may not feel a cultural, language, or interactional connection with the subject being taught, and so are at an educational disadvantage (Gibbons, 2003). The "traditional" science teacher must employ methods that make course material meaningful and comprehensible to all students, including ELLs (Atwater, 1994; Chamot & O'Malley, 1994; Garcia, 1999; Lee, 2004).

Research has shown that the majority of science teachers are novices at teaching literacy and second language in the context of subject matter (Stoddart , Pinal, Latzke, & Canaday, 2002). The increased diversity of U. S. society and low English literacy among ELLs creates a challenge for science teachers who have had little to no formal training on meeting the needs of ELLs (Lee, 2004), especially in urban schools where there are other variables such as school climate and teacher quality that impact student learning and are more prevalent (Ruby, 2006).

This chapter describes a research-grounded professional development model for K–12 science teachers, developed in response to the crisis in our schools resulting from the growing number of Hispanic ELLs with low literacy skills and little cultural connection to the science concepts they study.

TRANSFORMATIVE PROFESSIONAL DEVELOPMENT MODEL

The Transformative Professional Development (TPD) model is designed to specifically support urban science teachers of Hispanic ELLs in transforming their teaching of science to reflect a more appropriate approach for diverse student populations (Johnson & Marx, in press). The TPD model was created in response to the dearth of professional development program models that link science pedagogical content knowledge to enhanced pedagogy for ELL student literacy and cultural experiences, making instruction more relevant and meaningful (Lee, Luykx, Buxton, & Shaver, 2006). Research at all levels of training has indicated that only a few programs targeting "teaching for diversity" resulted in fundamental lasting change in teacher beliefs, perceptions, and teaching strategies, and only for a fraction of the teacher and pre-service teacher participants in the science professional development programs (Bryan & Atwater, 2002; Lee et al., 2006; Yerrick & Hoving, 2003). Research on the TPD program, therefore, had a dual purpose: (1) development and implementation of a model for combining science pedagogical content knowledge with literacy and cultural experiences, and (2) contribution of additional findings relating to teacher change and student learning relative to the science program implementation.

The TPD model emerged from, and is grounded in, the intersections of the following three bodies of research: culturally responsive teaching, effective science instructional practice, and literacy, home language, and culture.

Culturally Responsive Teaching

The first body of research is focused specifically on improving the educational experiences of ELLs and children of color in struggling urban schools through teaching in culturally responsive ways (e.g., Howard, 2001; Ladson-Billings, 1994; Lipman, 1995) and helping teachers to better recognize, understand, and build on the strengths their diverse students bring with them into the classroom (González , Moll, & Amanti, 2005; González, Moll, Tenery, et al., 1995). Findings from studies of teacher learning and change have revealed teacher beliefs as a key in determining whether or not practices change, as well as how changes are implemented and sustained (Anderson, 2002; Borko & Shavelson, 1990; Fullan, 2001; Krajcik, Blumenfeld, Marx, & Soloway, 1994; Loucks-Horsley & Matsumoto, 1999;

Van Driel, Beijaard, & Verloop, 2001). Teacher beliefs about Hispanic ELLs may contribute to the education gap in science for these students. There is a rich literature base related to low expectations of Hispanic ELLs and children of color overall (Valencia, 1997; Valenzuela, 1999) as well as the implications of teachers using "pedagogies of poverty," resulting in teacher-controlled instruction and lack of opportunities for student creativity and choice (Haberman, 1991). Research on successful teachers and schools consistently has encouraged classrooms where children take active roles in their learning through inquiry, cooperation, and leadership (Anyon, 2001; Flick & Lederman, 2006; Haberman, 1995; Ladson-Billings, 1994).

Effective Science Instructional Practice

The second area of the research literature is focused on effective science instructional practice and the experiences and support teachers need in order for change to occur (Johnson, Kahle, & Fargo, 2007a; National Research Council, 1996). Science teacher quality is a concern across the United States (Johnson, 2007). Many science teachers do not feel prepared in regard to having adequate content knowledge and/or the instructional skills to teach science effectively (Crawford, 2000; Johnson, 2007; Keys & Bryan, 2000; Wright & Wright, 2000). Few teachers feel prepared to teach students who have cultural and ethnic backgrounds different from their own (Bryan & Atwater, 2002; Fuller, 1994; Rodriguez & Kitchen, 2005). Hispanic ELLs are a specific subgroup that is prominent among those teachers consider as having different cultural and ethnic background because of its rapid growth in numbers (McCandless, Rossi, & Doherty, 1997). Effective science teaching, through the use of inquiry as a central teaching strategy, is a powerful context for Hispanic ELLs, as it enables the development of English language and literacy skills along with science concepts (Stoddart, et al., 2002). Academic discourse, social discourse, and cultural discourse are developed and constantly refined for all students through scientific educational practices such as inquiry and cooperative learning (Lee & Fradd, 1998). Effective science practices, including discussing observations, solving problems, and communicating findings, lend themselves to the development of English language proficiency (Chamot & O'Malley, 1994).

Although research findings demonstrate the positive influence standards-based instruction (NSES) can have on learning, many science teachers have been reluctant to implement required changes in teaching practices (Anderson, 2002; Keys & Bryan, 2001), as revealed by a synthesis of the National Science Foundation-supported Local Systemic Change programs, which revealed that only 14% of science lessons observed across the nation were of high quality, as defined by "providing students an opportunity to learn important science concepts" (Banilower, Smith, Pasley, & Weiss, 2006, p. 376).

According to Berns and Swanson (2000), teachers are not prepared to use student-centered instructional practices, as "this area is often neglected by teacher preparation programs, and given little attention in the professional development offerings by school districts across the nation" (p.2). Bateman (1990) argued that the problem with inquiry called for in standards-based instruction is that "not enough teachers use it, not enough understand the power inherent in inquiry, not enough see their job as other than transmitting information" (p. XV). Many current science teachers, unfortunately, struggle to teach science effectively and have not received the support needed to transform their instruction. There have been two approaches to professional development for science teachers; targeting of individual teachers and targeting of collective groups from individual schools. Desimone, Porter, Garet, Yoon, and Birman (2002) argued that "collective participation of groups of teachers from the same school, department, or grade level" should be the focus of professional development, "as opposed to the participation of individual teachers from many schools" (p. 83). The research base on systemic change in the teaching of science has supported the school-based collective approach (Czerniak, Beltyukova, Struble, Haney, & Lumpe, 2005; Johnson, Kahle, & Fargo, 2007b; Kahle, Meece, & Scantlebury, 2000). The Transformative Professional Development model described here is based upon the school as unit of implementation.

Literacy, Home Language, and Culture

The third area of the research literature is focused on the components of the Instructional Congruence Model, which include connections between science instruction, literacy, home language, and culture so that science can be delivered in ways compatible with a student's native culture and language (Lee, 2004). The Instructional Congruence Model's framework mirrors the National Science Education Standards' (NSES) emphasis on the need to learn important science concepts, engage in inquiry, engage in discourse, and provoke scientific thoughts (NRC, 1996). Some specific strategies highlighted in the Instructional Congruence Model include helping students learn science concepts and vocabulary; engaging students in science investigations cooperatively; developing science-related thinking skills; and encouraging students to talk about science with other students, their parents, and their teachers (Lee, 2004). Research results indicate that these strategies, included in standards-based instruction, hold significant promise (Anderson, 2002; Chang & Mao, 1998; Von Secker & Lissitz, 1999) including the ability to reduce and/or eliminate achievement gaps in science (Johnson et al., 2007a; Kahle et al., 2000).

Home language and culture are critical components in science instruction that enable Hispanic ELLs to be successful through bringing personal, real-world meaning to concepts that initially may appear abstract to all students. Culturally relevant pedagogy, according to Ladson-Billings (1995) includes three components: "an ability to develop students academically," "a willingness to nurture and support cultural competence," and "the development of a sociopolitical or critical consciousness" (pp. 475–76). Culturally relevant pedagogy allows students to maintain cultural integrity while succeeding academically (Ladson-Billings, 1995). In the TPD model, student home language and culture are seen as strengths and are purposefully and explicitly included in science instruction. The resulting TPD model is responsive to school climate, teacher needs, and teacher beliefs, with the intention of promoting change in urban science teacher practice, reflecting a more relevant pedagogy for Hispanic ELLs.

The TPD model has three main components grounded in research: (1) intensive, sustained, whole-school efforts focused on the development of student conceptual understanding through culturally relevant science and effective teaching methods that incorporate literacy and language strategies; (2) a focus on building relationships between teachers and teachers, teachers and students, and teachers and university faculty members; and (3) the creation of a positive school and classroom climate through procedures and routines for participating in science class and high expectations for success. A key component of the program is that its structure and purpose continue to emerge as it is facilitated based on the input of participating teachers (Settlage & Meadows, 2002). Intended outcomes of the TPD program include the following: (a) increased use of effective instruction and improved student learning, (b) effective learning and working environments for students and teachers, and (c) shared vision and concern for students and colleagues through strengthened relationships.

IMPLEMENTATION OF THE TRANSFORMATIVE PROFESSIONAL DEVELOPMENT MODEL

Setting

The TPD model was developed through the Institute of Education Sciences (IES) Teacher Quality Research in Science and Mathematics program grant funding for the project, "Utah's Improving Teacher Quality Initiative," also known as the UTQ program. The UTQ program took place from July 2005 to August 2008 and was a three-year, sustained, collaborative pro-

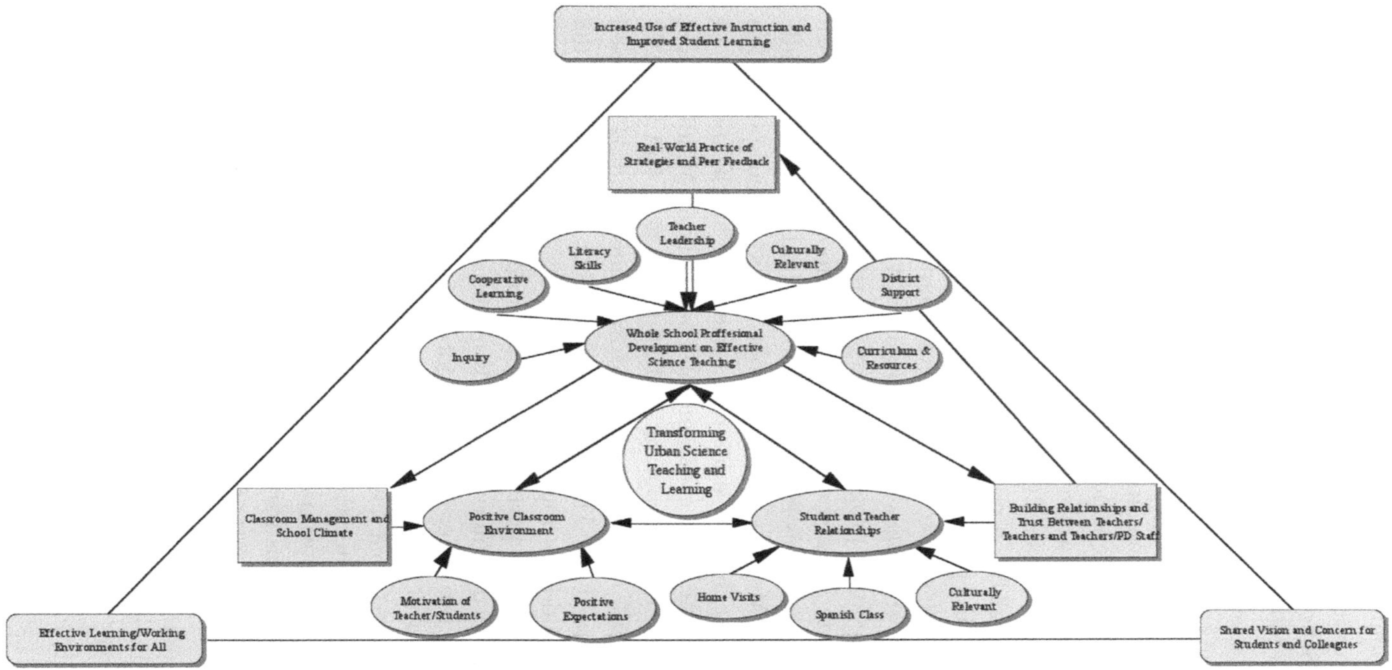

Figure 10.1 Model of transformative professional development (TPD).
The TPD model has the goal in the center "Transforming Urban Science Teaching and Learning". The three main components of TPD are: whole school professional development on effective science teaching, positive classroom environments, and student and teacher relationships. Anticipated foci of each of the three components are listed as inputs to each component in the small circles. Unanticipated foci which emerged from teacher needs revealed in professional development sessions are shown in rectangles as inputs to each component. The three main anticipated outcomes from TPD are found in the three boxes which enclose the model on the outside perimeter.
Source: From "Transformative Professional Development: A Model for Urban Science Education Reform," by C. C. Johnson and S. Marx (2009). *Journal of Science Teacher Education, 20*(2). Reprinted with permission of the authors.

fessional development program that involved all four middle schools in one district—two intervention middle schools that used the TPD model (Bryce and Zion Middle Schools [pseudonyms]) and two control middle schools. The first summer (2005), teachers from the intervention group attended a two-week summer session, followed by eight released days each school year (approximately one each month) for three years.

The school district had a 39% Hispanic student population overall at the middle-school level at baseline, which grew to 58% by the end of the study. Only 27% of the total student population in the district attained mastery on the State Criterion Referenced Test administered annually in science at baseline. Of its student body, 68% were on the free lunch program for those whose families fall below poverty level, and 49% were considered proficient in science, having passed the state science assessment in 2003. Bryce Middle School had five science teachers. Its students were 49% Hispanic. Zion Middle School's students were 64% Hispanic. Within its student body, 66% were on the free lunch program, and 16% were considered proficient in science on the state assessment. Zion Middle School had three science teachers. The control schools included Meadow Middle and Hill Middle schools. At baseline, Meadow Middle School had a student population that was 32% Hispanic, 56% of students received free lunch, and 54% of students passed the state science assessment. Four Meadow Middle School teachers participated in this program. Hill Middle School had 26% Hispanic enrollment, 69% of students received free lunch, and 34% passed the state science assessment. There were four teacher participants in the program from Hill Middle School.

Fifteen science teachers participated in the study. Of these teachers, eight were involved in the intervention TPD model and seven were in the control group at the two other middle schools in the same district. Teaching experience for participants in the intervention and control groups was very similar and ranged from less than 1 year to over 30 years. Two of the five male teachers in the intervention group were fluent Spanish speakers, including one male teacher who was Hispanic and the other who was White and had learned Spanish as part of a missionary experience. Six female participants were in the intervention group. There were two male and four female participants in the control group. None of those participants were fluent in Spanish and all spoke English as their primary language. Seven members of the intervention group were Caucasian and all were native English speakers. Six members of the control group were Caucasian and one was Native American.

Methods

A mixed methods approach was used to capture the rich data available from the TPD program. The data collected for the intervention schools included teacher classroom observations, teacher and student interviews, teacher and student questionnaires, field notes from released day activities, and teacher focus group sessions. Control school teachers were not interviewed or communicated with during the program other than through the completion of teacher surveys. These surveys indicated a prevalence of teacher-centered instructional strategies and that the use of inquiry and other engaging strategies such as cooperative learning were not a priority for them. Control school teachers also were observed teaching across the year with the intervention teachers.

The Local Systemic Change (LSC) Classroom Observation Protocol (Horizon Research, 2005) was used for classroom visits to evaluate teacher effectiveness in this study. One rater observed all teachers and a second rater observed the same teachers during the same class period with the first rater at least twice each school year. Inter-rater agreement reached acceptable levels as median weighted Kappa coefficients for items within each teacher quality domain were: Design = .94, Implementation = .86, Classroom Culture = .85, and Math/Science Content = .93.

In this study, it was important to capture the impact of the TPD program on students' attitudes, perceptions, and perceived abilities to do science. Students completed the *Discovery Model Schools Science Questionnaire* (DMSSQ) each year (Damnjanovic, 1998). This questionnaire was created by Discovery for use in the State Systemic Initiative, funded by the National Science Foundation. Internal consistency reliability for the DMSSQ was .88 Cronbach's alpha coefficient.

For qualitative data collection at the student level, six students were randomly selected to participate in a case study on the influence of science teaching practices on Hispanic ELL science self-efficacy from Bryce Middle, a TPD intervention school. Student participants did not know their teacher was participating in a program to improve their effectiveness and were very candid when speaking about teachers in the program.

Student learning was measured for all students enrolled in the four participating schools each year. Growth was measured through a pre/post science assessment that was grade-level specific (6th, 7th, 8th). The science assessments were developed through selecting 15 released items from the state science assessment and constructing two formats (pre and post) for each grade level. The assessments were validated by science education experts and the internal consistency reliability from the previous use of items was high (ranging from 0.93 to 0.88 using Cronbach's alpha coefficient). The assessment was scored on a 15-point scale.

Program

The TPD program activities are designed to reveal teacher and student needs that become the focus of the program. In July, 2005, teachers began the intensive, sustained, professional development program by attending a two-week summer workshop that introduced them to their new science curriculum, which consisted of two new curriculum units per teacher. All grade levels at the middle schools in this state teach integrated science content with life, physical, and Earth science taught in a spiraling manner, meaning that concepts are introduced in earlier grades and are expanded upon in later grades. Eighth-grade teachers received curriculum units on biography of the Earth and force and motion. Seventh-grade teachers received human body and properties of matter units, and sixth-grade teachers received space science and heat, light, and sound units. Two modules (18 weeks of instruction) of the Premiere Science curriculum were purchased from Frey Scientific to align with the science content mandated by the state for each grade level. Teachers were provided an overview of the modules during the first summer. The expectation was that they would implement the modules "as is" during year one and then modify them in the second summer to include culturally relevant strategies for implementation in year two. The primary focus of the summer session was an introduction to effective science teaching strategies. The NSES were a topic of discourse as teachers learned about the use of inquiry and participated in inquiry investigations in the role of learner. The secondary focus was an overview of culturally relevant teaching strategies and principles. For example, teachers engaged in book study using *Learning and Not Learning English* by Guadalupe Valdez to better understand the struggles their Hispanic ELLs experienced in school. A final focus of the TPD summer program was a brief introduction to integrating literacy skills such as reading comprehension and writing into the teaching of science.

Eight released day sessions across the academic year involved teachers in working on refining cooperative learning structures and classroom management strategies to better facilitate student learning. Teachers also completed a thirty-hour class in which they learned conversational Spanish. Teachers, in addition, reflected upon and made modifications to the new science curriculum (Premier Science mentioned previously) in order to better meet the diverse needs of their students. Teachers were provided with two curriculum units that only covered one third of the content standards they taught each year due to the purposeful decision to give teachers a place to start with some curriculum with which to begin the year. The long-term goal was to help teachers develop their own supplemental curriculum using the purchased curriculum as a guide. Some sessions focused on developing, identifying, and modifying current teacher resources to compile grade-level curriculum emphasizing inquiry instruction.

During the second summer session of three days, the collective group of seventh- and eighth-grade teachers decided to design and implement a common set of expectations and procedures for science class. All of the teachers spent time drafting "welcome back" letters for incoming students, which were mailed to parents along with a questionnaire used to gather data about how to better support students to be successful.

Released day sessions in years two and three worked to build upon student relationships through a deeper exploration of designing culturally relevant lessons and curriculum, conducting home visits with parents and students, deepening content knowledge in earth science, integrating technology such as classroom response systems, further refining instruction through the use of student-driven station investigations, and developing a new student motivation/incentive program to reach the reluctant students in their classrooms. Earth science was chosen as a focus area due to the state content standards which included Earth science in about 75% of sixth-grade curriculum and 25% of eighth-grade curriculum. Peer teacher observations and some targeted teacher mentoring also took place in year two. One released day was spent going out into the community and talking with business owners about a newly developed incentive program and partnerships with the school. Teachers had developed a program that used science bucks to reward students' classroom performance and meeting of teacher expectations. Students were rewarded for completing assignments, performing well on assessments, and being a good citizen with science bucks and at the end of the semester they could spend the bucks to bid on donated items at a school auction.

TRANSFORMATIVE PROFESSIONAL DEVELOPMENT FINDINGS

Teacher Change

Science teacher quality or effectiveness has been tied to student learning and achievement in science (Czerniak et al., 2005; Johnson, et al., 2007a; Kahle, et al., 2000). The TPD program study, therefore, looked specifically at teacher change through the use of a classroom observation protocol for the eight classroom visits each year.

Results of the analysis of classroom observation data collected through the LSC protocol indicated the TPD group experienced more growth in teacher quality across the three-year study on the 5-point scale (see Table 10.1). Specifically, in the four domains of the LSC protocol, TPD teachers had an 18% increase in Design of Lesson, a 13% increase in Implementation of Lesson, 8% increase in Culture of Classroom, and 15% increase in Content Knowledge. The control group, in contrast, experienced a decrease in all areas: Design (19% decrease), Implementation (35% decrease), Culture (28% decrease), and Content (28% decrease).

At the onset of this study, most of the participating teachers had very structured, teacher-controlled classrooms. During the first summer session, collaborative relationships between teachers, at grade level, school level, and across schools, were fostered and began to grow, according to teachers in their focus groups and individual interviews. Teachers slowly became more comfortable with allowing students to become more active learners. Teachers engaged in opportunities to observe other teachers' classrooms, even across schools in the cases of Bryce and Zion. Peer observations resulted in individualized feedback on problems such as classroom management and

TABLE 10.1 LSC Protocol Teacher Quality Change by Year and Overall

	TPD Group (Intervention)	Control Group	Difference
Fall 2005	2.99	3.35	
Spring 2006	3.57	3.48	
Year One Raw	0.58	0.13	.45
Year One %	0.19	0.04	
Fall 2006	3.71	3.28	
Spring 2007	3.41	3.33	
Year Two Raw	–.30	0.05	–.35
Year Two %	–.08	0.01	–.09
Fall 2007	3.08	2.89	
Spring 2008	3.44	2.46	
Year Three Raw	0.36	–.43	.79
Year Three %	0.12	–.15	.27
Overall Raw	0.45	–.89	1.34
Overall %	0.15	–.27	.42
Overall t-statistic	1.46	2.73	
Overall *p*-value	.217	.053	
Overall *d*	.65	0.96	

alternative instructional strategies, which included both constructive comments and positive reinforcement from their peer group. Teacher groups at each school formed strong professional learning communities and had figured out ways in which they could collaborate to enhance the impact of their science instruction and classroom management. Teachers noted in monthly released day meetings, as well as in interviews and informal communications, including email, that student engagement, performance, and attitudes were positively influenced by the student incentive program (science bucks). Teachers noted the additional time to collaborate with and learn from others was a very beneficial component of the program.

During the three year implementation of the TPD program in this district, there was tremendous conflict at the district level with the policy for closing of schools, dramatic shifting of district boundaries, an increase in class size to over 40 students in several classes, rotation of the building administration, and other activities that had a direct impact on school climate and classroom instruction. It was troubling to see the decline in teacher quality in the control schools across the study but, at the same time, promising to see that the teachers with project support were able to stay the course and continue to improve over the three-year period. All teachers in the TPD group who were low in the beginning moved from the "Ineffective Instruction" category to at least "Beginning Effective Instruction" or from other levels to "Effective Instruction" across the three-year program. The TPD group also experienced an overall 10% growth in effectiveness. Individual teachers in the control group who had been at the "Effective Instruction" saw their individual quality decline.

Those teachers experiencing implementation of the TPD model refined their teaching through the incorporation of new strategies. Teachers shared during release day focus groups that the most important aspect of the program was the internal support networks that were established through relationships with their colleagues, which enabled them to navigate the terrain of turbulence in this urban school district without allowing student learning of science to be negatively impacted. The professional learning community and mutual respect generated through the TPD program provided support for teachers as they worked through issues that urban schools sometimes experience, such as a heavy focus on assessment and lack of consistent support for initiatives.

By comparison, the control schools did not have the benefit of collaborative time, professional development, or curricular resources during the period of this program. Teachers at the control schools were observed monthly and it was noted that the school climate was challenging, instructional strategies were primarily teacher-centered, and little use of home language and/or culturally relevant strategies were recorded.

Student Attitudes toward Science and Self-Efficacy

The DMSSQ was used for the TPD program and had been used in the Ohio State Systemic Initiative previously (Damnjanovic, 1998).The DMSSQ revealed that teacher participation in the TPD program had a positive influence on student attitudes toward science. Students in the TPD intervention group improved their attitudes toward science overall ($t = 4.19$, d = 0.24), and specifically the attitudinal component of relating to doing science with friends improved ($t = 2.95$, d = .17), over the course of the three-year program. The attitudes of the students in the control schools declined overall ($t = -0.04$, d = 0), and relating to doing science with friends ($t = 4.53$, d = 0.24).

Interestingly, students were very cognizant in their interviews with program personnel of their particular teacher's lack of effective strategies and were able to clearly articulate views of teachers that aligned with our teacher quality findings. It also was revealed that teachers who used more effective science instructional practices and who made science more culturally relevant for students had a positive impact on student science self-efficacy. This was found among both students who entered science class liking science and with those who did not care for the subject. Students with less effective teachers who did not encourage discourse and inclusion of home language and culture had lower science self-efficacy. Teachers in the TPD model program who effectively implemented strategies from the program had a positive influence on their students' science self-efficacy (Jennings, 2007).

STUDENT INCREASE IN SCIENCE LEARNING

Student science scores were analyzed to determine if intervention schools (Bryce and Zion) experienced statistically significant gains from pre- to post-test in year 1 and year 2. Baseline (fall, 2005) student science scores were compared to scores at the end of year 1 (spring, 2006). Science scores collected at the beginning of year 2 (fall, 2006) were compared to those from the end of year 2 (spring, 2007). The comparisons were done by computing a series of two linear mixed-effects (LME) models. Student science scores served as the outcome variable, with Time (beginning versus end of each year), Group (Treatment versus Control), and the interaction between Time and Group serving as predictors in the LME models.

In year 1, the control group slightly outperformed the TPD group, gaining 2.40 points, compared to 1.55 points. In years 2 and 3, the TPD group gains more than doubled those at the control school (see Table 10.2).

The impact of professional development programs on student learning and achievement in science has been hard to document, with only a few emerging studies demonstrating a link over time between teacher partici-

TABLE 10.2 Student Science Performance on Pre/Post Assessment by Year

	TPD Group (Treatment)	Control Group	Difference
Fall 2005	5.27	5.82	
Spring 2006	6.82	8.22	
Year One Raw	1.55	2.40	–0.85
Year One %	0.29	0.41	–.21
Fall 2006	5.01	5.92	
Spring 2007	8.87	7.54	
Year Two Raw	3.86	1.62	2.24
Year Two %	0.77	0.27	.500
Fall 2007	4.14	4.15	
Spring 2008	6.45	4.57	
Year Three Raw	2.31	0.42	1.89
Year Three %	0.56	0.10	.460

pation in a program and student learning (Czerniak, et al., 2005; Johnson, Fargo, & Kahle, 2008; Kimble, Yager, & Yager, 2006). One reason this is hard to do is that it takes time for teachers to feel comfortable with new strategies and implement them effectively. Our findings support previous work indicating that it takes some time, possibly over a year, for teacher change in practice to be reflected in student achievement. Such change requires a shift in learning strategies for students as they begin to experience student-centered environments and get away from memorization and traditional textbook-driven instruction (Anderson, 2002; Johnson, 2007; Keys & Bryan, 2000). It is often the case, unfortunately, that school districts and teachers do not give new innovations the time needed to begin to make an impact before they move on to another strategy, thereby leading to the assumption that the current program is not working. The findings in this study show that if teachers are given the time and support needed to learn new strategies, implement new curriculum, and enable students to thrive in student-centered classrooms, a positive impact can be realized in student learning of science.

CHAPTER SUMMARY AND IMPLICATIONS

In this chapter, we discussed the literature indicating the lack of preparation found among most science teachers when they have to meet the needs of the increasingly diverse U.S. student population they are discovering in their classroom, with specific attention to Hispanic ELLs. It simply is not enough to focus on the preparation of pre-service teachers when attempting to meet the needs of linguistic minority students. We must begin to provide support and strategies for in-service teachers. The majority of Hispanic ELLs live in urban areas and attend schools that have their own dilemmas and dynamics that negatively impact school climate, teacher morale, resources, and expectations (Ruby, 2006). The TPD study presented in this chapter demonstrated that the TPD model was successful in improving the performance of intervention teachers and their students during the study, as indicated by the differences between intervention teachers and their classes in scores and improvement versus the declines noted among control teachers and classes. These findings demonstrate the continued growing need for programs that support teachers in making necessary improvements and changes relative to the dynamics of the science classroom. Professional development is a life-long learning process. A system of continual learning support needs to be in place for teachers, especially in urban settings.

When exploring externally funded projects and their impact on the educational system, it is important to keep in mind the nature of funded research projects such as the one presented in this chapter. The TPD program was funded by the Institute of Education Sciences as a three-year, $913,000 research study to develop and implement a professional development program with the goal of improving science teacher quality and ELL students' learning of science. Further implementation of this model requires funding for released time for teachers, as well as facilitators and funding for curriculum. This could be accomplished in a scaled-down version using existing district-level resources and professional development time. Large scale funding enables us to explore innovative models. School districts, however, in some instances have become accustomed to science programs being externally funded and provide fewer internal resources toward improving the science program. Hopefully, through the emerging focus on global competitiveness and STEM (science, technology, engineering, and mathematics) careers, more resources will be allocated toward improving science instruction and learning for all.

In this study of the implementation of the TPD model, we identified four critical components of this professional development program: (1) a school-based focus on developing professional learning communities and support systems within teacher groups; (2) integration and emphasis on culturally relevant science, effective instructional methods, and focus on literacy and language; (3) purposeful building of relationships and creation of positive climates; and (4) allowing time for teachers and students to adjust and thrive in the new learning environment. There is a need for further research exploring these critical components and the relationship between these types of professional development programs and teacher change and, ultimately, ELL student learning of science.

REFERENCES

Anderson, R. D. (2002). Reforming science teaching: What research says about inquiry. *Journal of Science Teacher Education, 13*(1), 1–12.

Anyon, J. (2001). Inner cities, affluent suburbs, and unequal educational opportunity. In J. Banks & C. Banks (Eds.), *Multicultural education: Issues and perspectives* (4th ed., pp. 85–102). New York: John Wiley & Sons, Inc.

Atwater, M. M. (1994). Research on cultural diversity in the classroom. In. D.L. Gabel (Ed.), *Handbook of research on science teaching and learning* (pp. 558–576). New York: Macmillan.

Banilower, E. R., Smith, P. S., Pasley, J. D. & Weiss, I. R. (2006). The status of K–12 science teaching in the United States: Results from a national observation survey. In D. Sunal & E. Wright (Eds.), *The impact of state and national standards on K–12 teaching.* Greenwich, CT: Information Age Publishing.

Bateman, W.L. (1990). *Open to question: The art of teaching and learning by inquiry.* San Francisco, CA: Josey-Bass.

Berns, B. B. & Swanson, J. (2000, April). *Middle school science: Working in a confused context.* Paper presented at the annual meeting of the American Educational Research Association, New Orleans, LA. (ERIC Document Reproduction Service No. ED444944).

Bryan, L. A. & Atwater, M. M. (2002). Teacher beliefs and cultural models: A challenge for science teacher preparation programs. *Science Education, 86,* 821–839.

Borko, H. & Shavelson, R.J. (1990). Teacher decision making. In B. F. Jones & L. Idol (Eds.), *Dimensions of thinking and cognitive instruction* (pp. 311–346). Hillsdale, NJ: Lawrence Erlbaum Associates.

Business Roundtable. (2005). *Tapping America's potential: The education for innovative initiative.* Retrieved July 30, 2007, from http://www.businessroundtable.org/pdf/20050727002TAPStatement.pdf.

Chamot, A. & O'Malley, M. (1994). *The CALLA handbook.* Reading, MA: Addison-Wesley Publishing Company.

Chang, C. & Mao, S. (1998). *The effects of an inquiry-based instructional method on earth science students' achievement.* Paper presented at the annual meeting of the National Association for Research in Science Teaching, San Diego, CA. (ERIC Document Reproduction Service No. ED 418858).

Committee on Prospering in the Global Economy of the 21st Century. (2007). *Rising above the gathering storm.* Washington, DC: National Academies Press. Retrieved July 11, 2007 from http://www.nap.edu/catalog/11463.html.

Crawford, B. (2000). Embracing the essence of inquiry: New roles for science teachers. *Journal of Research in Science Teaching, 37*(9), 916–937.

Czerniak, C. M., Beltyukova, S., Struble, J., Haney, J. J., & Lumpe, A. T. (2005). Do you see what I see? The relationship between a professional development model and student achievement. In R. E. Yager (Ed.), *Exemplary science in grades 5–8: Standards-based success stories* (pp. 13–43). Arlington, VA: NSTA Press.

Damnjanovic, A. (1998). Ohio statewide systemic initiative (SSI) factors associated with urban middle school science achievement: Differences by student sex and race. *Journal of Women and Minorities in Science and Engineering, 4*(2–3), 217–233.

Desimone, L., Porter, A. C., Birman, B. F., Garet, M. S., & Yoon, K. S. (2002). How do district management and implementation strategies relate to the quality of the professional development that districts provide to teachers? *Teachers College Record, 104*(7), 1265–1312.

Flick, L., & Lederman, N. (Eds.). (2006). *Scientific inquiry and nature of science: Implications for teaching, learning, and teacher education.* Dordrecht, The Netherlands: Kluwer Academic Publishers.

Fullan, M. (2001). *The new meaning of educational change.* New York: Teachers College Press.

Fuller, M. L. (1994). The monocultural graduate in the multicultural environment: A challenge for teacher educators. *Journal of Teacher Education, 45*(4), 269–277.

Garcia, E. (1999). *Student cultural diversity: Understanding and meeting the challenge* (2nd ed.) Boston, MA: Houghton Mifflin.

Garcia, E. (2002). *Student cultural diversity: Understanding and meeting the challenge.* (3rd ed.) Boston, MA: Houghton Mifflin.

Gibbons, B.A. (2003). Supporting elementary science education for English learners: A constructivist evaluation instrument. *The Journal of Educational Research, 96*(6), 371–380.

González, N., Moll, L. & Amanti, C. (2005). *Funds of knowledge: Theorizing practices in households, communities, and classrooms.* Mahwah, NJ: Lawrence Erlbaum Associates.

González, N., Moll, L., Tenery, M., Rivera, A., Rendon, P., Gonzales, R., & Amanti, C. (1995). Funds of knowledge for teaching in Latino households. *Urban Education, 29*(4), 443–470.

Haberman, M. (1991). The pedagogy of poverty versus good teaching. *Phi Delta Kappan, 73*(4), 290–294.

Haberman, M. (1995). *Star teachers of children in poverty.* Indianapolis, IN: Kappa Delta Pi.

Horizon Research, Inc. (2005). *Local systemic change classroom observation protocol.* Retrieved June 15, 2005 from http://www.horizon-research.com.

Howard, T. (2001). Telling their side of the story: African American students' perceptions of culturally relevant teaching. *The Urban Review, 33*(2), 131–149.

Jennings, V. (2007). The inner scope: Exploring the influence of classroom teachers' sense of efficacy on Hispanic students' science efficacy at one urban middle school. Unpublished dissertation.

Johnson, C.C. (2007). Whole school collaborative professional development and science teacher change: Signs of success. *Journal of Science Teacher Education. 18*(4), 629–662.

Johnson, C. C., Fargo, J. D., & Kahle, J. B. (2008). The cumulative and residual impact of a systemic reform program on teacher change and student learning of science. Manuscript submitted for publication.

Johnson, C.C., Kahle, J.B. & Fargo, J. (2007a). Effective teaching results in increased science achievement for all students. *Science Education. 91,* (3), 371–383.

Johnson, C. C., Kahle, J. B., & Fargo, J. (2007b). A study of sustained, whole-school, professional development on student achievement in science. *Journal of Research in Science Teaching. 44,* (6), 775–786.

Johnson, C. C., & Marx, S. (2009). Transformative professional development: A model for urban science education reform. *Journal of Science Teacher Education, 20*(2), 113–134.

Kahle, J. B., Meece, J., & Scantlebury, K. (2000). Urban African-American middle school science students: Does standards-based teaching make a difference. *Journal of Research in Science Teaching, 37*(9), 1019–1041.

Keys, C. W., & Bryan, L. A. (2000). Co-constructing inquiry-based science with teachers: Essential research for lasting reform. *Journal of Research in Science Teaching, 38*(6), 631–645.

Kimble, L. L., Yager, R. E., & Yager, S. O. (2006). Success of a professional development model in assisting teachers to change their teaching to match the more emphasis conditions urged in the *National Science Education Standards. Journal of Science Teacher Education, 17*(3), 309–322.

Krajcik, J. S., Blumenfeld, P. C., Marx, R. W., & Soloway, E. (1994). A collaborative model for helping middle grade science teachers learn project-based instruction. *The Elementary School Journal, 94*(5), 483–496.

Ladson–Billings, G. (1994). *The dream keepers: Successful teachers of African American children.* San Francisco: Jossey-Bass Publishers.

Ladson-Billings, G. (1995). Toward a theory of culturally relevant pedagogy. *American Education Research Journal, 32*(3), 465–491.

Laird, J., DeBell, M., Kienzl, G. & Chapman, C. (2007). *Dropout rates in the United States: 2005* (NCES 2007-059). U.S. Department of Education. Washington, DC: National Center for Education Statistics. Retrieved December 12, 2007 from http://nces.ed.gov/pubsearch.

Lee, O. (2004). Teacher change in beliefs and practices in science and literacy instruction with English language learners. *Journal of Research in Science Teaching 41*(1), 65–93.

Lee, O. & Fradd, S. (1998). Science for all, including students from non-English-language backgrounds. *Educational Researcher, 27*(4), 12–21.

Lee, O., Luykx, L, Buxton, C. & Shaver, A. (2006). The challenge of altering elementary school teachers' beliefs and practices regarding linguistic and cultural diversity in science instruction. Manuscript submitted for publication.

Lipman, P. (1995). "Bringing out the best in them": The contribution of culturally relevant teachers to educational reform. *Theory into Practice, 34*(3) 202–208.

Loucks-Horsley, S., & Matsumoto, C. (1999). Research on professional development for teachers of mathematics and science: The state of the scene. *School Science and Mathematics, 99*(5), 258–271.

McCandless, E., Rossi, R., & Doherty, S. (1997). *Are limited English proficiency (LEP) students being taught by teachers with LEP training?* (NCES 97-907). U.S. Department of Education, National Center for Education Statistics. Washington, DC: US Government Printing Office.

National Center for Education Statistics. (2006). *Public Elementary and Secondary Students, Staff, Schools, and School Districts: School Year 2003–04* (NCES 2006–307). Washington, DC: Author.

National Research Council. (1996). *National science education standards*. Washington, DC: National Academy Press.

Pew Hispanic Center. (November 11, 2005). *Latinos more likely than Blacks, Whites to attend the largest public high schools, according to new Pew Hispanic Center report.* Retrieved June 21, 2006, from http://pewhispanic.org/newsroom/releases/release.php?ReleaseID = 36.

Rodríguez, A. & Kitchen, R. S. (Eds.). (2005). *Preparing prospective mathematics and science teachers to teach for diversity: Promising strategies for transformative action.* Mahwah, NJ: Erlbaum.

Ruby, A. (2006). Improving science achievement at high-poverty urban middle schools. *Science Education, 90*(6), 1005–1027.

Settlage, J. & Meadows, L. (2002). Standards based reform and its unintended consequences: Implications for science education within America's urban schools. *Journal of Research in Science Teaching, 39*(2), 114–127.

Stoddart, T., Pinal, A., Latzke, M. & Canaday, D. (2002). Integrating inquiry science and language development for English language learners. *Journal of Research in Science Teaching, 39*(8), 664–687.

Valencia, R. R. (Ed.). (1997). *The evolution of deficit thinking: Educational thought and practice.* London: Falmer Press.

Valenzuela, A. (1999). *Subtractive schooling.* Albany: State University of New York Press.

VanDriel, J. H., Beijaard, D., & Verloop, N. (2001). Professional development and reform in science education: The role of teachers' practical knowledge. *Journal of Research in Science Teaching, 38*(2), 137–158.

Von Secker, C. E. & Lissitz, R. W. (1999). Estimating the impact of instructional practices on student achievement in science. *Journal of Research in Science Teaching, 36*(10), 1110–1126.

Wright J. C. & Wright C. S. (2000). A commentary on the profound changes envisioned by the national science standards. *Teachers College Record, 100*(1), 122–143.

Yerrick, R. K. & Hoving, T. J. (2003). One foot on the dock and one foot on the boat: Differences among preservice science teachers' interpretations of field-based science methods in culturally diverse contexts. *Science Education, 87*(3), 390–418.

CHAPTER 11

SCIENCE AS SPRINGBOARD

Promoting Science Achievement and Aspiration among Latino English Language Learners in the Secondary School

Bernadette Musetti and Sara Tolbert

ABSTRACT

Latino students, particularly English Language Learners (ELLs), are severely underrepresented in the sciences at the post-secondary level. These students often receive mixed messages regarding post-secondary education. Those messages encourage students to work hard, to become college eligible, and to be confident about college admissions, funding, and higher education outcomes, while often simultaneously implying that they are not welcome, will not be admitted to higher education institutions, and indeed are taking slots away from more "deserving" students, especially in science. This chapter describes outcomes for student participants in a school–university partnership summer science enrichment program for ELLs. Student outcomes included increased interest in and motivation toward the pursuit of post-secondary education, increased interest in science, and development of academic language.

Teaching Science with Hispanic ELLs in K–16 Classrooms, pages 253–272

INTRODUCTION

The dynamics of a new, flat, global economy are such that social and economic intergenerational mobility are increasingly contingent on post-secondary educational achievement, on high levels of education, and on multiple professional credentials (Fry, 2005, 2004, 2002; Chapa & De La Rosa, 2004; Swail, Cabrera, & Lee, 2004; The National Center for Public Policy and Higher Education, 2004; United States President's Advisory Commission on Educational Excellence for Hispanic Americans, 2002, 2000,1996; Council of Economic Advisors, 2000; White House Initiative on Educational Excellence for Hispanic Americans, 2000, 1999, 1998). Such dynamics, combined with shifting national demographics that position Latinos as the largest non-dominant community, have created national concern at the policy level about the factors influencing secondary students' views of the opportunities they have to participate in higher education and especially in higher education as a science major.

BACKGROUND TO THE STUDY

Across the U.S.A., there are approximately 40 million Latinos, forming 15% of the U.S. population, but this group is growing at three times the rate of other groups. There are over 11 million Latinos of school age and almost half of all Latino students are ELLs, among whom over 70% are Spanish speakers. The 2003 Census Bureau and National Center for Education Statistics (NCES) data indicate that 43% of all Latinos have less than a high school diploma (Chapa & De La Rosa, 2004).

In the last decade, Latinos became the minority group with the most rapid rate of population growth in the U.S.A. By 2050, if current growth rates continue, one of three residents in the U.S.A. is expected to be Latino (U.S. Census Bureau, 2008). Growth rates for Latinos currently are most significant in areas unaccustomed to serving them, such as Southern and Midwestern states (Schemo, 2002). Such drastic increases in Latino student populations mean that states, institutions of higher education, schools, administrators, and teachers must adapt to this changing demographic environment in order to meet the educational and social needs of all students, where those students are increasingly Latino.

Latinos, in addition, are particularly underrepresented in the sciences. Latino students are significantly less likely than their Anglo counterparts to pursue advanced coursework or degrees in science (Commission on Professionals in Science and Technology [CPST], 2007; NCES, 2007). Only 7.3% of bachelor's degrees, 4.3% of master's degrees, and 2.7% of doctoral degrees in science and engineering are awarded to Latinos (CPST, 2007). Additionally, Latinos who enroll in science and engineering degree programs demonstrate higher levels of attrition than White students in such programs (Vela, 2003). Latinos disproportionately attend poorly-funded and under-resourced schools that lack strong science and mathematics programs and the highly qualified teachers necessary to prepare students for admission to and success in college science and mathematics degree programs (Chavez, Soriano, & Oliverez, 2007; Noguera, 2006; Vela, 2003). Enrichment programs for “at risk” Latino youth, furthermore, often are geared toward remediation and not designed to prepare them to pursue post-secondary coursework in science or engineering (Vela, 2003). The achievement scores of Latino students in science reflect this trend. The most recent National Assessment of Educational Progress (NAEP) conducted in 2005 indicates that significant gaps persist between the science achievement of Anglo European students and Latinos (NCES, 2006). Although the achievement gap between Anglo European and Latino students at grade 4 grew smaller as measured by NAEP science scores in 2005 versus those in 2000, score gaps were unchanged for Latino versus non-Hispanic white students at grades 8 and 12. Latino students are disproportionately represented in high poverty schools, while low-income students, as measured by their eligibility for free and reduced lunch, continue to score much lower than students who are ineligible (NCES, 2006, 2008). Score gaps for ELLs show improvements between the 1996 NAEP and 2005 NAEP science assessments for fourth and eighth grade, although no gains were documented for ELLs at the 12th-grade level, and gaps are still significant at all three grade levels..

While K–12 educators across the nation are struggling to meet the needs of a rapidly changing student population, many of whom are ELLs, an increasing number of policy-makers and educators have realized that concern for the education of Latinos is a social and economic imperative. To that end, university–school partnerships targeting Latino achievement in schools and in the transition to higher education have begun to emerge. There are several large-scale initiatives nationwide designed to improve science education for minority students and ELLs and to increase minority enrollment in the sciences, with the aim of creating pipeline programs from K–12 to higher education. Many of these programs support educationally underserved and underrepresented students in the sciences. One of the most innovative, studied, and efficacious of these is the multi-state initiative MESA: Math, Engineering and Science Achievement (MESA), which provides rigorous academic development including math and science, individualized academic planning, study skills training, peer group learning, career exploration, parent involvement, professional development, transfer assistance, and special orientation classes and services for higher education students. MESA works with students at three levels; K–12, community college, and four-year institutions. California MESA has the goal of graduating 20,000 new engineers over the next decade in that state alone (MESA, 2008).

Another successful research-based program that promoted science education for ELLs was Language Acquisition through Science Education in Rural Schools (LASERS), a National Science Foundation (NSF) Local Systemic Change Project (Stoddart, 2008; Stoddart & Tolbert, 2008; Tolbert & Stoddart, 2008; Stoddart, Pinal, Latzke, & Canaday, 2002). This program was designed to help teachers learn to use inquiry science instruction as a medium for facilitating English language development. The science–language integration method was designed to help facilitate language learning through participation in a contextualized science activity and was based on the idea that the teaching of science cannot be separate from the teaching of language (see also the chapter by Stoddart, Solis, Tolbert, and Bravo in this book). Inquiry science instruction involves students in experimentation and exploration of scientific phenomena that are directly linked to concrete, hands-on activities (Stoddart et al., 2002). During the LASERS study, elementary school teachers from rural school districts heavily populated with migrant farm workers in central California participated in a five-week Summer School Academy to receive training in the science–language integration method. Elementary teachers taught science during a summer school program to 500 ELLs in the mornings and then participated in action research-based professional development sessions in the afternoon. In Action Research Teams with staff developers and professional development coaches, teachers discussed videotaped clips from their morning classes and evaluated student work samples in order to develop effective techniques designed to facilitate both academic language and science learning through science–language integration. Teachers then implemented what they had learned from the Summer School Academy with their ELL students during the following school year. Both science and language achievement gains for ELLs in bilingual and English-only classrooms were significant.

In the study described in this chapter, we embedded key elements of programs like MESA and LASERS, where high-school ELLs placed as "at risk" within the educational system took part in "Steps to College," which is an intensive summer science enrichment program held on a college campus. This program was an attempt to address the needs of one group of Latino ELLs by creating a pre-collegiate program that could provide the kinds of supports and experiences that students with high social capital usually possess and which contribute to these students being more competitive from the standpoint of university admissions. Steps to College, as described here, was implemented in Georgia, a state which is a poignant example of the "new Latino south" (Wortham, Murillo, & Hamann, 2002).

THEORETICAL FRAMEWORK

Social Capital

On a national level, arguments for providing greater access and opportunity to traditionally underperforming students typically include national competitiveness and workforce demands. In the recent past, we have seen many scholars and business leaders writing and speaking to academic and business audiences about encouraging greater minority, specifically Latino, achievement in science, technology, engineering, and mathematics, where this underrepresentation is referred to as "a silent crisis" (Litow, 2008). It is true that the nature of scientific discovery and innovation demands a large pool of intellectual capital and, as a nation, we currently are failing to integrate the significant intellectual resources of cultural and linguistic minority students, which affects not only these students, but the nation as whole.

Failure to close Hispanics' education and language gaps risks compromising their ability to both contribute to and share in national prosperity. How these risks and opportunities play out over the decades ahead will define not only the kind of future Hispanics will inherit, but also the economic and social contours of the United States in the 21st century (National Research Council, 2006). Although economic arguments for better serving Latino and ELL students are compelling, our primary organizing principle for this work is around educational equity. Increased global competitiveness should not determine whether or not all students are provided access to quality education, although this is clearly an obvious and favorable outcome. We understand equitable access to higher education, instead, as a civil rights issue. In this study, we focus our attention on the relationship between social capital and student achievement, where it is understood that the more social capital students bring with them to school, the more likely they are to succeed in school. Students with less social capital, conversely, are not as likely to succeed in school, where immigrant students and ELLs, whose populations are predominantly Latino, are among those students most underserved and least likely to succeed within our educational system (Chavez et al., 2007; National Council of La Raza, 2007; Noguera, 2006; Pew Research Center, 2005; Chapa & De La Rosa, 2004). Research has shown that Latino students aspire to pursue higher education, but that "blocked opportunities" rather than a lack of motivation can help explain their disproportionately low enrollment in colleges and universities (Kao & Tienda, 1998). These blocked opportunities include inequitable access to rigorous coursework and information about college admissions requirements, as well as the support of a college network, including college-bound peers and effective high school counselors. This is especially critical since many low-income Latino students are the first in their families who plan to attend college (Chavez et al., 2007; Kao & Tienda, 1998).

Social Constructivism

This research with the Steps program took place within a sociocultural approach to teaching and learning (Vygotsky, 1978; Lave & Wenger, 1991; Rogoff, 2003) wherein school success and failure are viewed to a great extent as socially constructed, and within a Vygotskian framework in which human cultural development is a product of social interaction, which fundamentally involves a transformation of interpersonal processes into intrapersonal processes. Children learn tools for thinking, which are the mediating devices their cultures provide, through interactions with more skilled partners in what Vygotsky called a *zone of proximal development* (1978). Our understanding of Vygotsky allowed us to conceive of student achievement at the secondary level as a product of social interaction. The goal of the program described here was to engage students socially and academically, recognizing that these are inextricably linked, and to allow students to operate within their zones of proximal development and achieve outcomes beyond what they could do without such assistance.

In order for students to be motivated to want to learn academic English and succeed in school, however, they must be engaged in collaborative relationships with educators who view their native languages and cultures as assets, such that learning English is additive, rather than subtractive (Noguera, 2006; Valdes, 2001; Valenzuela, 1999; Erickson, 1986). Cummins (2000) reminds us that underachievement among language minority students is not caused by a lack of fluency in English, but rather it is the result of specific interactions in school that lead these students to mentally withdraw from academic effort. For Cummins (1998, 1996), the zone of proximal development for language minority students is as much a zone of identity development as it is a zone of academic or cognitive development. Minority and bilingual students become disengaged when their identities are threatened and their experience of school is subtractive, rather than additive. Many minority students, therefore, are said to "achieve" failure as the only way to maintain their linguistic and cultural identities within an educational environment that reinforces the notion that the sociocultural and linguistic resources students bring are deficient and must be remediated or replaced (Valenzuela, 1999). Research has shown, however, that when students are not only encouraged to maintain their cultural and linguistic identities, but also given access to the social capital required to succeed academically within the dominant society, they demonstrate increased participation and achievement in school (Carter, 2005; Gibson, 1988). The Steps program described here is one example of how schooling can be additive and socially construct success among linguistic minority students through the subject area of science.

Science and English Language Learners

Although science provides an ideal medium for academic language development, ELLs largely are deprived of opportunities to participate in advanced science courses, since fluency in English often is mistakenly viewed as a prerequisite for success in science (Lee & Luykx, 2006; Stoddart et al., 2002). Students commonly perceive scientists as white males (Solomon, 1993). Both restricted access to science courses and a culturally reinforced perception of scientists as white males contribute to students' identity development in terms of whether or not they perceive themselves as capable of doing science or becoming scientists. In most science courses in which language minority and other students are enrolled, abstract science concepts typically are presented in a decontextualized manner, whereby students are expected to decipher the complex language of science and distill advanced science concepts through transmission of material from text or teacher to student or through "cookbook labs" that do little to authentically contextualize new material in a way that makes it linguistically accessible or socially relevant (Stoddart et al., 2002). Research has shown, however, that contextualized science instruction integrating inquiry science and the cultural and linguistic resources of language minority students can effect significant gains in both language development and science understanding (Lee, Deaktor, Hart, Cuevas, & Enders, 2005; Lee & Luykx, 2006; Stoddart et al., 2002; Moje, Collazo, Carillo, & Marx, 2001; Warren, Ballenger, Ogonowski, Rosebery, & Hudicourt-Barnes, 2001). Contextualizing science involves numerous components: "scaffolding" language and concepts for students, linking students' cultural and linguistic backgrounds on the topic under investigation, using what students know experientially and linguistically to build new understanding, allowing students to discuss concepts and procedures in their primary language and work collaboratively across flexible groups, adapting materials to make them more comprehensible, simplifying verbal and text-based linguistic input (e.g., simpler syntax, slower rate of speech, anticipation and reading guides), modeling, emphasizing key vocabulary, allowing for multiple opportunities to practice and apply what is being learned, and allowing students to show what they know in multiple formats. These and other components of what constitutes an effective modified, "sheltered," or highly contextualized and comprehensible lesson have been delineated in the Sheltered Instruction Observation Protocol as developed by Echevarria, Vogt, and Short (2008) and in the Center for

Research on Education, Diversity, & Excellence (CREDE) Five Standards of Effective Pedagogy (Dalton, 1998).

STUDY

The main instructional goals of the Steps to College program were to (1) offer current and former high-school English speakers of other languages (ESOL) students an experience of a college campus and promote a college-going mentality, (2) promote knowledge of and interest in science through teaching an engaging, relevant, rigorous, science-based curriculum for which students would earn academic credit, and (3) develop academic English through, and as the result of, science. As such, the study addressed the following research questions:

1. What outcomes are evidenced among student participants in the Steps to College intensive summer science enrichment program?
2. Specifically, does participation in the summer science program promote students' higher education aspirations, development of academic English, and interest in science?

Design and Methods

The month-long Steps to College program required students to attend class and participate in related activities for 19 full instructional days, totaling 90 hours of instruction. Electronic, web-based, pre-and post-surveys were administered on the first and last days of the program and included questions related to students' higher education aspirations and interest in science. Surveys were analyzed using quantitative methods by an outside evaluator, showing the percentage of difference for each item from the program's beginning to its end. The students also were asked to provide written responses to a series of questions evaluating their overall experience with the program and the effectiveness of the program in meeting stated objectives. Ten students and three teachers also were interviewed and videotaped individually on the last day of the program. Data included three videotaped focus groups of 22 participating students (7–8 students per focus group) at the culmination of the program. In focus groups, students discussed prompts such as, "What do you most value from the Steps experience?" The remaining 33 students answered the focus group questions in writing only, without discussion, during the last class. Student work samples—including science journals and final reports—were collected and analyzed to evaluate science language development and students' ability to write academic English on science topics. The assistant principal, the main administrator from the school involved in the program, was interviewed on several occasions during and after the program. Selected parents and teachers (both teachers from the program and other teachers from the students' school) were interviewed at the culmination of the summer program. Discourse and narrative analyses were used to analyze all data with exception of the survey data, and are reported here using qualitative and descriptive means. Teacher-designed materials and lessons also were analyzed using components of the SIOP protocol for effective lesson planning for highly salient features, including linking to student background to assist students in building new knowledge and creating meaningful activities that promote high-order thinking and academic language development.

Program Participants

Fifty-five students in grades 9–12, all current and former ESOL, self selected to participate in a four-week, intensive summer program held on a public university campus. Interested students were required to attend a mandatory informational meeting with a parent or guardian prior to enrolling in the program. All expenses, including transportation, materials, meals, and fees, were covered by the program. All interested students were admitted. Student participants were from the same large urban high school in the metropolitan Atlanta area. The program was offered for academic credit.

Program Curriculum and Instructional Strategies

The summer enrichment program was created with attention to our theoretical assumptions: (1) our students wanted to go to college and lacked not the desire nor ability, but the preparation, information, and access to resources needed for a college-bound trajectory; (2) knowledge is socially constructed most optimally within a student's zone of proximal development, and (3) science is an ideal medium for academic language development, a primary instructional goal of the program. The enrichment curriculum was aligned with both the district science standards and World Class Instructional Design and Assessment (WIDA) standards for ELLs, which include language development standards linked to the content area of science. Students were placed into three types of grouping arrangements at different times: whole group, by English language proficiency, and in heterogeneously mixed groups across grades and English language proficiency levels. The curriculum was built around the theme of water, and the program used the slogan "Water is life" to contextualize it to the local environment and ecology of the area in which the study took place. The students lived in an area close to a large lake and reservoir, upon which the state and its local communities, businesses, and farms depend for most of their water supply. The lake is a very popular recreational site for residents living in the vicinity and throughout the state. Lessons and related activities focused on issues related to the health of the lake and the sustainability of the water it provides for the state's residents. Lesson objectives were designed to engage students in science learning related to hydrology, wastewater, and drinking water treatment, the water cycle, water contamination and testing, and water chemistry. Students also learned about related concepts such as osmosis and diffusion. Students participated in several types of activities: lectures, science labs, field excursions, interacting with guest speakers, college and university visits, a college fair, and a final ceremony during which they presented their research projects. These activities often were focused on the water theme, and thus the field excursions included trips to the world's largest aquarium, a local dam that produces hydroelectric power and houses a fish hatchery, and the local water treatment plant. The teachers selected all were experienced science educators and held teaching credentials in science, with additional endorsements in teaching English as an additional language (ESOL). Importantly, all had taught science to students in very diverse classrooms and had experience in "sheltering" or otherwise modifying and contextualizing instruction so that all students could access the content.

FINDINGS AND IMPLICATIONS

Data evidenced that the instructional program simultaneously promoted higher education aspirations, academic English language development, and science knowledge and interest. The program also had dramatic impacts on students' perceived efficacy and identities. Each of these findings is discussed below.

Higher Education Aspirations

An especially telling and dramatic outcome from the program was reported by the assistant principal at the school after visiting the program and speaking with each student individually. He reported subsequently being inundated with requests from more than half of the students who wanted to change their schedules for the upcoming semester and, in effect, "track themselves up" into more rigorous classes, including science classes. After fourteen days in the program, the assistant principal sent the following e-mail to the program organizers:

> I have been very busy these past 2 days answering scheduling questions and moving those kids up to the honors and AP levels! They are very motivated and have a new found confidence that wasn't evident during the initial registration process in the spring. Before they thought they couldn't make it in the upper level courses, now they expect more from themselves and are willing to try.

The pre- and post-surveys showed that students reported an increase in planning to attend a four-year higher education institution from a pre-program level of 51% to a post-program level of 92%. All students reported wanting to continue in the program in the future, if possible. Students also reported wanting to visit more college campuses in the coming year.

The 55 students who were in the Steps program are much more likely to make it to higher education as a result of the information and confidence they gained during the program regarding what it takes, academically, to be able to go to college and how to pay for college, regardless of their documentation status. For the vast majority of the students in the program, they would be the first person in their family to participate in higher education. This program, although small (55 students), is a significant case in terms of demonstrating how, in one month's time, with the right supports, experiences, and information sharing, Latino students become determined and more likely to attend college. This change is important because the educational trajectories of Latinos within post-secondary education ultimately will determine the path this nation will follow into the millennium (Pew Research Center, 2005; Chapa & De La Rosa, 2004; United States, 2002, 1996; White House Initiative on Educational Excellence for Hispanic Americans, 1999).

Identity Development and Self Efficacy

An important, yet unintended, outcome from the program was students' developing identities as leaders. A core group of students from the program returned to their school in the fall and began a club to continue the work of Steps and to teach other students what they had learned (unbeknownst to the Steps program faculty or organizers). The students tellingly named their club FLOW: Future Leaders of the World. They offered one another tutoring support and planned university Shadow Days and other events, including community service. Through peer mediation they helped one another to achieve at higher levels and in more rigorous classes, including more rigorous Advanced Placement (AP) science classes.

While students overwhelmingly reported field excursions (e.g., to Lake Lanier and the Georgia Aquarium) as the most enjoyable aspect of the program, those program aspects they reported to be of most value to them were: (1) information about how to go to and pay for higher education and (2) messages they received about their ability and efficacy in terms of participating in higher education. Students changed their aspirations about higher education as well as their understanding of what is required in order to be accepted into various kinds of higher education institutions. They reported wanting to take Honors- and AP-level courses, improve their grade point averages, get involved in extracurricular activities, and, most importantly, go to college. Many students responded that the most valuable information they gained from the program was that Latinos can go to college. One particularly poignant response to the question, "What do you most value from the Steps experience?" was the following reply: "There are people other than my mom who want me to go to college."

Throughout each focus group session, students' responses demonstrated that they had been impacted by an "at risk" identity. Before attending the Steps to College Program, many of the students thought that they could not go to college for various reasons and often cited among them grades or financial hardship. Some students specifically noted the negative impacts of an "at risk" institutional identity on Latino students. As one student noted in the focus group interview:

> A lot of students think that it is not possible to go to college and so they are afraid, uh, like me, I was afraid, I thought, I thought we [Latinos] could go, it's just that other people didn't.... We really believed what the other people say, that it's too expensive, it's so hard to get into college, so we cannot.

Negative beliefs such as those expressed by this student stem from external societal pressures and input. These students were exposed to negative discourse from multiple sources regarding their "at risk" identities prior to attending the program—sources that could have included school teachers, various media, or other influences. Through the Steps to College Program, importantly, these students learned the logistical steps of how to go to college. For example, students learned about the coursework required for acceptance and other admissions-related factors that might affect their admissions status, such as service and extracurricular activities. They also learned how to apply for financial aid and were given opportunities to meet and talk with admissions officers from various colleges. During the focus group sessions, students' attitudes were positive and hopeful about facing the challenges that lay ahead in terms of going to college.

Academic Language Development through Science

Our experience is that students in linguistically diverse classrooms are able to successfully learn and use the academic language of science when it is taught in a highly engaging, contextualized manner and applied in meaningful ways. Our study confirms what many other researchers have concluded: science is a powerful vehicle for learning English because the two processes are complementary, even synergistic (Douglas, Klentschy, Worth, & Binder, 2006; Fathman & Crowther, 2006; Stoddart et al., 2002), especially when taught using a functional approach (Dobb, 2004). There was a concerted effort by all teachers in the program to create meaningful opportunities for critical and academic reading and writing. For example, students read newspaper articles about the current drought and related "tri-state water wars" occurring in Georgia during the time of this study. After visiting Lake Lanier, they learned about the Lanier watershed and how it had been affected by the drought. They used the Internet to research information about how lake water levels are monitored by the U.S. Army Corps of Engineers and learned about elevation, discharge, inflow, and rainfall. Each day following a field trip, students were also required to write about what they had learned during their excursion. Students wrote about how water at a sewage plant was treated and released, how osmosis was used in the process of water purification for public drinking water, and how fish populations were tracked and monitored. Students focused on higher-order skills and on functional uses of language to write across academic text types and genres, as evidenced in their final work products for the program. During the first week, students were told that they would be required to present a final project related to the program theme, "Water is life." Students were given time during each non-field trip day to research their topic and prepare their presentation. In order to allow for differentiation in the products across language proficiency levels, students had the choice to display their information in the form of (a) a poster, (b) a tri-fold brochure, or (c) a written report. Many students chose to augment their final products with visuals such as photos and Power Point and to display and share these electronically.

There was an emphasis on language development from the word to the discourse level. Each day began with a "word of the day" focused on academic language related to the day's lesson (such as *hydrology*), a saying for the day focused on colloquial or idiomatic language (such as "It's water under the bridge"), and an inspirational quote of the day (such as Nelson Mandela's "Education is the most powerful weapon which you can use to change the world"). Teachers focused on academic language development from a functional perspective, with a focus on what we use language to do and say in science, and made efforts to culturally and linguistically contextualize the science concepts and language. Students, for example, analyzed and compared water samples, described causes and effects of the 2006 drought in the state and much of the country, enumerated the steps in the process of how cacti are used to filter water in Mexico, categorized watersheds across the state by water quality and stream health, and summarized the effects of global climate change on oceans and water levels worldwide.

Within the program there was an attempt to deconstruct and reconstruct new vocabulary and make science learning relevant by linking it to what students already knew. Students learned common Latin prefixes and suffixes to help distill meanings in new science words. The Latin roots often were recognizable to Spanish-speaking students as cognates, often similar to more "everyday" words in their first language, where 90% of program participants were native Spanish speakers. Students learned new words by analyzing both content and structure, as when they studied the divisions of biology, for example, zoology, cytology, and parasitology. Finally, teachers in the program made a concerted effort to approach science teaching using a constructivist approach, whereby students constructed their own meanings for new ideas, words, and concepts, as exemplified in the following graphic organizer from a program lesson on tonicity:

Word	Prefix Means	Root Means	My Scientist Definition	My "Word on the Street" Definition	What it Reminds Me of
Hyper-tonic	Hyper: over; above	Tonic: related to normal function of tissue or organ	Water leaves cell	*Sha-rink! SUCKS water out* from the cell	Eating too much popcorn—it makes me really thirsty!
Hypo-tonic	Hypo: under; less	Tonic: related to normal function of tissue or organ	Water enters cell	*Bloating! More water outside the cell membrane, so the cell gets so full of water*	Squeezing a balloon and one side gets "mad" fat!

Students studied cell behavior in hypertonic solutions and hypotonic solutions through participation in an inquiry activity in which they observed the effects of hardboiled eggs in isotonic, hypertonic, and hypotonic solutions, using common household items such as vinegar, rubbing alcohol, detergent, and yarn. All students had some familiarity with these common household materials. During the inquiry activity, students formulated their own hypotheses and recorded their observations and conclusions, working in groups to make sense of what they observed. Using a constructivist approach, students were not introduced formally to the vocabulary related to the lesson (*hypertonic, hypotonic, isotonic, solution, osmosis*) until after they had participated in the inquiry. In this way, all students had prior and direct experience with the concepts to which they could then relate the new vocabulary.

Interest in Science

The data revealed that students developed interest in science while also learning a great deal of science across various learning and teaching contexts, including interactive science lectures and guided readings, university science labs, field excursions/studies, and through doing web-based and other research. Students reported being most engaged during field experiences where science was "hands-on," for example, at a popular lake and reservoir where they collected and analyzed water samples. Analyses of student writings revealed that they overwhelmingly stated that the most important thing they learned was related to the critical role of water in our lives and how important it is to protect and conserve water. Many students reported becoming more interested in chemistry as a result of the program, and the majority of students reported a greater appreciation for the need to study science for their future careers.

There was a conscious attempt by the organizers of the program to have students work with scientists who shared their cultural backgrounds. The percentage of students who said they knew a scientist with a cultural or ethnic background similar to their own increased from the 13% at the program's beginning to 34% at its end. As reflected in the pre- and post-responses, students who said they were good in science declined from 68% to 62%. This decline likely was the result of students becoming more realistic about what it means to be "good" in science, having met many "real" scientists. The program promoted students' identities as "scientists" by an increase from the beginning (28%) to end of the program (50%) in students answering yes to the question "Are you a scientist?" Overall, student interest in science was very high throughout and increased slightly from 89% at the beginning of the program to 91% by the end of the program. A relatively large percentage (74%) also said that they would be interested in taking science courses in college.

CONCLUSIONS

In the Steps to College program, we worked to frame summer school as privilege, not punishment, and as enrichment, not remediation. In the program, teaching was characterized by teachers linking to prior knowledge and what students knew in their primary language, with a curriculum that was rigorous, relevant, experiential, and hands-on. Cognitively demanding work was contextualized for students and thus scaffolded or "sheltered," allowing all students to access the science content despite their English proficiency levels. Other key characteristics of the teaching–learning context were high levels of personal attention and peer mentoring, modeling, collaborative work, and positive affect. Students were given choices whenever possible, in an attempt to increase interest and relevance, while allowing for differentiation of learning. Finally, data evidenced that teachers had very high expectations of students and were operating from an ability-centered perspective regarding their students. Students' perceptions of themselves and their aspirations changed as well. With Steps to College teachers as curricular innovators and agents for social justice, the subtle and overt messages students received about their educational rights and abilities constantly were empowering.

Teacher selection is a key variable in the success of such a program and of its students. Teachers selected must be pedagogically and attitudinally prepared to effectively teach rigorous content to students who are "at risk" within our educational system—in this case, immigrant, predominantly Latino, current and former ESOL students. As with many such outreach endeavors, two perennial issues emerged with the culmination of the program: how to sustain it and how to replicate or grow it. The real challenge facing us, however, is to make students' ongoing experiences of school additive, enriching, and an exciting fulfillment of their potential, including their potential as scientists. Our collective goal must be to have schools where all students have access to high-quality science classes taught by science teachers who are pedagogical as well as subject area experts, and who have an ability-centered perspective of all students where success is routinely constructed in diverse classrooms. Although powerful, we cannot rely on an "add-on" or enrichment approach to raise the achievement levels of ELLs in science. As science teachers, rather, we must recognize our responsibility to all of our students. Such responsibility means that we must work individually and collaboratively to make our classes accessible to all of our students, including the rapidly growing numbers of ELLs, as they are the future scientists of our nation, as well as the "Future Leaders of the World," as our students had come to see themselves.

While research on interventions such as this one will continue to be pertinent in improving minority students' access to participation in the sciences and enrollment in higher education, further research is needed to more specifically determine the support that minority students, particularly ELLs, need and receive once they make the decision to pursue a college degree. Increasing aspirations to participate in higher education are only a starting point. We must continue to find ways in which to assist students as they navigate the admissions process, as well as provide ample support, such as writing workshops, support networks, degree program advising, and science tutoring, once they enter an institution of higher education.

REFERENCES

Carter, P. (2005). *Keepin' it real: School success beyond Black and White.* New York: Oxford University Press.

Chapa, J., & De La Rosa, B. (2004). Latino population growth, socioeconomic and demographic characteristics, and implications for educational attainment. *Education and Urban Society, 36*(2), 130–149.

Chavez, M., Soriano, M., & Oliverez, P. (2007). Undocumented students' access to college: The American dream denied. *Latino Studies, 5,* 254–263.

Commission on Professionals in Science and Technology (2007). CPST Professional Women and Minorities: A Total Human Resources Data Compendium.

Council of Economic Advisors. (2000). *Educational attainment and success in the new economy: An analysis of challenges for improving Hispanic students' achievement.* Washington, DC: Executive Office of the President.

Cummins, J. (1996). *Negotiating identities: Education for empowerment in a diverse society.* Ontario, CA: California Association of Bilingual Education.

Cummins, J. (1998, March 29–31). *Linguistic and cognitive issues in learning to read in a second language.* Paper presented at Reading and English Language Learner Forum, Sacramento, CA.

Cummins, J. (2000). *Language, power and pedagogy: Bilingual children in the crossfire.* Clevedon, England: Multilingual Matters.

Dalton, S. (1998). *Pedagogy matters: Standards for effective teaching practice. Research Report No. 4.* University of California, Santa Cruz: Center for Research on Education, Diversity, & Excellence.

Dobb, F. (2004). *Essential elements of science instruction for English learners, 2nd ed.* Los Angeles, CA: California Science Project.

Douglas, R., Klentschy, M., Worth, K., & Binder, W. (2006) (Eds.). *Linking science and literacy in the K–8 classroom.* Arlington, Virginia: NSTA Press.

Echevarria, J., Vogt, M., E. and Short, D. (2008). Making content comprehensible for English Language Learners: The SIOP model. 3rd edition. Boston: Allyn & Bacon.

Erickson, F. (1986). Culture difference and science education. *The Urban Review, 18*(2), 117–124.

Fathman, A. K., & Crowther, D. T. (Eds.) (2006). *Science for English language learners: K–12 classroom strategies.* Arlington, Virginia: NSTA Press.

Fry, R. (2002). *Latinos in higher education: Many enroll, too few graduate.* Washington, DC: Pew Hispanic Center.

Fry, R. (2004). *Latino youth finishing college: The role of selective pathways.* Washington, DC: Pew Hispanic Center.

Fry, R. (2005). *Recent changes in the entry of Hispanic and White youth into college.* Washington, DC: Pew Hispanic Center.

Gibson, M. (1988). *Accommodation without assimilation.* Ithaca, NY: Cornell University Press.

Kao, G. & Tienda, M. (1998). Educational aspirations of minority youth. *American Journal of Education, 106*(3), 349–384.

Lave, J., & Wenger, E. (1991). *Situated learning: Legitimate peripheral participation.* Cambridge, UK: Cambridge University Press.

Lee, O., Deaktor, R. A., Hart, J. E., Cuevas, P., & Enders, C. (2005). An instructional intervention's impact on the science and literacy achievement of culturally and linguistically diverse elementary students. *Journal of Research in Science Teaching, 42*(8), 857–887.

Lee, O., & Luykx, A. (2006). *Science education and student diversity: Synthesis and research agenda.* Cambridge, UK: Cambridge University Press.

Litow, S. (2008). A silent crisis: The underrepresentation of Latinos in STEM careers. *Education Week.* Retrieved August 6, 2008 from www.edweek.org/go/litow.

MESA Connect Newsletter (2008). Retrieved February 21, 2009 from http://mesa.ucop.edu/news/connect_archive/08_winter_directorcorner.html.

Moje, E., Collazo, T., Carillo, R., & Marx, R. (2001). "Maestro, what is quality?": Language, literacy, and discourse in project-based science. *Journal of Research in Science Teaching, 38,* 469–498.

National Center for Education Statistics (2006). *The Nation's Report Card: Science 2005 (NCES 2006-466).* U.S. Department of Education. Washington, DC: U.S. Government Printing Office.

National Center for Education Statistics (2007). *The condition of education 2007 (NCES 2007–064) Special Analysis.* U.S. Department of Education. Washington, DC: U.S. Government Printing Office.

National Center for Education Statistics (2008). *The condition of education 2008 (NCES 2008-031).* U.S. Department of Education. Washington, DC: U.S. Government Printing Office.

National Council of La Raza. (2007). *Hispanic education in the United States.* Washington DC: Author.

National Research Council, (2006). *Multiple original, uncertain destinies: Hispanics and the American Future.* M. Tienda and F. Mitchell, (Eds.). Washington, DC: National Academies Press.

Noguera, P. (2006). Latino youth: Immigration, education, and the future. *Latino Studies, 4,* 313–320.

Pew Research Center. (2005). *Hispanics: A people in motion.* Washington, DC: Author.

Rogoff, B. (2003). *The cultural nature of human development.* Oxford, UK: Oxford University Press.

Schemo, D. J. (2002). National briefing/Education: Latino student population rises. *The New York Times.* Retrieved September 30, 2008 from http://query.nytimes.com/gst/fullpage.html?res = 9E03E2DD153FF933A15755C0A9649C8B63.

Solomon, J. (1993). *Teaching science, technology, and society.* Milton Keynes, UK: Open University Press.

Stoddart, T. (2008, March). *Integrating science and language pedagogy: Cases of teaching and learning.* Paper presented at the annual meeting of the American Educational Research Association, New York, NY.

Stoddart, T., Pinal, A., Latzke, M., & Canaday, D. (2002). Integrating inquiry science and language development for English language learners. *Journal of Research in Science Teaching, 39,* 664–687.

Stoddart, T., & Tolbert, S. (2008, April). *Preparing elementary school teachers to integrate inquiry science instruction and language development for English Language Learners.* Paper presented at the annual meeting of the National Association of Research in Science Teaching, Baltimore, MD.

Swail, W., Cabrera, A., & Lee, C. (2004). *Latino youth and the pathway to college.* Washington, DC: Pew Hispanic Center.

Tolbert, S. & Stoddart, T. (2008, March). *Academic language development through inquiry science: A comparison of results from a science–language integration program in English-only and bilingual elementary classrooms.* Paper presented at the annual meeting of the American Educational Research Association, New York, NY.

The National Center for Public Policy and Higher Education. (2004). *The educational pipeline: Big investment, big returns.* San Jose, CA: The National Center for Public Policy and Higher Education.

United States. President's Advisory Commission on Educational Excellence for Hispanic Americans. (1996). *Our nation on the fault line: Hispanic American education.* Washington, DC: Author.

United States. President's Advisory Commission on Educational Excellence for Hispanic Americans. (2000). *Creating the will: Hispanics achieving educational excellence. A report to the President of the United States, the Secretary of Education, and the nation.* Washington, DC: Author.

United States. President's Advisory Commission on Educational Excellence for Hispanic Americans. (2002). *The road to a college diploma: The complex reality of raising educational achievement for Hispanics in the United States. An interim report.* Washington, DC: White House Initiative on Educational Excellence for Hispanic Americans.

U.S. Census Bureau (2008). An older and more diverse nation by midcentury. Retrieved November 11, 2008 from http://www.census.gov/Press-Release/www/releases/archives/population/012496.html.

Valdes, G. (2001). *Learning and not learning English: Latino students in American schools.* New York: Teachers College Press.

Valenzuela, A. (1999). *Subtractive schooling: U.S.–Mexican youth and the politics of caring.* Albany: State University of New York Press.

Vela, C. (2003). Obstacles impacting Latino representation in science and engineering. Center for the Advancement of Hispanics in Science and Engineering (CAHSEE). Retrieved November 11, 2008, from http://www.cahsee.org/about/obstacles.asp.htm

Vygotsky, L. S. (1978). *Mind in society: The development of higher psychological processes* (M. Cole, V. John-Steiner, S. Scribner & E. Souberman, Trans.). Cambridge: Harvard University Press.

Warren, B., Ballenger, C., Ogonowski, M., Rosebery, A., & Hudicourt-Barnes, J. (2001). Rethinking diversity in learning science: The logic of everyday sense making. *Journal of Research in Science Teaching, 38*, 529–552.

White House Initiative on Educational Excellence for Hispanic Americans. (1998). *Latinos in education: Early childhood, elementary, secondary, undergraduate, graduate.* Washington, DC: U.S. Department of Education.

White House Initiative on Educational Excellence for Hispanic Americans. (1999). *Latinos in education.* Washington, DC: U.S. Department of Education.

White House Initiative on Educational Excellence for Hispanic Americans. (2000). *What works for Latino youth* (second edition). Washington, DC: U.S. Department of Education.

Wortham, S., Murillo, E. G., & Hamann, E. T. (2002). *Education in the new Latino diaspora: Policy and the politics of identity.* Westport, CT: Ablex.

CHAPTER 12

A FRAMEWORK TO SUPPORT HISPANIC UNDERGRADUATE WOMEN IN STEM MAJORS

Barbara A. Burke and Dennis W. Sunal

ABSTRACT

Much effort has gone into preparing and supporting young women in their K–12 years to matriculate into institutions of higher education as majors in science, technology, engineering, and mathematics (STEM). Fewer resources have been utilized to sustain them through graduation. In this chapter, the study of an intervention model, with the goal of increasing the graduation rate of women in mathematics and the computational/physical sciences, and its implementation at a large public urban Hispanic-serving institution (HSI) is described. In the study there were 52 participants, 14 of whom were Hispanic; all had achieved junior/senior status in one of the targeted disciplines when they entered the program. This intervention model provides a framework for support that includes mentoring, supportive networking, and developing a personal professional identity through participation in research projects and attending/presenting at professional conferences. Based on the results of this study, factors are identified that are associated with sustaining academic progress for the participants, with special emphasis on those of Hispanic heritage and their distinctive socio-cultural issues. Findings are based on quantitative and qualitative data gathered via focus group interviews and analysis of study

Teaching Science with Hispanic ELLs in K–16 Classrooms, pages 273–312

and faculty materials and artifacts. Results include positive findings in the areas of: providing students with meaningful and real life problem situations; building long-term, deep student and faculty interactions; expanding students' perceptions of the discipline and career; improving students' confidence; providing time for professional socialization; providing experiences in challenging research; professional writing and professional presentations; and creating a learning community of peers. The factors provide a framework that can be tested at other similar institutions serving diverse student populations.

INTRODUCTION

Based on recent National Science Foundation (NSF) statistics, one might infer that gender parity has been reached in mathematics and the physical/computer sciences, since women and men have the same, albeit low (<40%), graduation rates in these fields. The actual number of women graduates, however, is about half that of men and, for women of color, that number drops to about one-tenth. In addition, women make up only about 30% of the workforce in these fields (NSF, 2002, 2004).

Donna Shalala gave voice to the national concern about gender disparities in the STEM workforce (National Academy of Sciences, 2006):

> Women... must deal with lifelong questioning of their ability in science and mathematics and their commitment to a career.... Women scientists and engineers with minority racial and ethnic backgrounds are virtually absent from the nation's leading science and engineering departments. This needless waste of the nation's scientific talent must end... we urgently need to make full use of all our talent to maintain our nation's leadership.... (p. xii)

Since the mid-nineties, the "leaky pipeline" analogy has been used to describe the loss of women, to a much greater extent than men, at every juncture of the STEM educational system (Sonnert, 1995). This analogy, even though it aptly illustrates the situation, does not explain how and why this disparity has come about nor its persistence. What are the barriers? Are they the same for all women or do they vary with race, ethnicity, and first-generation and/or socioeconomic status? In this chapter we focus on these questions, their relationship to the retention and graduation of undergraduate Hispanic women in the fields of mathematics and the physical/computer sciences, and strategies that can sustain them through their junior/senior years.

BACKGROUND

The "pipeline model does not capture the complexity of the educational and career processes of becoming a scientist" posit Xie and Shauman (2003, p. 8). Their study investigates the changing personal, cultural, and academic experiences that affect women, compared to men, as they progress through the STEM educational system and on into careers. They found that a student's high-school experience determines whether or not she majors in a STEM field in college. It is also true that another pathway (often not taken into account) to these majors occurs when a woman changes into them from a non-STEM major. Individual or familial factors were not found to be a significant influence in persistence and graduation. There was no correlation between mathematical achievement and coursework performance and successful graduation in STEM majors. The effects of race, ethnicity, or first generation and/or socioeconomic status on women's decisions to persist in STEM were not addressed in the study.

The culture/atmosphere of established, male-dominated STEM programs often is not welcoming and is perhaps even somewhat hostile to women undergraduates. A study by Karen Tonso documented the adjustment of undergraduate women and men to inequities that were part of the culture/atmosphere in an institution where support for change in the culture/atmosphere was not apparent (2007). A recent longitudinal study (Fox, Sonnert, & Nikiforova, 2009) identified successful women's support programs, using as a measure of success the increase in graduation rate of women in STEM fields. The most successful were those focusing on institutional, systemic change through identifying problems and working with administrators to find solutions. These programs also had strong, individualized student support (bridging programs, living/learning residence halls, mentoring) but not as their main focus. These programs "adapted" the institutional environment, rather than "adopting" it (Fox et al., 2009). Institutional changes must occur not only in the lecture classroom, but also within the research environment, offering a more collaborative approach, more female role models, greater access to career advisement and career development through research opportunities, and travel to professional conferences. The Louis Stokes Alliance for Minority Participation (LSAMP) is a nationwide initiative supported by the NSF that has been successful in helping universities institutionalize changes supportive both to women and to those traditionally underrepresented in STEM (Burke, 2007).

A retrospective study of chemists found that, as undergraduates, women were less likely than men to have received mentoring, information about careers and graduate school, or a research experience (Nolan, Buckner, Marzabadi, & Kuck, 2008). Two other recent studies of women mathematics majors show that stereotyping can have negative effects and perhaps contribute to fewer women choosing mathematics as a major (Kiefer & Sekaquaptewa, 2007; Rodd & Bartholomew, 2006). De Palma (2001) believes the low numbers of women pursuing computer science is due to the way in which it is taught, whereas Huang (2003) cites the lack of sufficient numbers of women, personal issues, and the potential loss of confidence as factors to be considered. Others have explored the relationship between gender and computer games and how that may have a negative effect on women's interest and retention in computer science (Martinson, 2006). Greater strides have been made in physics to acknowledge that changes, especially in teaching, need to be made and institutionalized (Fencl & Scheel, 2006; McCullough, 2007; Schneider, 2001).

Hispanic women often are the first in their families to attend college, which can compound the gender issues they encounter in the STEM disciplines. The studies described above address gender issues, and only peripherally are race, ethnicity, and first-generation and/or socioeconomic status considered. Studies using longitudinal data show that 33% of Hispanic students who enrolled in college graduated, compared with 60% of White students. Women make up over 50% of the Hispanic post-secondary population. Many Hispanics begin their post-secondary educational journey in community colleges, attend part-time, or delay entry (Fry, 2002; Swail, Cabrera, Lee, & Williams, 2005). These factors can exacerbate differences in preparation, academic and career advisement, participation in the major, and academic expectations.

Hispanics place a strong emphasis on family relationships (*familisimo*) and respect (*respeto*), which are both hierarchical in nature and usually result in greater restrictions on female behaviors, as compared to their male siblings. It is unclear, however, to what extent Hispanic family culture influences daughters as they matriculate through college (Raffaelli & Ontanai, 2004). A study by Susan Sy and Aerika Brittian indicates that Hispanic women take family obligations in stride: it is just the way life is, even though these obligations can impede their academic progress at college (2008).

First-generation college students have no familial history of the college experience and are very often "on their own" as they navigate their way through the community college or university. Without adequate advising and mentoring, they base their expectations and their choices on the limited world from which they come. During the collegiate experience, a world of wider possibilities arises that can lead to conflicts between the world in which they were brought up and the world they now know. These students can, at least to some extent, bridge these two worlds by participating in programs addressing their academic, social, cultural, and emotional needs (Olive, 2008). For Hispanic students at a community college (Miami-Dade), Leonard Bliss and Janice Sandiford found students' study behaviors, which arise from their individual culture, can be correlated with collegiate success (2004). First-generation Latina women seek their friends and families, who hadn't attended college, for support as they cope with barriers particular to them, such as low socioeconomic status and cultural and gender stereotyping. It would, therefore, be advantageous for colleges and universities to more effectively communicate with families, especially parents, to understand higher education's structure, expectations, and how their children can successfully navigate that environment (Gloria, Castellanos, & Orozco, 2005).

Both ethnicity and gender inequities contribute to the growing shortages in the California STEM workforce (Offenstein & Sulock, 2009). A significantly lower percentage of Latinos than Asian and Whites was awarded Bachelor's degrees in 2007, and only 15% of that number was awarded in STEM fields. Women in STEM fields also earned a smaller proportion of degrees than in other non-STEM fields (Offenstein, & Sulock, 2009).

Much of the literature on the history of women in science was synthesized by Suzanne Le-May Sheffield (2006), who observed that women at the university level feel isolated and often feel "uncomfortable in the social situations in which networking connections... are made.... Instead they have often been left to struggle alone..." (p. 162). She concluded, "At both the undergraduate and graduate levels, women can benefit remarkably from access to support networks for women, study groups, mentoring programs, and the choice of attending all-women tutorials and seminars" (p. 192). Organizations such as the American Women in Science Network (AWIS) promote these types of activities through local university chapters, conferences, publications, and on-line resources and information. Not all universities, however, have chapters, and thus more individual, concerted efforts on the local level are needed. Even such efforts are not enough. "Creating a woman-friendly university is a daunting task because of the historical, statistical, and cultural traditions that have built the university to fit male needs, developmental stages, and interests." (Rosser, 2006, p. 70) The NSF, through its ADVANCE program, is focusing on institutional change that increases the number of women in university leadership positions and thereby effects a sustainable increase in women in STEM (Bystydzienski & Bird, 2006).

During the critical junior and senior years in college, students often switch out of science, mathematics, and computer science majors, leading to significantly decreasing numbers each year in course enrollments in these fields. Factors leading to switching include the economic need to work and commuting time, along with a loss of interest in the major, and/or the very high curricular demands of these majors (Seymour & Hewitt, 1997). The commonalities among the switchers and non-switchers led Elaine Seymour and Nancy M. Hewitt to "posit that problems which arise from the structure of the educational experience and the culture of the discipline... make a greater contribution to S.M.E. attrition than the individual inadequacies of the students or the appeal of other majors" (p. 392). A loss of confidence also can be a significant factor in women dropping out of STEM (Huang, 2003). Developmental relationships, especially in their crucial junior and senior years, are needed to enable women to maintain/increase their self-esteem (Downing, Crosby, & Blake-Beard, 2005).

Seymour and Hewitt (1997) also found that undergraduates have a stereotypical idea of scientific personalities and potential scientific careers. To dispel these stereotypes, undergraduate women need to be involved in professional activities such as research and attendance at professional meetings; these activities have been shown to help motivate and retain students (Doyle, 2000; Light, 2001; Sands, 2001; Tsui, 2007). Indeed, participation in "authentic" research, that is, working collaboratively with the faculty mentor to ultimately make an original contribution to the discipline (Pemberton, 2003; Hunter, Laursen, & Seymour, 2006), provides undergraduate women entree into the scientific community. Gafney (2005) posits that, in order for students to have a successful, meaningful research experience, their advisor needs to provide professional and personal mentoring. Students themselves acknowledge the importance of undergraduate research in "becoming a scientist," especially with respect to experiences enabling them to carry out research, learn more about what research requires, gain more skills, and more precisely define their own career plans (Hunter et al., 2006). Hunter et al. also "define UR [undergraduate research] as a powerful affective, behavioral, and personal discovery experience whose dimensions have profound significance for their emergent adult identity, sense of career directions, and intellectual and professional development" (p. 68).

As we pointed out earlier, the educational pipeline is not graduating women in STEM at a rate that indicates parity in the near future, despite massive investment in gender studies aimed at recruiting women into these programs (Bystydzienski & Bird, 2006; Chubin, May, & Babco, 2005). Between 1993 and 2003, 211 grants were issued under the NSF Diversity in Science and Education program. Despite the fact that most of these studies focused on improving the inclination of young women to pursue a career in STEM fields, enrollment numbers for women in several areas actually declined (Bystydzienski & Bird, 2006; Mervis, 2003).

Few research studies have been concerned with effective strategies in and outside the classroom for sustaining junior/senior undergraduate women in the STEM pipeline. The remainder of this chapter focuses on the results of our research at a large urban Hispanic-serving institution (HSI) on the west coast. Our study involved junior/senior undergraduate women majoring in mathematics and the physical and computer sciences. The results of our research identified effective intervention strategies that can sustain women through graduation. Our study, we believe, advances a critical understanding of the characteristics and needs of STEM undergraduate women, especially those of Hispanic heritage.

PROGRAM DESCRIPTION

The purpose of the research conducted with the Women's Educational Equity Act (WEEA) program described here was to examine the factors associated with sustaining academic progress for undergraduate women and the effects of an intervention program model. This program model was designed to increase the graduation rate of women in STEM, especially those majoring in mathematics, computer science, and the physical sciences. The program created experiences enabling upper division women to build a support network that focused more precisely on their needs. Based on the results of this study, the barriers encountered by Hispanic women in these majors can be compared and contrasted with those of students from a variety of other ethnic and socioeconomic backgrounds.

The public polytechnic HSI university in which the model was implemented awards Bachelor's and Master's degrees. Ninety percent of its students reside within 15 miles and are reflective of the state's racial and ethnic makeup. A majority of the undergraduates are minority, which includes Latina, Asian, African American, and European ethnicities. More than 50% of the students are receiving financial aid and about 14% are immigrants or refugees.

At this university, 48% of all first-time freshmen (FTF) graduate. Of the FTF who enter the university in STEM disciplines, 28.7% persist and graduate as STEM majors within six years. The system average is 25.0% (Consortium for Student Retention Data Exchange [CSRDE], 2008). Baseline data collected for upper-division women in the targeted majors from 2000, 2001, and 2002 show that, after reaching junior/senior status, 54% graduated in those majors, 2% changed to engineering, 3% changed to biology, 16% changed to non-STEM majors, 10% left the university because of a low grade point average, and 15% left the university in good academic standing. Eighty-eight percent of the women who graduated did so within one to three years of attaining junior/senior status. (Some transfer students were already at senior status when they transferred).

In the College of Science at this university, women make up less than 23% of the undergraduate students and 27% of the faculty in the targeted areas. About 54% of upper division women in the targeted majors are transfer students. For the 2005–2008 cohorts of upper-division women in the targeted majors in this study, the percent of Hispanic women ranged from 19% to 33%, with an average of 27%.

The WEEA/Mentoring and Professional Development Equity (MPE) Project at the higher education institution under study provides an intervention model aimed at increasing women's graduation rate in the targeted discipline areas. The MPE Research Apprentice Program (RAP) was designed to create a variety of professional experiences for upper-division women that built an experience base and a support network focusing more precisely on their needs (Braxton, Hirschy, & McClendon, 2004). The goal was to achieve greater educational equity while also promoting equity for women who suffer multiple forms of discrimination, namely, sex, race, ethnic origin, and limited English proficiency (Li, 2002). The expected outcomes for the RAP program included increased confidence in one's ability to "do" science, learning to communicate science to science professionals, becoming familiar with degree requirements for various scientific and/or professional careers, and learning to network in a professional setting,

As the upper-division women participants develop relationships with professionals, they more accurately learn what the careers in their majors encompass. They also begin to develop their own "personal professional identity" (Johnson & Ridley, 2004, p. 128). Participants have many opportunities to interact and network with women role models, that is, peers further along in their academic program, professors in their disciplines, as well as professionals in industry. Mentoring and professional development activities include formal research symposiums, informal discussions of issues about women in science, planning and conducting research projects, attendance and participation in professional conferences, community-building social events, luncheons, and meetings between students and faculty and professionals in the work world.

Students in RAP work one-on-one with their faculty mentor as they plan and carry out a discipline-based research project; for many students this becomes their senior project. It helps them understand a specific real-world application of their major as well as learn what it means to be a professional scientist. At the end of each quarter, their scientific communication is a written progress report to their research mentor as well as to the WEEA RAP director. In a few cases, their work has resulted in publication in a professional journal or presentation at a professional scientific meeting. These experiences enable the participants to begin their transition from "apprentice" to "professional."

All RAP students present their research progress at the annual College of Science Research Symposium held in May. They learn how to prepare either a poster or oral presentation. For many students this is the first time they have participated in a professional conference. This enables them to gain experience in professionally interacting with other scientists. All RAP students are encouraged to attend other national or regional professional conferences in their disciplines. Most often, faculty mentors also attend these conferences and facilitate students' navigation of the conference as well as networking. These are important experiences that will help them develop as a STEM professional.

Every quarter, the WEEA program holds a community-building "social" to which all women in the targeted fields are invited through flyers posted around campus and on the WEEA website. Personal invitations are sent to junior/senior women undergraduates, women faculty members, WEEA RAP mentors, and other faculty in each of the targeted departments. During the social, students and faculty interact informally in small groups; time is allotted for informal reports by RAP students and mentors. In addition, other students and faculty are encouraged to participate through questions and comments related to the WEEA RAP program and other issues of women in science. Faculty disseminate information about summer programs, discipline seminars, upcoming professional conferences, and other professional development opportunities. Smaller discipline luncheons are held once per year. These opportunities allow women students and faculty in their major to form a stronger learning community within a specific discipline.

Through these activities, junior/senior undergraduate women gain valuable experience and self-confidence that will enable them to graduate and become part of the STEM workforce.

PROCEDURES

The WEEA MPE Project involved a mentoring and professional development equity program situated at an HSI of higher education. The program, RAP, was designed at the university under study to create a replicable intervention model to increase the number of women undergraduate students graduating in the targeted academic discipline areas of physics, chemistry, computer science, and mathematics. Through effective mentoring and professional development, the goal was to increase the number of women eligible to participate in the professional workforce in the targeted STEM disciplines. These are disciplines with low graduation rates among minority women, particularly Hispanic women. The focus was on the critical junior and senior years of these women university students because the transition point from lower to upper division status is when students often switch from or drop out of STEM academic majors (Seymour & Hewitt, 1997). The four disciplines of physics, chemistry, computer science, and mathematics experience very high switching and drop-out.

This research project, conducted concurrently during each year of MPE RAP, utilized data collected using multiple formats and multiple times. The research objectives were:

1. Determine factors associated with sustaining academic progress for undergraduate women in the targeted areas.

2. Examine evidence regarding perceived outcomes and extent of effectiveness of the MPE/RAP in meeting its goal of sustaining academic progress for undergraduate women in the targeted areas.
3. Identify intervention program model components that have been most successful in sustaining academic progress in the targeted academic areas.

A mixed design with qualitative and quantitative data collected investigated the magnitude and range of impact that the program and specific elements of the program had on undergraduate minority women in mathematics and computer/physical science fields. The research attempted to identify and explain factors influencing women in the targeted majors to continue toward graduating or leave without graduating.

Data sources were MPE/RAP faculty mentors, women program undergraduate participants, student records from the targeted academic programs, other MPE/RAP personnel, observations of participant activities during the program, and artifacts used and/or developed during the program (Sunal, Ogletree, & Burke, 2006, 2007; Sunal, Steele, & Burke, 2008). Data were collected over the time period the MPE project was in operation. Each year of the MPE project, about 160 upper-division women were in the targeted majors and about 65 of those comprised a new cohort. During the entire project, data were collected from a total of 52 undergraduate women and 32 faculty mentors participated in RAP.

The study used a variety of tools, including individual interviews; focus group interviews; observational notes from program events; program documents and products produced; and survey questionnaires, including a RAP student application form and essay and STEM population surveys. Individual and focus group interviews were structured around determining basic demographic data of all participants, objectives and schedule of MPE project events, and perceptions and understandings of the program's progress.

Focus group interviews were completed during site visits each year of the MPE project. The students were interviewed in multiple focus-group settings each year. The interviews were conducted in separate focus groups of three to five, since the participants changed, with new students and faculty replacing graduated students and some new faculty selected to match new students joining the program. About 29 faculty (including several who were not RAP mentors) and 32 students participated in focus groups over the four years of the MPE Project. The focus group data collection instruments and processes used enabled the exploration and clarification of understandings focused, in part, on six areas.

1. range of ideas or feelings participants have about the discipline and about their participation in it as a professional

2. differences in perspectives between participants, non-participants, drop-outs, mentors, and project staff
3. factors influencing opinions, behaviors, and motivation among participants
4. ideas and perceptions emerging from each group using the saturation technique
5. information to help interpret the quantitative data gathered through surveys
6. language, personal culture, and social interaction used by the participants to help interpret the qualitative data

The basic content for the interview questions was written by individuals external to the project operation and were developed from the grant documents. Evaluation research guidelines were followed and then validated using MPE faculty personnel at the HSI institution. Focus group data collection was conducted with groups until a saturation point was reached for each group. The resulting questions, the *Student Focus Group Interview Protocol* (WEEA-SFGIP), around which the undergraduate focus groups were conducted, are found in Figure 12.1. The faculty focus group questions, the *Mentor Focus Group Interview Protocol* (WEEA-MFGIP), found in Figure 12.2, were developed, validated, and used with faculty RAP mentors.

Additional extensive probing questions were asked to clarify participants' responses. The interviews took place over a range of 45 to 120 minutes. The interviewer-researcher audio taped all conversations and took extensive notes on the students' responses with the agreement of the participants. All tapes were transcribed. The transcribed tapes and extensive interviewer notes were analyzed using the goal and purpose of the grant to guide the procedure (Krueger & Casey, 2000).

Surveys were constructed and administered to monitor MPE and full RAP participant's interests prior to joining the program and to the STEM student population at the HSI institution. The RAP student applications, 2005–2009, included an abstract of the proposed project, a letter of support from the research mentor, and GPA. Each student also wrote a personal statement in which she described her long-term goals and how the proposed RAP project fit those goals.

The analysis procedure, driven by the purpose of the project, was systematic, sequential, verifiable, and a continuing and recursive process based on the transcripts and checked by returning to sections of the detailed notes, audio tapes, and planned question statements. The collected data relating to the factors and intervention elements influencing women in the targeted discipline majors to graduate were then analyzed. In this case study, an inductive data analysis was conducted in order to discern themes, patterns, and assertions from the varied sources of data collected. The focus of the

Date of Focus Group ________________ Location ______________________________

Number of participants in focus group __________

Range of Disciplines in RAP Program: ____

RAP activities completed to date: ____

____

____

1. Describe your current academic area of study here at ______________. Describe your current career goal as the result of your studies here at ______________.
2. Describe your understanding of RAP. What are RAP's goals?
3. Are there comments and ideas you wish to discuss about the RAP, your academic area, or the life events you find yourself in that relate to the goals of RAP?
4. Describe the variety of types of experiences you have had with RAP the past year. Which were most beneficial to you? Why?
5. Describe your experiences as a researcher in your discipline. How would you describe them—satisfying, interesting, challenging, or ___________? What were the most beneficial aspects about the research experiences?
6. What impact if any have the MPE/RAP experiences had on staying in your major? In RAP? Your career goal you started the RAP with?
7. What have been some supportive events/people and difficulties/barriers you faced here at ________________ that helped/forced you to reconsider (and stay on course or change) your academic program of study?
8. What have been some supportive events/people and difficulties/barriers you faced at home, in your personal life, or elsewhere that helped/forced you to reconsider (and stay on course or change) your academic program of study?
9. How important were the following in creating your current academic program of study and your career goal? (e.g., parents, relatives, peers, role model at home, public school teachers, instructors, courses, role model at _____________, RAP Advisor, RAP research mentor, others of importance)
10. Describe your current plans for your future following your Bachelor's degree here at ______________________.
11. Is the RAP program meeting its goals? Describe your attitude about RAP at this time?
12. Other comments and ideas you wish to discuss about the program, the academic area, or the life events you find yourself in that relate to the goals of RAP.

Figure 12.1 Student focus group interview protocol (WEEA-SFGIP) (RAP Participants).

qualitative research analysis was understanding and interpretation (Merriam, 1998). The data were coded and analyzed as an ongoing process. The initial categories informed the direction of subsequent data collection. During the analysis, weight was given to frequency of similar responses,

Date of Focus Group ________________ Location ______________________________

Number of participants in focus group __________

Range of Disciplines in RAP Program: __________________________________

RAP activities completed to date: ______________________________________

____

____

1. Describe your current discipline of teaching/mentoring here at ____________.
2. What is the process by which undergraduate women form and enact their STEM career/life goals?
3. Describe your understanding of RAP. What are RAP's goals?
4. Why did you become a research mentor in the MPE project? Has the reason changed over time?
5. Are there comments and ideas you wish to discuss about the RAP, your academic area, or the life events you find yourself in that relate to the goals of RAP?
6. Describe the variety of types of experiences you have had with RAP the past year. Which were most beneficial to the women students, to you? Why?
7. Describe your experiences as a researcher/mentor in the MPE RAP project. How would you describe them for yourself, for the women participants—satisfying, interesting, challenging, or?
8. What were the most beneficial (supportive) aspects about the research/mentor experiences? What have been some difficulties (barriers) you faced as a research/mentor?
9. What impact, if any, have the RAP project experiences had on sustaining women undergraduate students in the targeted majors? On you staying as a researcher/mentor in the RAP program?
10. What have been some supportive events/people and difficulties/barriers RAP undergraduate women face here at ______________ that helped/forced them to reconsider (and stay on course or change) their academic program of study?
11. What have been some supportive events/people and difficulties/barriers RAP undergraduate women face at home in their personal life, or elsewhere that helped/forced them to reconsider (and stay on course or change) their academic program of study?
12. What have been some difficulties (barriers) RAP undergraduate women face here at ________________, at home, or in their personal lives that forced them to reconsider (and stay on course or change) their program of study discipline or career goals?
13. How important for RAP undergraduate women were the following in creating staying on with their current discipline of study and their career goal? (E.g., parents, relatives, peers, role models at home, public school teachers, instructors at _____________, courses at _______________, role models at ___________, RAP Advisor, RAP research mentor)
14. Is the RAP program meeting its goals? Describe your attitude about RAP at this time?
15. Other comments and ideas you wish to discuss about the program, the academic area, or the events you find yourself in that relate to the goals of RAP.

Figure 12.2 Mentor focus group interview protocol (WEEA-MFGIP) (RAP Mentors).

specificity of the responses, and extensiveness—how many different people reported similar responses (Krueger & Casey, 2000). Bias was reduced by having two independent reviewers analyze the data and compare results. As the data were collected, they were coded and sorted in an iterative process and examined for patterns (Bogdan & Biklen, 1992). Categories of social phenomena as described by Lofland (1971) include acts, activities, meanings, participation, relationships, and setting. These categories and themes, guided by a symbolic interactionist framework and sensitivity to socio-cultural factors, were analyzed for interpretive meaning. Triangulation of data was established using multiple data sources (Merriam, 1998).

RESULTS

Five areas resulting from data analysis are discussed: (1) baseline data; (2) ways in which the principal participants involved perceived, understood, and valued the goals and activities implemented by the program; (3) relevant factors related to this attitude, and conditions and circumstances that influenced that attitude; (4) factors associated with sustaining academic progress for undergraduate women in the targeted discipline areas; and (5) effects of an intervention program model designed to increase the number of undergraduate college women eligible to join the professional workforce or enter professional schools and/or graduate programs in the targeted STEM majors.

Baseline and General Results

For the period 2000–2008, the total number of junior/senior women in the targeted majors each fall averaged 152, within which each new cohort of juniors/seniors averaged 64. Baseline graduation data were obtained using three cohorts of junior/senior women from 2000–2002 to ensure that nearly all members of these cohorts graduated. The data show that 54% graduated, and of the 46% who left the targeted majors, 2% changed to engineering, 3% to biology, 16% to other colleges (not engineering) within the university, and 25% left the university (15% in good standing and 10% on probation or under disqualification). About 88% of those who graduated did so within one to three years of reaching junior/senior status.

Of the 52 undergraduate students participating as research apprentices, 63% graduated by September 2009, 33% are on track to graduate, one left the university, and one changed to a non-STEM major. All but three RAP students who graduated are either employed in industry or are in a graduate/professional school. The RAP graduation rate is almost 10% higher than the baseline (54%). Fifty-eight percent of the RAP participants were transfer students. Twenty RAP students were chemistry majors, 10 computer science majors, 19 mathematics majors, and 3 physics majors. The demographics of the RAP students were: 42% Asian, 27% Hispanic, 25% White, 3.9% Black, and 1.9% Other. Most RAP students presented posters on their research at the annual College of Science Research Symposium.

Student RAP, 2005–2009, applications provided insights into how these young women viewed themselves both as persons and as scientists. The students shared experiences that were influential in their life, as an immigrant, a transfer student, and/or a family member. Many also spoke about their enthusiasm for their major and about their goal of serving the community through their professional work.

Of the 56 women who participated in the social community building/networking events and/or travel, but were not in the RAP program, 21 graduated, 20 are on track to graduate, seven left the targeted majors, and eight attendees were not in the targeted STEM majors. It is noteworthy that 19 RAP students were introduced to RAP through these events and/or travel. One can, therefore, argue that these social interactions and professional opportunities have the potential to produce a significant "ripple" effect.

The research experience was a major component of the MPE RAP project. During the RAP activities all undergraduate students were required to plan, conduct, and present a research project in their STEM major. The research conducted was both basic and applied, and included laboratory, computational, and educational research. Many involved research on real-world applications (see Table 12.1)

Thirty-six of the RAP participants made at least one oral or poster presentation at local/regional professional conferences that included California State Polytechnic University at Pomona's College of Science Annual Research Conference, the Southern California Undergraduate Research Conference, and California State University's Program for Education and Research in Biotechnology. Twenty-four RAP participants attended professional conferences in their subject areas, such as those of the American Chemical Society, Mathematics Association of America, National Science Teachers Association, Frontiers in Education, and Grace Hopper Conference. Ten RAP participants worked as interns at companies such as the Jet Propulsion Laboratories and Metropolitan Water District or carried out research projects during the summer at universities such as University of California at Berkeley and Arizona State University.

TABLE 12.1 MPE RAP Undergraduate Student Research Project Examples

Discipline	Selected Project Topics
Chemistry	• To study the secondary structure of proteins using Circular Dichroism • To develop multimedia tutorials for general chemistry • To use Boronate Affinity Chromatography to study the binding capacity of a cross linked agarose column • To synthesize 1-phenylcyclophexene-3-one using retrosynthetic techniques • To use computational modeling and simulations to study electron transfer in biological molecules
Computer Science	• To develop 3D simulation of the movement of rail cars to determine factors needed for successful calculations for the simulation • To design a user friendly interface and supporting features of automatic query • To develop a GUI using Visual C++ that will create a 3D graphics model of a building • To develop a robot team that is heterogeneous and still capable of accomplishing an effective hunt • To study mathematical concepts (powers modulo m, successive squaring) in Number Theory that are related to computer science and network security
Mathematics	• To analyze a mathematical model for the genetic evolution of autism • To study the properties of Cayley graphs for various values of n and when S has only generators • To study knot theory and its potential applications in the real world • To develop assessment instruments to study the self-reporting differences in men and women when they write a "math biography for themselves" • To investigate the effect that changes in mosquito birth and death parameters have on population level and perform numerical simulations
Physics	• To study the composition of plasma in order to study the synergistic effects of radiation, ions, electrons and reactive radicals on selected microorganisms • To investigate Respiratory Sinus Arrythmia by measuring successive times between heart beats and breathing rates for subjects • To construct computer models that demonstrate system behavior according to the laws of physics using Vpython, a 3D animation program

An analysis of the 33 WEEA research apprentices who graduated found 22 of them in graduate or professional schools, and 8 are working in STEM fields in industry.

Undergraduate Student and Mentor Results

Based on the evidence collected, the results are reported using significant categories that emerged from the analysis (Burke, Sunal, & Ogltree,

2008; Burke, Sunal, & Steele, 2009). The categories (see Table 12.2) emerging from the data sources included (1) factors associated with sustaining academic progress for undergraduate women, (2) perceived outcomes of RAP as an intervention, and (3) effective components of RAP as an intervention. One additional category emerging from faculty mentor interview data alone was the process and effects of faculty mentoring.

TABLE 12.2 Sustaining Academic Progress: Categories Related to Student Responses

Factors Associated with Sustaining Academic Progress Leading to Graduation in Target Science Fields	Outcomes of the Research Apprentice Program	Effective Components of the Research Apprentice Program
Daily life	Student self-esteem/personal efficacy; provided positive element of resume to have been a part of RAP conducting research	Real-world research projects
Culture and family support, the role of women in society	Independence and empowerment; provided confidence and encouragement to further their education	Presenting at conferences and symposia
Socialization; lack of other women students in the targeted STEM fields	Students gained self-concept; they want the mentors to be proud of them, so they attempt challenges that they would not have done and they work harder	Regular meetings or luncheons for women, peers, and role models
Self-esteem/Self-efficacy, feeling confident with oneself	Camaraderie and socialization	Having a mentor for support and a role model
Individual and personality differences as personal barriers for completing a degree	Provided students with real-world experiences through the research, conferences, and presentations	
Gender behavior differences	Provided connections with HSI university professors and professors at other universities; readily available academic and professional recommendations	
Health issues	Improved professional communication skills and networking among peers	

Factors Associated with Sustaining Academic Progress Leading to Graduation in Target Science Fields

Undergraduate RAP students were interviewed about possible factors that influence women to continue in their science major, participate in the RAP mentoring program, change out of the target science majors, or drop out of school before graduating. The students in each of the focus groups during the MPE RAP project discussed and proposed several reasons that might cause women to drop out of their science, mathematics, or computer majors late in their program, before graduation.

The most significant and extensive factors that all students reported concerned their daily life (family responsibilities, living at home, commuting time to college, and economic need to work). The HSI university in this study is a commuter school whose students may have to commute and hour or more each way. Students living on campus said that if they had to commute, it would be extremely difficult to complete the program in terms of time and cost. Other students, especially Hispanics who lived at home, expressed concern over women having more responsibilities at home than men: "I know a lot of people [peer friends] are married or have children, so, maybe the women have a harder time because they are expected to do certain things at home." "I live with my only brother and I have more things to do than him with respect to chores." The students reported several experiences where home chores (parents' or personal home) were either expected or required and required several hours daily: washing clothes for the family, cleaning house, preparing daily meals for the family, babysitting, and many other duties in caring for siblings or their own children.

The necessity of working while attending the university was a concern in all focus groups and was given as the most common reason delaying Hispanic women from graduating in a timely manner. Several students in different focus groups reported "All of us work." A poll taken in each group found that 83 percent of the women worked each quarter in which they were in RAP, about half worked full-time, the rest part-time. Almost all worked full-time before joining RAP. Working, commuting, and family responsibilities deterred many women who tried to finish their science or mathematics program major.

A majority of the students cited culture and family support as a major contributing barrier to women's success. One Hispanic student felt her family did not understand how hard the program was and put her down instead of being more supportive: "They don't understand that it is a lot of work, and they don't understand that you are in school for this amount of time and you have to put this much time into it." "Women our age, in the twenties and thirties, are supposed to be married, and many women fall into that trap and they don't finish their programs." Similarly, another said that her father did not support her in mathematics at all because he did not consider it a major for women. Another said she had family support, but they still doubted whether or not she, as a woman, could make it. She stated that lack of support makes women feel less confident, weak, and causes them to possibly back out of school. This was a particularly strong theme for Hispanic RAP students. In general, family issues, money issues, and wanting to have a family constituted reasons for not completing a degree.

The ways in which society and specific cultures view the role of women in society emerged as a significant factor preventing women from graduating. Several comments were made in each focus group concerning this issue, with statements about their college majors being "male dominated": "People think it's a guy role and society is playing a part." These beliefs affected the women's family members and almost always were negative and discouraging. One student said, "It's a family thing." Another said, "My father didn't want me to continue in college because it's not part of our culture." Still another stated that "women are not taught to use pressure [be aggressive, as in choosing a male-dominated professional role] because it is not feminine." This theme was more prevalent with Hispanic RAP students.

The women in these focus groups reported not being able to socialize with other females, and the lack of other women in these male-dominated fields (socialization) is a factor influencing females' lack of completion of these majors. One stated, "When I started, I felt intimidated and lonely because there were not many women in the class. When I joined this program, I felt that I was not the only woman." Others confirmed that it can be intimidating to be in a class with all men. One remarked, "I don't know a lot of women in my program . . . there are so few. I know of only one other woman in my major courses this quarter." Another said, "There just aren't that many women. I know of only two others in my courses." Other comments, especially by Hispanic RAP students, were, "I think that possibly, for the females, it may be hard because there are not many females to speak to, and maybe they are too shy to go up to other students and speak to them." "If you are not outgoing and you don't have as many other females in your classes, you have to make an extra effort to talk to other people." "We need more help with speaking and getting more social." Not all of the women expressed such feelings, but the difficulties identified were most extreme in computer science courses, followed by physics and chemistry.

Students' self-concept or self-esteem/self-efficacy was another factor that came up repeatedly. Many of these students have faced situations in which they questioned their self-efficacy.

> I always loved math, but when I was in engineering I guess I didn't know a lot of the mechanical parts of the engine. I was so lost and most of the people in there knew what they were talking about. I had to spend more time on actually looking up what something was, when most of the people already knew what it was. That's why I switched out of engineering into math.

Other adults have affected students' self-efficacy. One student stated, "The counselor told me not to get into calculus and that it was too tough." A chemistry student had a college instructor who said, "You shouldn't be here. Chemistry is not a major for a woman." One student reported trying to overcome self-concept issues by repeatedly telling herself, "I am capable." Another student (Hispanic) felt lucky to have peer friends in the mathematics department and thought "a lot of females aren't aware of the many opportunities that are strictly for females."

Individual and personality differences were mentioned as personal barriers for completing a degree. One student remarked that even some men are not as successful as others because they choose not to work at it. Another stated that some students do not hear the voice, "'You can do it, you can do it!' . . . That is the difference between the women who are successful in this science field and the women who are not successful. Most women that are successful say, 'I can do it!'" Another declared that if someone had negative feelings toward her in regard to being able to succeed, it gave her even more reason to keep going. "I am a really stubborn person. I want to show people that I can do it. I want to succeed and show the world that it's not true."

Several of the focus groups discussed differences in gender behavior. When asked why more women were not enrolled in the hard sciences, students suggested that women think it is masculine and that men know more than women. "The guys can get really geeky on you, and so it is like, 'Wow, they know so much.' I don't know as much compared to them." One Hispanic RAP student stated society creates a "social thing" because women are supposed to do certain things: "Even if you are not thinking about it, I think subconsciously you are." She also stated that her brother never had to wash dishes and had more study time. Groups concluded that women have to try harder because men are "more into it." One student stated, "It's like men get more gratification from knowledge than women do. The more they know, the more masculine they feel." A Hispanic student remarked that men and women think differently: "I think men can concentrate on one thing. For example, as a woman I like plants. I like cooking, and I like to style my hair." She stated that when men take physics, that is all they think about, and they don't surround their lives by more things.

Two experiences were commonly reported concerning health issues for women. Students stated women believed they should not be exposed to a lot of chemicals or that they had been told that they should not be exposed to a lot of chemicals. One stated, "Chemicals interfere with hormones" and "you just have to be cautious." The students went further to state that their RAP mentors worked with chemicals safely, even though they worked with chemicals extensively and had families. The mentors may have been the first tangible/concrete role models for these chemistry majors.

Perceived Outcomes of the Research Apprentice Program as an Intervention

All of the students in the focus groups were positive about their experiences in the program, considering many aspects of the mentoring program to be effective.

Students' self-esteem and self-efficacy were enhanced as a result of this mentoring program. The RAP faculty mentor provided encouragement, motivation, and support. One student stated,

> I would be lost if I didn't have my mentor.... If I have a problem; if I cannot make something more, I go to him and he tries to help me and sometimes he opens my eyes for me, and I'm like, 'Oh, I know how to do this now.' The support is great. It is personal support.

The program was viewed as empowering these women in several ways. "It made me realize that I could do more, I wasn't sure what to do with my major. Now I know I can do research. I can work for a laboratory and in different areas using skills, not just teaching other people." "It expands our horizons." One Hispanic student remarked, "It makes me feel glad in a way that I am doing something meaningful in my field." Comments indicated the women, especially those who were Hispanic, have gotten over the fear of presenting in small groups and in public settings and have confidence when speaking to peers, other adults, and professionals in the field. The program not only helped the students become more interested, more motivated, and more confident in their major field, but lowered their frustration level.

Several students mentioned that the program helped them to become independent. Several stated that the mentors set high expectations for them and constantly pushed them, which encouraged them to try harder: "You know, I'm not sure I can do it, but I know I have support. So, I'm going to try." The students, especially Hispanics, wanted the faculty mentors to be proud of them, so they attempted challenges that they would not have undertaken otherwise, and they worked harder. One stated that the mentor always had ambitious ideas about what to do with her project, but said, "I'm not sure how to do it, but I'm going to try very hard to give her what she wants." Another remarked that the mentor gave her guidelines, but then expected her to work by herself. She stated that it gave her confidence to be in the laboratory working by herself and not constantly worrying if she could do it or not: "I do know what I'm doing and it gives me the motivation and confidence that you are doing it by yourself. That is neat!"

The camaraderie and socialization with other women and with professors played an important role in making this an effective program. The women mentioned working with others, working in groups, and being able to meet other women in the targeted major fields as being extremely important to them. Numerous comments, especially by Hispanic women, were made about the value of being with and being accepted by others. The program and the faculty mentors involved were described as a constituting safe environment. One student stated that the program motivated her to be a part of the "science community on campus." Another said, "I definitely think working in a group is a lot stronger because we can share our ideas.... It is really a good way to meet people." Another commented, "I think it's good to have women meeting other women. I've met a lot more females through the program that I would have never known." Another stated she was "much more comfortable working with somebody that I've had practice with, and being able to be more personal and having that person close by. I can even call them at home and, if I have a problem, they will help me." Many generalized these perceptions, but one person actually said, "You are not so alone. The importance is not just having friends or peers, men and women, but having peers and being accepted by others like you, other women in the same scholarly field. In a sense, it becomes a safe learning community."

In addition to academic and professional socialization efforts, the program included community-building events (e.g., luncheons, ice cream and pizza socials). The MPE program encouraged women to bring a friend to these events. The students felt this was a way for more women to find out about RAP. They liked the socials with other departments because they were able to interact with more women and had a reason to talk with them. The students particularly liked the smaller departmental luncheons because they were more likely to see those people on a regular basis. One student said the luncheons provide a time during which she could meet with her mentor and talk about things other than work. During the luncheons, she got to know her faculty mentor: "There is more of a connection, you know, because we are in a relaxed environment because there is no work to talk about, which is great."

These female students felt the RAP program provided them with real-world professional experiences through the research, conferences, and presentations in which they participated, and they built a closer connection not only with their own mentor professor, but also with professors at other universities. Each student was involved in a research project with her individual academic advisor. The students presented posters at a local university-level symposium, and two (both Hispanic) presented at national conferences during the first year of the project. The women stated that their professional communication skills had improved because of their participation at professional conferences and presentations.

On a practical note, getting to know a professor better meant there was someone to write academic and professional recommendations for them. One student commented,

> In the past when I was applying for anything, I'd always get hung up on the fact that I need to approach somebody in the faculty for a recommendation, and I don't feel that I know anybody there well enough or they know me well enough to give me a really good recommendation.

She went on to discuss how supportive the faculty mentors had been in helping her to make applications for higher-level opportunities: "It allows me to apply for programs that in the past I would have had problems applying for, because I need to find people to write me the recommendations." Another student mentioned how working on a real research project will help the students get a job because such experience "looks good on a resume."

One student described working on real research projects as "a challenge, but it is very, very valuable." Since her project worked with other people's data, it was important to figure out how to make the project work without starting over, even though she would have liked to do so. She found it a challenge, but knew it was important to change only what was necessary. The situation was a learning opportunity for her that she felt put her ahead of others looking for jobs. She noted that these experiences are not found in typical classroom coursework. Another said, "The research is so much greater... an experience than courses alone." One student stated that the research experience has helped strengthen some of her skills: "It has helped me strengthen things that I don't get as much practice on—in writing abstracts." Another commented, "I think it is a good opportunity to get a woman educated and get some experience before they even graduate, because when we graduate and get further along in our fields, we are going to have to do this." One student said that the "trouble-shooting was the most valuable." Most, including Hispanics, felt they were given experiences working with hands-on and real-life problems. One woman commented that they had never had hands-on research before: "It is a way for us to do our research, and you get to know what research is from first hand experience." Another stated, "The RAP program provides a platform for us to do more independent research."

Lastly, these women thought they were being prepared for the future by having the opportunities afforded them by RAP. One student said it was helpful to have a professor "willing to take time out and instruct you, and help guide you through this learning process." Another comment was, "It gives us the opportunity to work one-on-one with instructors that are more knowledgeable, that can give us guidance." Another said, "Being able to work with a professor really helps out." Some Hispanic (and other) students reported they would not have known what they wanted to do, or where to go for help without this program, so it has helped prepare them for graduate school and professional work. The women now realized there were opportunities out there for them. It was best summed up with the statement, "Now that you've opened up the research part of it, you can actually see that there are opportunities."

The students, in general, expressed the perception that RAP had benefited them and their education in many other ways and had given them the confidence and the encouragement needed to further their education. About half of the women were considering continuing their education through graduate studies. They recommended RAP be available for freshman women to help encourage them and thought it would be a good experience for all college undergraduates. "It's a really good program, actually, no matter what level you are at," one student remarked. A second student stated, "It has had a pretty major impact on me."

Effective Components of the Research Apprentice Program as an Intervention

The undergraduate women in the focus groups identified four specific program elements they thought were most effective: (1) researching with real-world research projects, (2) presenting at a symposium, (3) regular meetings or luncheons for women, and (4) having a mentor for support. These elements are best described as a process providing a positive socialization program that enhances their professional careers in physics, chemistry, mathematics, and computer science.

Real-World Research Projects. All of the women were involved in real world research projects that helped improve their self-confidence and communication skills. Several mentioned the positive growth they experienced conducting research, and planning and preparing for their poster presentations for the symposium. They liked the challenge of writing abstracts and presenting. One student stated that she enjoyed the seminar given by the project on writing abstracts because it was one of the "hardest things for me to do." A Hispanic student talked about the difficulty she had working on her mathematics research because there was not much information available. Then, when she was finished, she had to present her research and poster at a national meeting.

Presenting at a Symposium. The women in the focus groups said presentations made at the HSI university symposia and at professional conferences were not only exciting, but very good practice for professional careers. Several reported the local symposium was important and meaningful. When asked what was most beneficial, one said, "My big one was the symposium. That was awesome because I got to present and you get to gloat a little over what you did." Another stated, "I agree that the symposium was the best." One felt it made her confident enough to be able to present again, even though she had always been shy about talking in front of people: "I was a bit shaky, but I'm also excited because this is a real thing that you are going to be doing. This is a very big deal for me." A Hispanic RAP liked the experience not only because she presented her research, but because she met other mathematics professors and discussed their projects with them. She found out about programs at other universities and saw many women there in mathematics: "It was a very huge experience for me." Several of the undergraduates reported attending and/or presenting experiences at national and regional conferences.

Regularly-Held Forums for Women. All of the focus groups identified the luncheons and program meetings as providing a unique experience for them to get to know peers, other women role models, their own mentors, and other professionals on a more personal level. Networking and becoming part of larger existing networks was seen as a critical aspect of RAP. These activities were cited as building confidence, providing information in private conversations, and building a circle of friends—a learning community.

Having a Mentor for Support and a Role Model. The faculty mentors were discussed as very positive role models in the program, providing an organized personal support system for their RAP students. The students, especially Hispanics, described this role model as unique and necessary, in that they saw few and personally knew no other professional in their field before they entered RAP.

Process and Effects of Mentoring: Categories Related to Mentor Responses

Sustaining Academic Progress Leading to Graduation in Target Academic Fields

The mentors discussed reasons why more women are not in physics, chemistry, mathematics, or computer science and why more women are not finishing these undergraduate degrees, identifying seven contributing categories or factors as the major reasons why female students do not graduate from the university (see Table 12.3): family and work issues, societal issues, students not knowing their academic and career options, self-efficacy and self-conception issues, struggling with upper-level classes, social and communication issues, and the lack of the ability to think abstractly.

Many of the women had a family and a job that required them to work too many hours, according to statements made in the mentor focus groups. Some of these female students were pressured by their spouses to go to work and "bring in an income." A mentor said some women just "don't get that support" from their families. With younger students, a mentor noted, "If they get married, they drop school." One mentor talked about a student whose father-in-law discouraged her from going to school because her husband worked and she needed to stay home with the baby: "In those relation-

TABLE 12.3 The Process of Mentoring: Categories Related to Mentor Responses

Sustaining Women RAP Participants' Academic Progress Leading to Graduation in Target Science Fields	Mentoring Process	Effective Elements of MPE/RAP
Family and work issues	Rationale for being a mentor varies	Research performed by the students
Social and cultural issues	Several factors inhibit mentor effectiveness (lack of time, heavy teaching load, orientation to skills needed in the mentoring process)	Conferences and meetings students attend and make professional presentations at because of their research
Students not knowing their academic and career options or requirements necessary for pursuing the field of study	Lack of incentives such as release time limits effectiveness	Building of a learning community among students and between students and faculty
Self-efficacy and self perfection issues	Professional schedule provides little room for work with mentee	Increased the participation of women in research programs in the fields at HSI university
Social and communication issues; not enough women students in the these male-dominated classes; women did not see many women faculty (role models in their academic fields) handling a career and a family	Few women in Departments; few chemistry, physics, mathematics, or computer science faculty to take on mentoring junior-level students	Symposiums, luncheons, other social events, and one-on-one meetings are important for RAP students and bring in and affect other women (and, in some cases, men) not in the program
Struggling with their upper-level classes	Rationale for mentoring: mutual interest and motivation in solving problems in a discipline; personal rewards and making a difference in another person; professional accomplishment in being a positive role model; challenge of research, professional writing, and making professional presentations	Motivating students by improving students' self-confidence
Other factors: Time management, Cultural issues, Students taking a heavy class load, Students were in the academic programs only as a place holder for another academic program		Challenge and higher expectations of professional writing and making professional presentations

ships, there is probably more support for the male possibly to continue his education and that it's not necessarily as supported for the woman to continue her education." In another case cited during the mentor focus group sessions, one female student dropped out because she had to help her sister and her sister's children, and did not have enough money to provide support and put herself through college. One mentor stated, "all the students have to work full-time to earn enough money to support themselves, and even during regular class quarters, students still must work part-time.... It is a major factor that most female students cannot overcome." Mentors thought students should actually see a financial benefit for putting in the time here in a science major at the university.

Social issues arise from society's perception of women's role. This perception can cause undergraduate women to face many problems that men do not face. Representative mentor statements were: "There are still expectations that the woman has to do certain things more than men in relationships," and "It is just inconceivable that the woman could be the primary bread winner of the house." Some professional men/professors were described as "having something against" women in science and told the females they "should not be in this class" and needed to go "get married and have a family." Another mentor stated, "There are still so many people that think that mathematics is a man's field." Several mentors stated that the higher the degree goes, the more it is male-dominated.

A few professors felt another barrier was that female students do not know of possible academic and career options. It is important to "know what options are out there and that those options are obtainable." One mentor wanted to help students consider graduate school because most have never even thought about it.

> RAP students had no exposure to graduate school before entering the program. We want them to realize what their options are and to realize all the resources that they have to talk to people. Some students think they will never get a job in their field.

Self-efficacy and self-perception of female students was discussed as another barrier. Two of the mentors expressed concern over a lack of confidence in students among instructors. One mentor, citing research, suggested that "women internalize problems, and men will either blame the instructor or just be OK with the problems. If the guys fail a class, they just retake it. The women begin to feel that maybe I'm not good at this." In another situation, a mentor described a female student who broke down in tears because she did not like engaging with people she did not know: "They do not realize that there is nothing wrong with not knowing."

The mentors stated that there were not enough women students, a lack of a quorum, in the male-dominated classes in the targeted disciplines at the university, and that students did not see many women faculty (lack of role models in their academic fields). Not only were the mentors concerned about the number of women in these fields, but several expressed concerns about the lack of women as faculty members. They described mentoring as difficult due to the intensity and time commitments required. One mentor returned from a professional conference in her discipline where there were only three women out of 150 people in attendance. Because there are not many females, the communication between women in professional training and working as professionals is lacking in the targeted STEM areas. "Communication is a problem especially when we are trying to recruit women. If there were more women, it might be more comfortable." Several mentors described the problem as difficult and to be solved only by increasing the number of women in college majors: "Few women seem to enter into groups with men at college." "Sometimes a relationship with males develops, and then it ends up being a problem." "Women feel isolated, and that might be the reason they might not continue in the program." "Women are underrepresented in their majors. [In physics,] it is something like 10% of the PhDs are given to the women, and so it could only help having more women."

The mentors also were concerned that some women were struggling with their upper-level classes because they were lacking in experiences with abstract thinking. One mentor stated, "There are a few excellent women students, but there are a few who really need help." Another stated, "Many of the students drop because they are not doing well in their coursework." The mentors agreed that some students did not realize, for example, that a lot of mathematics was needed for computer science. Because "they did not take the right math in the freshman and sophomore years, in the computer science program the students decided to go into different majors," or "they just don't have computer experience." In the junior and senior years, more abstract concepts are introduced in the courses, so the courses are fundamentally different. One mentor suggested, "When the subject moves into a more abstract level, maybe that is preventing students from staying." Another reported, "It's on the junior level that it really gets very hard. I think the subject matter is harder than just the year before." The consensus was that students must have a strong interest in their major field to do well.

The faculty mentors also briefly discussed additional negative factors impacting female students as including being unable to pass the prerequisites, time management, cultural issues (e.g., speaking in class, speaking to a professor, etc.), and students taking too many classes during a quarter while working full-time.

Finally, there were examples of instances cited in most of the fields in which some students were in programs as a place holder, with little intention of graduation from the program. Chemistry majors, for example, are not required to get a Bachelor's degree to go to pharmacy school. This contributes to students leaving the program before they reach a B.S. degree in chemistry. One mentor stated, "If a decent portion of them have no intention of getting a Bachelor's degree or are kind of camping out as a chemistry major until they get accepted in pharmacy, and then they are gone, that is not a failure on our part if that's their goal." It was suggested that chemistry (and other fields) should follow students' progress and studies in order to see where they end up in school.

Mentoring

Even though mentors discussed the enjoyment of mentoring, they commented about the lack of time available to mentor students. All mentor focus groups acknowledged time as the factor causing the biggest problems for mentoring. One older mentor felt that age might be an issue because she thought the students might want a younger mentor. One female mentor felt it was very difficult to "pull women from the department together. Some are busy working, and some you don't see, and there are so few numbers in a department." A female mentor wondered if male faculty members felt as strongly about helping women in science, and if not, then there were not enough women faculty to handle an expanded program. One of the mentors expressed this concern: "I don't know what would happen once I won't be able to do this anymore, because it is very hard to do with all our teaching roles and everything. I feel that I'm approaching my end, and I'm not sure how many years I'll be able to maintain [this demanding schedule]." On numerous occasions, different mentors reported that faculty were busy, overbooked, had too heavy of a workload, and needed more time to be able to continue or expand mentoring students. One stated, "I am really busy, and it is hard to devote any time to a research student." Another said, "There are lots of benefits for the students, but there is really nothing there for the faculty, and with our very busy schedule, we could have been doing a much better job if we would have had the time."

Several faculty members in the focus groups were mentoring in the RAP program for their second year. One professor said he had not sought out mentorship in the program, but was asked to do so by the coordinator. "I have no misgivings about doing it, but it is not something I chose to do." This professor thought the program was beneficial. Another mentor stated that she liked working with female students and actively sought out students to work on research projects. "I want to at least work with them and see what a new generation of female students are thinking."

Reasons for Mentoring

Faculty mentors gave numerous reasons for choosing to mentor women in the WEEA/MPE student program: "I love being able to work with someone who is actually interested in math. It's really rewarding to see who is actually interested and feel that you are actually doing something that makes a difference." "This is the thing I enjoy the most, doing students' projects. I enjoy being creative with the students on these projects." "I like just doing physics, and where else can I do physics but on the student's projects? I like seeing the students grow. I try to find a project that I think is going to be fun for me too. It is often the most enjoyable part of the week." "I get more adrenaline if I do something with the students. I enjoy having as many as I can." Being a positive role model for the students also was important. The following additional comments were made concerning the reasons to mentor: "It is so easy for us to have a positive impact on the students." "What I get out of it is the enjoyment and fun that I see the student get when she sees the light turn on." "For me, that's the enjoyment I get out of it is them learning and gaining confidence." "We are not doing it for the money."

Effective Elements of MPE/RAP

The faculty mentors agreed on the effectiveness of the program, mentioning many positive aspects. The program was effective because of the research that the students did and the conferences and meetings students related to their research. The factor mentioned most was building of a learning community among students and between students and faculty.

The women students were involved in real research they completed with the guidance of a mentor and then by themselves. One student already had published two papers. The program increased the participation of women in research programs in mathematics and other fields. The RAP program also provided allocation for research in terms of providing some supply money. The students have the opportunity to present their research at the college symposium that provides professional experience in a less threatening atmosphere. Students also have attended and presented at national meetings. One of the female math students in the RAP program won an award in a national competition.

The symposia, luncheons, socials, and one-on-one meetings with their students all build a sense of community that these faculty members believed was the most effective part of the program. One mentor stated, "I think the sense of community and bringing students together, even the students who are not directly part of the program coming to the luncheon, the ice cream social, or various activities, is a valuable experience for both RAP students and non-RAP students." Another faculty member agreed and said there is a trickle-down effect because the women bring their friends to the socials.

> There is actually a real positive effect to more than just the research mentees or the people who are actually in the program. It has had a positive impact on other women, but that has been through friendships. So bringing those people together to foster those friendships is important. So it will be research mentors, the apprentices, and then the apprentices are inviting friends so that we can just spread the word about this experience.

The mentors liked and wanted more community-building activities so these women could get in touch with each other and find out what others were doing. The program allowed students to get close to faculty members, which helped students see them as more like themselves.

Mentors' Understanding of the Effectiveness of MPE/RAP

Faculty mentors thought the project was effective in several important areas. The interaction between faculty and students, showing the students options for the future, improving students' confidence, and providing time for socialization are a few of the many benefits and effectiveness identified for the MPE project. But, most effective was providing students with meaningful and real-life situations.

SUMMARY OF INVESTIGATION OF MODEL INTERVENTION PROGRAM

Triangulation resulted in a convergence on several key outcomes. The undergraduate women, faculty mentors, and results from other data sources demonstrated a similar understanding of the problems related to the goals of the research project:

1. Determine factors associated with sustaining academic progress for undergraduate women in the targeted areas.
2. Examine evidence regarding perceived outcomes and extent of effectiveness of the MPE/RAP in meeting its goal of sustaining academic progress for undergraduate women in the targeted areas.
3. Identify intervention program model components most successful in sustaining academic progress in the targeted academic areas.

Summary of Factors Associated with Sustaining Academic Progress for Undergraduate Women in the Targeted Academic Disciplines

The data supported results converging on several important outcomes, with student, faculty mentor, and artifact data demonstrating similar results. Factors associated with sustaining academic progress for undergraduate women included:

Culture and Family

- Culture issues related to the specific participant focused on stereotypical views of roles women play as influenced by culture and being members of a family and societal group
- Family, parental, and sibling support
- Personal and family financial issues need to be resolved regarding students' need to work, cost of education, and family responsibilities (economics)

Personal Difference and Skills

- Individual self-efficacy and self-esteem and their role in the discipline
- Time management in organizing multiple, complex, challenging tasks including commuting time to the university

Role Models

- Presence of women career models in the targeted academic and professional fields
- Amount of communication with the few women students in male-dominated classes, women faculty, and women working as professionals in the targeted academic fields
- Presence of role models of women successfully participating in university education or handling a career and a family in the targeted academic fields

Gender Concerns

- Health issues
- Gender behavior differences in course work (competition, relationships, and responsibility)

University Processes

- Unsupportive policies or faculty
- Level of student interest in the academic field
- Amount of knowledge and confusing or complex requirements necessary for pursuing a field of study
- Level of class load and difficulty level associated with junior courses, inability to pass or do well in prerequisite courses

Summary of Perceived Outcomes of the Research Apprentice Program as an Intervention

The data analyses supported the positive impact of the program's components, with student, faculty mentor, and artifact data demonstrating similar results. Effective supports associated with sustaining academic progress for undergraduate women included:

- Self-esteem and self-efficacy
- Independence and self-sufficiency
- Camaraderie and socialization with other women and with professors
- Real-world research and professional experiences
- Preparation for the future by having these opportunities

The RAP intervention model facilitated significant growth and level of achievement for the undergraduate women students involved in the Program. Students and mentors were positive in their description of RAP. The effects of the program for only partially supported women and other women involved in part of the program functions were inconclusive.

Summary of Effective Components of the Research Apprentice Program as an Intervention

Components of the RAP model perceived as most successful in sustaining academic progress in the targeted academic areas were:

- Conducting real-world research as undergraduates
- Attending and making presentations at national professional conferences and regional and campus events related to students' research

- Building a learning community among students and between students and faculty (Brumm & Mickelson 2002)
- Increasing the participation of women students and faculty in research programs and visible, ubiquitous role models
- Having a mentor for support and a role model
- Establishing and increasing networking opportunities and bringing in and affecting other women not directly involved in the program
- Interacting regularly with a mentor for support and as a role model
- Combining the program elements provided a synergetic effect on women participants in accomplishing MPE/RAP goals.
- The intervention program created significant change, concrete and psychological, for women to sustain efforts to complete their academic goals for graduation in one of the targeted STEM disciplines.

Summary of the Process and Effects of Mentoring: Categories Related to Mentor Responses

Analysis of mentors' interviews identified conditions affecting faculty mentor participation in the MPE Program and reasons why mentors take on the task of mentoring undergraduate women students in the targeted academic fields. Issues and concerns for development other similar programs, sustaining a RAP program model, or future expansion of the RAP model were centered on:

- extensive time required for mentoring
- age difference between mentor and mentee
- faculty and professional schedule and commitments providing little time for work with mentee
- few women faculty in academic departments available to take on mentoring junior level students

Reasons why mentors take on the task of mentoring women students in their fields were given as:

- mutual interest and motivation in solving problems in a discipline
- personal rewards in making a difference in another person
- professional accomplishment in being a positive role model
- challenge of research, professional writing, and making professional presentations

CONCLUSIONS

Based on the results of the analysis of the data sources, it was concluded that the RAP intervention was making a significant impact on the lives of the undergraduate women in sustaining academic progress. The effects of the intervention program model indicated it was a cost-effective use of resources. The factors associated with those intervention components perceived as effective by the women participants were similar to their mentors' operational understanding. Both groups perceived that positive and effective elements now operating in the program were:

- Transfer theory to real-world research and practice—providing students with meaningful and real-life problem situations
- Student and faculty interaction—long-term, deeper interaction than occurs with regular students
- Expanding students' perceptions of the discipline and career—showing options for the future
- Motivating students—improving students' confidence,
- Planned social events—providing time for socialization
- Challenging and higher expectations—of research, professional writing, and making professional presentations
- Encouraging friendship and working with other women peers in university activities and settings—creating a learning community

FINAL THOUGHTS

Efforts and resources have focused on making women aware, recruiting them, and providing diverse support for them so they may be able to begin academic undergraduate majors in STEM areas. Less effort and fewer resources have been invested in sustaining women after they begin these academic majors. Women in the targeted areas are not graduating in numbers that indicate parity now or in the near future. Baseline data for the HSI university reported here show that women in the targeted majors, once they reach junior/senior status, graduate at a lower rate (54%) than men (63%). The implications of this difference become clear when one considers the relative numbers. Women made up less than one-fourth of the total population (in baseline years 2000, 2001) and only about 17% (77/444) of the graduates. This disparity, as a recent report indicated (Offenstein & Sulock, 2009), impacts the STEM workforce. These low numbers clearly demonstrate the great need for a structure to sustain all women—but especially those of color, whose numbers are smaller—in these targeted majors as they move toward graduation.

Few research studies have been concerned with effective strategies in and outside the classroom for sustaining women in these targeted areas. The significant effects of this intervention and analysis of elements of the model impact the current understanding of strategies that are important for sustaining all women in these undergraduate majors. This study advances a critical understanding of the characteristics and needs of women in undergraduate science if their academic progress is to be sustained (Tinto, 1993).

Our data support the prominent role, especially for Hispanic women, that culture, family, and gender play in their lives and their potential to be barriers. In an earlier work, it was not clear the extent to which these influenced Hispanic students' progress in college (Raffaelli & Ontai, 2004). Another study (Sy & Brittian, 2008) noted that Hispanic women take family obligations in stride, even though these obligations can impede their academic progress. Since all but one of the Hispanic RAP students in this study have graduated in the targeted majors or are on track to graduate, we conclude that our intervention strategies have minimized these barriers and helped sustain their academic progress.

Hispanic RAP students, as well as most other RAP students, acknowledged both the importance and impact of the personal (mentor) and community-building (luncheons, informal social gatherings) activities and the professional development program components (research, presentations, meetings, travel). Collectively, these resulted in an increased graduation rate (63% versus 54%) for our RAP students as well as increased self-efficacy, independence, confidence, knowledge of careers, and motivation to continue in their STEM major. The intervention model at the HSI university demonstrates its effectiveness and agrees with earlier work (Seymour & Hewitt; 1997; Doyle, 2000; Light, 2001; Sands, 2001; Gafney, 2005; Hunter et al., 2007; Tsui, 2007).

Finally, based on our analysis of the results, we conclude that the MPE/RAP Project led to the development of an environment in which women in these targeted majors were not only engaged in professional activities, but also were comfortable networking with faculty, their peers, and other STEM professionals. This, along with the progress being made in the university's ADVANCE program, is the beginning of the transformation to a more "woman-friendly" (Rosser, 2006, p. 70) university. We are "adapting" our environment, rather than "adopting" it (Fox et al., 2009), thereby becoming more effective in graduating women in these targeted STEM fields.

The factors supporting the intervention components and strategies aimed at sustaining academic progress in this model can be tested in other settings and used to advance the success of undergraduate Hispanic women, as well as other underrepresented groups, at this and at other similar institutions serving diverse student populations. More broadly disseminated, these strategies should be tested and incorporated as part of intervention programs at other institutions to facilitate higher graduation rates of women in physics, chemistry, mathematics, and computer science.

ACKNOWLEDGMENT

This work is supported by a grant from the U. S. Department of Education (Women's Educational Equity Act, #U083A050034). Any opinions, findings, conclusions, or recommendations expressed in this publication are those of the authors and do not necessarily reflect the position, policy, or endorsement of the funding agency.

REFERENCES

Bliss, L. B., & Sandiford, J. R. (2004). Linking study behaviors and student culture to academic success among Hispanic students. *Community College Journal of Research and Practice, 28,* 81–295.

Bogdan, R. C., & Biklen, S. K. (1992). *Qualitative research for education: An introduction to theory and methods.* Boston: Allyn and Bacon.

Braxton, J. M., Hirschy, A. S., & McClendon, S. A. (2004). *Understanding and reducing college student departure. ASHE-ERIC Higher Education Report 30*(3).

Burke, B., Sunal, D., & Ogltree, G. (2008). *Women in undergraduate physics, chemistry, mathematics, and computer science: How can we sustain them through graduation?* Paper presented at the National Association for Research in Science Teaching, Baltimore, MD.

Burke, B. A., Sunal, D. W., & Steele, E. (2009). *Sustaining women in physics, chemistry, mathematics, and computer science through graduation.* Paper presented at the National Association for Research in Science Teaching, Anaheim, CA.

Burke, R. (2007). Women and minorities in STEM: A primer. In R. J. Burke & M. C. Mattis (Eds.) *Women and minorities in science, technology, engineering and mathematics: Upping the numbers* (pp. 3–27). Northhampton, MA: Edward Elgar Publishing, Inc.

Bystydzienski, J. M., & Bird, S. R. (Eds.) (2006). *Removing barriers: Women in academic science, technology, engineering, and mathematics.* Bloomington, IN: Indiana University Press.

Chubin, D. E., May, G. S., & Babco, E. L. (2005). Diversifying the engineering workforce, *Journal of Engineering Education, 84,* 73–86.

Consortium for Student Retention Data Exchange (CSRDE). (2008). *Data for California State University: Graduation rates by campus, ethnicity and gender.* Retrieved April 30, 2009 from http://www.asd.calstate.edu/csrde/index.shtml.

De Palma, P. (2001). Why women avoid computer science. *Communications of the ACM, 44*(6), 27–29.

Downing, R. A., Crosby, F. J., & Blake-Beard, S. (2005). The perceived importance of developmental relationships on women undergraduates' pursuit of science. *Psychology of Women Quarterly, 29*(4), 419–426.

Doyle, M. P. (Ed.), (2000). *Academic excellence: The role of research in the physical sciences at undergraduate institutions.* Tucson, AR: Research Corporation.

Fencl, H., & Scheel, K. R. (2006). Making sense of retention: An examination of undergraduate women's participation in physics courses. In J. M. Bystydzienski & S. R. Bird (Eds.). *Removing barriers: Women in academic science, technology, engineering, and mathematics,* pp. 287–302. Bloomington, IN: Indiana University Press.

Fox, M. F., Sonnert, G., & Nikiforova, I. (2007). Successful programs for undergraduate women in science and engineering: Adapting versus adopting the institutional environment. *Research in Higher Education, 50,* 333–353.

Fry, R. (2002). *Latinos in higher education: Many enroll, too few graduate.* Washington, DC: Pew Hispanic Center.

Gafney, L. (2005). The role of the research mentor/teacher: Student and faculty views. *Journal of College Science Teaching, 35*(1) 52–56.

Gloria, A. M., Castellanos, J., & Orozco, V. (2005). Perceived educational barriers, cultural fit, coping responses, and psychological well-being of Latina undergraduates. *Hispanic Journal of Behavioral Sciences, 27*(2), 161–183.

Huang, A. S. (2003). Confidence or arrogance? *AWIS Magazine, 32*(4), 6–9.

Hunter, A. B., Laursen, S. L., & Seymour, E. (2006). Becoming a scientist: The role of undergraduate research in sudents' cognitive, personal and professional development. *Science Education. 91,* 36–74.

Johnson, W. B., & Ridley, C. (2004). *The elements of mentoring.* New York: Palgrave Macmillan.

Kiefer, A. K., & Sekaquaptewa, D. (2007). Implicit stereotypes, gender identification, and math-related outcomes: A prospective study of female college students. *Psychological Science, 18*(1),13–18.

Krueger, R. A., & Casey, M. A. (2000). *Focus groups: A practical guide for applied research.* Thousand Oaks, CA: Sage Publications, Inc.

Li, L. (2002). Gender equity in science—who cares? *Journal of Chemical Education, 79*(4), 418–419.

Light, R. J. (2001). Making *the most of college: Students speak their minds.* Cambridge, MA: Harvard University Press.

Lofland, J. (1971). *Analyzing social settings: A guide to qualitative observation and analysis.* Belmont, CA: Wadsworth.

Martinson, A. M. (2006). Designing gender-sensitive computer games to close the gender gap in technology. In J. M. Bystydzienski & S. R. Bird (Eds.). *Removing barriers: Women in academic science, technology, engineering, and mathematics* (pp. 271–286). Bloomington, IN: Indiana University Press.

McCullough, L. (2007). Gender in the physics classroom. *The Physics Teacher, 45*(5), 316–317.

Merriam, S. B. (1998). *Qualitative research and case study applications in education.* San Francisco: Jossey-Bass Publishers.

Mervis, J. (2003). Report asks colleges to plug a leaky pipeline. *Science, 300,* 1353.

National Academy of Sciences. (2006). *Beyond bias and barriers: Fulfilling the potential of women in academic science and engineering.* Committee on Science, Engineering, and Public Policy/Committee on Maximizing the Potential of Women in

Academic Science and Engineering. Washington, DC: The National Academy of Sciences.

National Academy of Sciences. (2003). *Minorities in the chemical workforce: Diversity models that work.* A report by The Chemical Sciences Roundtable. Washington, DC: The National Academy of Sciences.

National Science Foundation. (2002). *Women, minorities, and persons with disabilities in science and engineering.* Retrieved April 20, 2009 from http://nsf.gov/statistics/women/

National Science Foundation. (2004). *Women, minorities, and persons with disabilities in science and engineering.* Retrieved April 30, 2009 from http://nsf.gov/statistics/women.

Newbart, D. (2009). DePaul is making science female-friendly. *Chicago Sun Times,* Retrieved January 5 from http://www.suntimes.com/news/education/1360643,CST-NWS-DEPAUL04.article .

Nolan, S., Buckner, J., Marzabadi, C., & Kuck, V. (2008). Training and mentoring of chemists: A study of gender disparity. *Sex Roles, 58*(3), 235–250.

Offenstein, J., & Sulock, N. (2009). *Technical difficulties: Meeting California's workforce needs in science, technology, engineering, and math (STEM) fields.* Sacramento, CA: Institute for Higher Education Leadership & Policy.

Olive , T. (2008). Desire for higher education in first-generation Hispanic college students enrolled in an academic support program: A phenomenological analysis. *Journal of Phenomenological Psychology, 39,* 81–110.

Pemberton, J. E., et.al. (2003). *Exploring the concept of undergraduate research centers. A report to the National Science Foundation.* Arlington, VA: National Science Foundation.

Raffaelli, M., & Ontai, L. L. (2004). Gender socialization in Latino/a families: Results from two retrospective studies. *Sex Roles, 50*(5/6), 287–299.

Rodd, M. & Bartholomew, H. (2006). Invisible and special: Young women's experiences as undergraduate mathematics students. *Gender and Education, 18*(1), 35–50.

Rosser, S. V. (2006). Using POWRE to ADVANCE: Institutional barriers identified by women scientists and engineers. In J. M. Bystydzienski & S. R. Bird (Eds.). *Removing barriers: Women in academic science, technology, engineering, and mathematics* (pp. 69–92). Bloomington, IN: Indiana University Press.

Sands, A. (2001). Never meant to survive, a black woman's journey: An interview with Evelynn Hammonds. In M. Wyer, M. Barbercheck, D. Geisman, H. O. Ozturk, & M. Wayne (Eds.). *Women, science, and technology: A reader in feminist science studies* (pp.17–25). New York: Routledge.

Schneider, M. B. (2001). Encouragement of women physics majors at Grinnell College: A case study. *The Physics Teacher, 39*(5), 280–281.

Seymour, E., & Hewitt, N. M. (1997). *Talking about leaving: Why undergraduates leave the sciences.* Boulder, Colorado: Westview Press.

Sheffield, S. Le-May. (2006). *Women and science: Social impact and interaction.* New Brunswick, NJ: Rutgers University Press.

Sonnert, G. (1995). *Who succeeds in science: The gender dimension,* New Brunswick, NJ: Rutgers University Press.

Sunal, D., Ogletree, G., & Burke, B. (2006). *Mentoring and Professional Development Equity (MPE) Project for the Women's Educational Equity Act Program: Year 1 Evaluation*. Tuscaloosa, AL: University of Alabama Office of Research on Teaching in the Disciplines.

Sunal, D., Ogletree, G., & Burke, B. (2007). *Mentoring and Professional Development Equity (MPE) Project for Women's Educational Equity Act Program: Year 2 Evaluation*. Tuscaloosa, AL: University of Alabama Office of Research on Teaching in the Disciplines.

Sunal, D., Steele, E., & Burke, B. (2008). *Mentoring and Professional Development Equity (MPE) Project for Women's Educational Equity Act Program: Year 3 Evaluation*. Tuscaloosa, AL: University of Alabama Office of Research on Teaching in the Disciplines.

Swail, W. S., Cabrera, A. F., Lee, C., & Williams, A. (2005). *Latino students & the educational pipeline*. Stafford, VA: The Educational Policy Institute.

Sy, S. R., & Brittian, A. (2008). The impact of family obligations on young women's decisions during the transition to college: A comparison of Latina, European American, and Asian American Students. *Sex Roles*, 58(9/10), 729–737.

Tinto, V. (1993). *Leaving college: Rethinking the causes and cures of student attrition*. Chicago, Illinois: The University of Chicago Press.

Tonso, K. L. (2007). *On the outskirts of engineering: How an engineering education culture moved women to the margins*. Paper presented at the American Educational Research Association Annual Meeting, Chicago, Illinois.

Tsui, L. (2007). Effective strategies to increase diversity in STEM fields: A review of the research literature. *The Journal of Negro Education, 76*(4), 555–581.

Xie, Y. & Shauman, K. (2003). *Women in science: Career processes and outcomes*. Cambridge, MA: Harvard University Press.

ABOUT THE AUTHORS

Jason T. Abbitt is an Assistant Professor of Instructional Design and Technology in the Department of Educational Psychology. Dr. Abbitt received his Ph.D. in Education and M.Ed. in Educational Technology from the University of Idaho and also holds a B.A. in English from Indiana University. Dr. Abbitt is currently the managing editor for the *Journal of Interactive Online Learning*. His current research interests include best practices for using technology in teaching and learning, the development of Internet-based learning and collaboration systems, and technology integration in STEM disciplines.

Daniel Bergman is an assistant professor and chair of Secondary Science Education at Wichita State University in Wichita, Kansas. Dr. Bergman has taught courses in secondary and elementary science methods; general secondary teaching methods; science, technology, and society; technology in the science classroom; and supervises various fieldwork experiences. Research interests include teacher self-evaluations with use of audio/video recordings, school-university partnerships, institutional constraints on research-based instruction, and the role of popular media in science and teacher education. His most recent publications include "Quality Questions" for *New Teacher Advocate* and "Bug Talk—A Learning Module on Insect Communication" for *Science Activities.*

Marco Bravo is assistant professor of education at San Francisco State University. His research focuses on language and literacy development with

Teaching Science with Hispanic ELLs in K–16 Classrooms, pages 313–319

English language learners. A particular focus of his work is the synergistic possibilities of literacy and science instruction.

Barbara A. Burke is professor of chemistry and director of Science Educational Enhancement Services (SEES) in the College of Science, California State Polytechnic University, Pomona. Her professional career encompasses teacher education (elementary & middle school), curriculum development, equity issues, and communication of science to the public. She developed/runs various programs for students traditionally underrepresented in STEM and is editor of an online column for the *Journal of Chemical Education* that features women/minority chemists. She's received awards for teaching, advising, and community service, including the NSF 2004 Presidential Award for Excellence in Science, Mathematics, and Engineering Mentoring. She can be contacted at baburke@csupomona.edu.

Cory A. Buxton is associate professor in the middle school education program in the College of Education at the University of Georgia. His research interests include the interplay of language, culture, and social justice through science curriculum, teacher professional development, and student and teacher reasoning.

Dr. Ann M. L. Cavallo earned a B.S. from Niagara University and two M.S. degrees and a Ph.D. in Science Education from Syracuse University. She holds teaching certificates in Biology, Chemistry, Earth Science, and General Science and was previously a middle- and high-school science teacher. Dr. Cavallo is Professor of Science Education and Associate Dean of Teacher Education at the University of Texas at Arlington. Her research investigates students' meaningful learning approaches and motivational factors and beliefs relative to their attainment of meaningful science understandings. Dr. Cavallo has more than 30 publications in refereed journals, a book and several book chapters, over 60 presentations at professional conferences, and significant grants to support her work. She may be contacted at cavallo@uta.edu.

Luciana C. de Oliveira is assistant professor of literacy and language education in the Department of Curriculum and Instruction at Purdue University. Dr. de Oliveira's research focuses on issues related to teaching English language learners at the K–12 level, including the role of language in learning the content areas (language arts, mathematics, science, and social studies). She has over fifteen years of English as an Additional Language teaching experience at various levels. In 2009, she was President of the Indiana Teachers of English to Speakers of Other Languages (INTESOL) professional association, an affiliate of the international TESOL organization.

Laureen A. Fregeau is currently Associate Professor of Educational Foundations at the University of South Alabama where she teaches multicultural and international/comparative education. She earned a B.S. in Biology/Chemistry and has worked with agricultural extension/research projects in Guatemala and Honduras. Her dissertation research examined how education influenced women's economic development in Costa Rica. She has consulted for the Ministry of Education in Nicaragua, assisting with the design of the post-Contra-war national education curriculum. With a Masters in ESL/Bilingual Education she taught ESOL at middle school through adult levels and ESL/Bilingual teacher preparation. She has conducted ESL teacher in-service training in southern Alabama and for Thomas Jefferson Institute in Mexico.

Dr. Patricia A. Gómez is Assistant Professor in the College of Education Department of Bilingual Education at Texas A&M University Kingsville. She holds degrees from Texas A&M University–Corpus Christi in Corpus Christi, Texas and Texas A&M University–Kingsville in Kingsville Texas. Dr. Gomez is a certified administrator and superintendent of schools. She has published in the areas of dual literacy curriculum, dual language program implementation, and science education for Ell's. Dr. Gomez has over 25years of teaching experience at the elementary, high-school, and college levels. She has taught in Spain, and in Texas.

Carla C. Johnson is an Associate Professor of Science Education and Director of the FUSION Center at the University of Cincinnati. Dr. Johnson's research is focused in three primary areas which include the dynamics of urban school reform; the impact of science professional development on teacher effectiveness and student learning; and barriers to teaching of authentic science K–20. Dr. Johnson has received funding to implement the Transformative Professional Development Model with ELL students at the elementary level from the Institute of Education Sciences. Dr. Johnson was the recipient of the Outstanding Early Career Scholar Award from the School Science and Mathematics Association in 2006 and currently serves on the Editorial Board of the *Journal of Research in Science Teaching* and the *Journal of Science Teacher Education.*

Teresa J. Kennedy has a Ph.D. in Education, with a concentration on international and bilingual/ESL science education, and also has a M.A. degree in the Spanish language. Her teaching experience includes 8 years at the secondary level, 7 years at the elementary/middle school level, and over 10 years teaching in higher education. She currently directs all international activities related to NASA's GLOBE Program, a science education program active in 110 countries. Dr. Kennedy has visited over 60 countries around the world through her work and educational experiences with ministries of education, public and private schools, as well as with formal and informal science education communities. Her current research *interests include online teaching and learning, content-based second language teaching and learning focusing on science and social studies, as well as brain research in relation to second language acquisition and bilingualism.*

Kerrie Kephart is Assistant Professor of Bilingual Education/ESL in the Teacher Education Department at the University of Texas at El Paso. She teaches courses in sheltered instructional methods, discourse analysis and academic writing for educators. Her research interests include advanced academic literacy, professional and disciplinary socialization, and second language learning. She is part of a cross-disciplinary team of researchers investigating innovative practices in engineering pedagogy and their effects on undergraduates' development of engineering literacy and discourse knowledge, a project funded by the National Science Foundation.

Okhee Lee is a professor of science education in the School of Education at the University of Miami. She has conducted extensive research and development in science teacher professional development focusing on student diversity, especially the needs of ELL students.

Robert D. Leier is currently Assistant Professor and program area coordinator for the Masters in ESOL Education Program at Auburn University. He has worked with reforestation and soil conservation projects in Guatemala and Honduras and was a field researcher for the Guatemalan "Biotopo" nature reserve system. His ethnographic dissertation research looked at how educational programs influenced Natural Resources development on the border of Panama and Costa Rica. This work was included as a chapter in the book, *Changing Tropical Forests.* He has consulted with the Ministry of Education in Nicaragua and recently taught graduate level courses on bilingual/bicultural education at the Thomas Jefferson Institute in Mexico.

Bernadette Musetti is a long time ESOL teacher and teacher educator. She served for many years at co-director of a center focused on professional learning to increase Latino student achievement. She is currently an Associate Professor of TESOL in the Department of Inclusive Education at Kennesaw State University.

Michael Odell has a Ph.D. in Curriculum Instruction/Science Education,and serves as the Roosth Chair of Science Education at the University of Texas–Tyler. His teaching experience includes 9 years at the secondary level, teaching middle- and high-school Earth and environmental science, and over 15 years teaching in higher education. Currently, he directs the Ingenuity Center, one of seven STEM centers funded by the Texas Education Agency. He has served as a National Space Grant Fellow at NASA Headquarters in Washington, D.C. (1995–97) and as a Faculty Fellow at NASA Kennedy Space Center (2003) in Florida. Dr. Odell is involved in research and development to vertically align high school and college curriculum to improve college readiness of high school students and improve pathways into the STEM disciplines.

Jorge L. Solís (Ph.D. in Language, Literacy, & Culture, University of California, Berkeley) is a post-doctoral researcher collaborating on two related projects developing effective ways to integrate science–language pedagogy in pre-service and diverse teacher education settings. His dissertation study concentrated on the transitional discursive contexts of English language learners in elementary and secondary school settings. Previously, he collaborated on the implementation of a more inclusive science curricular framework (Science Instruction For All) in 3rd- and 4th-grade classrooms funded by the National Science Foundation. Dr. Solís has been published in the *Journal of Linguistics and Education* and recently in a volume titled *Talking Science, Writing Science: The Work of Language in Multicultural Classrooms*, edited by Katherine Bruna and Kimberley Gomez.

Trish Stoddart is Professor of Education at the University of California, Santa Cruz and a Senior Scholar with the Center for Research on Education, Diversity and Excellence (CREDE) at the University of California, Berkeley. Her research focus is on the preparation of teachers to teach science in culturally and linguistically diverse classrooms and the relationship between teacher preparation, teaching practice and student learning outcomes. She is the author of over 70 research papers, monographs and book chapters and has been principal investigator on the NSF and USDOE funded LASERS, CCTD, and ESTELL science teacher education and diversity projects.

Cynthia Szymanski Sunal is Professor of Curriculum and Instruction at The University of Alabama and Director of the Office of Research on Teaching in the Disciplines. She primarily works with elementary and middle-school teachers. Among her publications are numerous books, journal articles, and monographs. She is Executive Editor of two journals and publishes a research series. She has been involved in several funded projects from the National Science Foundation, the Department of Energy, and other agencies. She may be contacted at cvsunal@bamaed.ua.edu.

Dennis W. Sunal holds a Ph.D. in science education, M.A. in Interdisciplinary Science, and a B.S. in Physics, all from the University of Michigan. He currently is a Professor of Science Education at the University of Alabama.

His university teaching experiences include undergraduate and graduate courses in physics, engineering, curriculum and instruction, and science education. He holds both Secondary, 6–12, and Elementary, K–6, teacher certification and has taught extensively on both levels. His research interests are in undergraduate science, pre-service teacher education, conceptual change in teachers and faculty, and web course design. He has been project director and co-director in numerous grants (e.g., NSF, NASA, Department of Education, USIA, and U.S. Department of Energy). Dr. Sunal has published numerous articles and chapters in refereed journals and books. Recent research presentations have been at the annual meetings of NARST, ASTE, NSTA, SCST, AACTE, and AERA. His published books include *Teaching Elementary and Middle School Science; Integrating Academic Units in the Elementary School Curriculum; Reform in Undergraduate Science Teaching for the 21st Century; The Impact of State and National Standards on K–12 Science Teaching;* and *The Impact of the Laboratory and Technology on Learning and Teaching Science K–16*. He may be contacted at dwsunal@bama.ua.edu.

Sara Tolbert is a former science and ESOL teacher, and is currently a doctoral candidate at the University of California, Santa Cruz, conducting research on the preparation of science teachers to teach in culturally and linguistically diverse classrooms. She received her Masters of Education at the University of Georgia in 2006, where she worked with the both the Center for Latino Achievement and Success in Education (CLASE) and the National Science Foundation [NSF]-funded project, Partnership for Reform in Science and Mathematics (PRISM). At UC-Santa Cruz, she is a research assistant and professional development coordinator for the NSF's Effective Science Teaching for English Language Learners (ESTELL).

Elsa Q. Villa is a doctoral student in Curriculum and Instruction at New Mexico State University (NMSU) and a full-time lecturer at The University of Texas at El Paso (UTEP) in the Department of Teacher Education. Ms. Villa teaches elementary mathematics and science methods courses and has research interests in teacher identity, inquiry, and science education. She has served as a co-principal investigator on collaborative projects with faculty in the UTEP College of Engineering including Affinity Research Group, an innovative model for engaging undergraduate and graduate students as co-researchers while learning and integrating the knowledge and skills required for research.

Emmett L. Wright, Ph.D., is Professor of Science/Environmental Education at Kansas State University. In recent years he served as the served as the Director of Research for the National Commission on Mathematics and Science Teaching for the 21st Century, and as a Program Director, TPC Section Head in the Division of Elementary, Secondary Education and Informal Education at the National Science Foundation and Deputy Director of Education and Science for the GLOBE Program. He was awarded a Fulbright Scholars position at the University of Malta during the 2006–07 academic year. He holds a Ph.D. from Pennsylvania State University (Science Education/Environmental Biology). Dr. Wright's research interests include decision-making attitudes, problem solving, misconceptions and scientific discrepant events, and international education. He has over 130 publications and has served as president of NARST and on the board of directors of NSTA and SCST. He has received major curriculum-development, teacher-education, and research grants, from EPA, NSF, U. S. Department of Education, U. S. Department of State, and U. S. Department of Energy. He may be contacted at birdhunt@ksu.edu.